3D Audio

T0133647

3D Audio offers a detailed perspective of this rapidly developing arena. Written by many of the world's leading researchers and practitioners, it draws from science, technologies, and creative practice to provide insight into cutting-edge research in 3D audio.

Through exploring the intersection of these fields, the reader will gain insight into a number of research areas and professional practice in 3D sonic space. As such, the book acts both as a primer that enables readers to gain an understanding of various aspects of 3D audio, and can inform students and audio enthusiasts, but its deep treatment of a diverse range of topics will also inform professional practitioners and academics beyond their core specialisms. The chapters cover areas such as Ambisonics, binaural technologies and approaches, psychoacoustics, 3D audio recording, composition for 3D space, 3D audio in live sound, broadcast, and movies – and more.

Overall, this book offers a definitive insight into an emerging sound world that is increasingly becoming part of our everyday lives.

Justin Paterson is Professor of Music Production at London College of Music in the University of West London, where he leads the MA Advanced Music Technology course. He has numerous research publications ranging through journal articles, conference presentations, and book chapters, and is the author of the *Drum Programming Handbook*. Justin is Co-chair of the Innovation in Music conference series and co-editor of its books. He is also an active music producer. His current research interests are 3D audio, interactive music, and haptic feedback. Together with Professor Rob Toulson and working with Warner Music Group to release prominent artists from their roster, he developed the variPlay interactive-music format, funded by the UK Arts and Humanities Research Council. His current research – funded by Innovate UK through its Audience of the Future programme with consortium partners including Generic Robotics and the Science Museum Group – is around the development of a novel music-production interface in mixed reality that utilizes haptic feedback.

Hyunkook Lee is Reader (i.e. Associate Professor) in Music Technology and Director of the Applied Psychoacoustics Laboratory (APL) at the University of Huddersfield, UK. His recent research has advanced understanding about the psychoacoustics of vertical stereophonic localization and spatial impression in 3D sound recording and reproduction. This has provided theoretical bases for the development of several 3D microphone arrays, including Schoeps ORTF-3D. His current research topics include the six-degrees-of-freedom perception and rendering of virtual acoustics, and the creation and evaluation of a multimodal immersive experience for extended reality applications. From 2006 to 2010, he was a Senior Research Engineer in audio R&D with LG Electronics, South Korea, where he participated in the standardizations of MPEG audio codecs and developed spatial audio algorithms for mobile devices. He received a bachelor's degree in music and sound recording (Tonmeister) from the University of Surrey, Guildford, UK, in 2002 and a PhD in spatial audio psychoacoustics from the Institute of Sound Recording (IoSR) at the University of Surrey in 2006. He is a Fellow of the Audio Engineering Society (AES) and Vice Chair of the AES High Resolution Audio Technical Committee.

Perspectives on Music Production

Series Editors

Russ Hepworth-Sawyer, *York St John University, UK*

Jay Hodgson, *Western University, Ontario, Canada*

Mark Marrington, *York St John University, UK*

This series collects detailed and experientially informed considerations of record production from a multitude of perspectives by authors working in a wide array of academic, creative and professional contexts. We solicit the perspectives of scholars of every disciplinary stripe, alongside recordists and recording musicians themselves, to provide a fully comprehensive analytic point-of-view on each component stage of music production. Each volume in the series thus focuses directly on a distinct stage of music production, from pre-production through recording (audio engineering), mixing and mastering, to marketing and promotion.

Cloud-Based Music Production

Sampling, Synthesis, and Hip-Hop

Matthew T. Shelvock

Gender in Music Production

Edited by Russ Hepworth-Sawyer, Jay Hodgson, Liesl King, and Mark Marrington

Mastering in Music

Edited by John-Paul Braddock, Jay Hodgson, Russ Hepworth-Sawyer, Matt Shelvock and Rob Toulson

Innovation in Music

Future Opportunities

Edited by Russ Hepworth-Sawyer, Justin Paterson, and Rob Toulson

Recording the Classical Guitar

Mark Marrington

The Creative Electronic Music Producer

Thomas Brett

3D Audio

Edited by Justin Paterson and Hyunkook Lee

For more information about this series, please visit: www.routledge.com/Perspectives-on-Music-Production/book-series/POMP

3D Audio

Edited by Justin Paterson and Hyunkook Lee

LONDON AND NEW YORK

First published 2022
by Routledge
2 Park Square, Milton Park, Abingdon, Oxon OX14 4RN

and by Routledge
605 Third Avenue, New York, NY 10158

Routledge is an imprint of the Taylor & Francis Group, an informa business

© 2022 selection and editorial matter, Justin Paterson and Hyunkook Lee; individual chapters, the contributors

British Library Cataloguing-in-Publication Data
A catalogue record for this book is available from the British Library

Library of Congress Cataloging-in-Publication Data
Names: Paterson, Justin, editor. | Lee, Hyunkook, editor.
Title: 3D audio / Justin Paterson and Hyunkook Lee.
Description: Abingdon, Oxon ; New York, NY : Routledge, 2021. |
Series: Perspectives on music production |
Includes bibliographical references and index.
Identifiers: LCCN 2021010452 (print) | LCCN 2021010453 (ebook) |
ISBN 9781138590038 (hbk) | ISBN 9781138590069 (pbk) |
ISBN 9780429491214 (ebk)
Subjects: LCSH: Surround-sound systems.
Classification: LCC TK7881.83 .A13 2021 (print) |
LCC TK7881.83 (ebook) | DDC 621.389/334–dc23
LC record available at https://lccn.loc.gov/2021010452
LC ebook record available at https://lccn.loc.gov/2021010453

ISBN: 978-1-138-59003-8 (hbk)
ISBN: 978-1-138-59006-9 (pbk)
ISBN: 978-0-429-49121-4 (ebk)

Cover Design: Justin Paterson
Photography: Elliot Smith (the APL listening room), Hyunkook Lee (the microphone arrays), Mark Wendl (the dummy head)
Location: Applied Psychoacoustics Lab (APL) at the University of Huddersfield

Typeset in Times New Roman
by Newgen Publishing UK

Contents

Figures

Tables

Notes on contributors

Cal Armstrong is an industry-focused R&D engineer specializing in the enhancement of spatial audio rendering and the personalization of sound source localization cues. During his PhD at the University of York, he developed techniques to improve the binaural reproduction accuracy of Ambisonic signals through head-related transfer functions (HRTFs) measurement and manipulation techniques. In 2018, he published the SADIE II database – a state-of-the-art collection of far-field human and dummy head HRTFs. He later went on to develop an alternative time-constrained near-field HRTF capture system suitable for more commercial applications. He has been involved in the standardization of the Immersive Voice and Audio Services codec for 3GPP and now works within the games industry. He is a keen musician with a particular interest in strength and fitness training.

Natasha Barrett is a composer and researcher specializing in acousmatic and live electroacoustic concert works, sound art, installations, and interactive music. Her inspiration comes from the natural and social world around us: the way it sounds and behaves, systems, processes, and resulting phenomena. These interests have led her to use cutting-edge audio technologies and exciting collaborations involving soloists, ensembles, visual artists, architects, and scientists.

The musical application of spatial audio has guided her work since the late 1990s. Since completing her PhD (City University, 1998), she has produced acousmatic compositions in formats spanning traditional stereo for sound diffusion performance, wave field synthesis, and seventh-order 3D Ambisonics.

Prominent among her 3D multi-media works are collaborations with the OpenEndedGroup and Marc Downie (USA), installations with Anthony Rowe (Squidsoup, UK), large sound-architectural projects with OCEAN Design Research Network (Norway), and sonification-based installations and compositions created in collaboration with geoscientists at the University of Oslo.

Her work is commissioned and performed throughout the world and she has received a solid list of international awards. These have included the Nordic Council Music Prize and the Concours International de Musique Electroacoustique Bourges Euphonie D'Or which, in the words of the IMEB, "represent particularly brilliant moments in the history of electroacoustic music."

She is also active in performance and education and is currently Professor of Composition at the Norwegian Academy for Music in Oslo.

Kevin Bolen supervises Skywalker Sound's interactive-audio department, which combines decades of cinematic audio experience with bleeding-edge technologies to create unforgettable immersive audio experiences, such as the Academy Award-winning *Carne y Arena*, the Peabody Award-winning *Queerskins, A Love Story*, and the Emmy-nominated *Star Wars: Vader Immortal*. Kevin's team collaborates with partners including Disney, Lucasfilm, Marvel, Legendary Entertainment, and ILMxLAB to extend the power of cinematic storytelling into location-based experiences and home entertainment alike.

James Cook is Lecturer in Early Music at the University of Edinburgh, where he also directs the BMus programme. He has previously taught at the Universities of Nottingham, Huddersfield, Cambridge, Bangor, and Sheffield. His research encompasses Medieval and Renaissance Music, both in its historical setting, and in its reuse in popular screen media, such as television, film, and video games. He publishes widely on both topics, most recently including his monograph *The Cyclic Mass: Anglo-Continental Exchange in the Fifteenth Century* and his co-edited collection *Recomposing the Past: Representations of Early Music on Stage and Screen*. He has been principal investigator on several funded projects on topics ranging from the use of virtual reality (VR) to reconstruct historical performance to the production of a prosopography of pre-reformation Scottish musicians.

James is a member of the British Academy's Early English Church Music Committee, scholar in residence for the internationally renowned ensemble The Binchois Consort, and has previously served both as Membership Officer and Treasurer of the Society for Renaissance Studies.

Etienne Corteel is Director of Education & Scientific Outreach, global, L-Acoustics. Governing the education and scientific outreach strategy, Etienne and his team are the interface between L-Acoustics and the scientific and education communities. Their mission is to develop an education programme tailor-made for the professional sound industry through research, technology, and analysis of field practices.

Etienne's extensive industry experience includes posts spanning from research scientist to chief technology officer. He has authored 70 scientific publications and 12 international patents in various audio fields, such as 3D audio and signal processing.

A native of Vernon, France, Etienne holds a masters and PhD in signal processing and acoustics from Paris's Sorbonne University, and an engineering degree from IMT Atlantique.

Paul Geluso's work focuses on the theoretical, practical, and artistic aspects of sound recording and music technology. As a recording engineer, producer, and musician, he has been credited on more than 200 commercially available titles, including Grammy and Latin Grammy-nominated recordings and award-winning films. Prior to joining the full-time faculty at New York University, he taught at the Peabody Institute of Johns Hopkins University, Bard College, and the State University of New York at Oneonta. He is an active member of the AES and has served as a voting member of the Recording Academy, Producers & Engineers Wing. His current research at New York University focuses on new ways to capture, mix, and process immersive audio for playback on binaural and multi-channel sound systems. He recently co-edited *Immersive Sound: The Art and Science of Binaural and Multi-Channel Audio* published by Focal Press-Routledge.

Gavin Kearney graduated from Dublin Institute of Technology in 2002 with an honours degree in electronic engineering, and has since obtained MSc and PhD degrees in audio signal processing from Trinity College Dublin. He joined the University of York as Lecturer in Sound Design in January 2011 and was appointed Associate Professor of Audio and Music Technology in 2016. He has written over 70 research articles and patents on different aspects of immersive and interactive audio, including real-time audio signal processing, Ambisonics, VR and augmented reality (AR), and recording and audio post-production technique development. He has undertaken innovative projects in collaboration with Mercedes-Benz Grand Prix, BBC, Dolby, Huawei, Abbey Road, and Google, among others. He is currently Vice Chair of the AES Audio for Games Technical Committee, as well as an active musician, sound engineer and producer of immersive audio experiences.

Sungyoung Kim received a BS degree from Sogang University, Korea, in 1996, and a Master of Music and PhD degree from McGill University, Canada, in 2006 and 2009, respectively. His professional work experience includes working as a recording/balance engineer at the Korea Broadcasting System (KBS) in South Korea and a research associate at Yamaha Corporation in Japan. He works at the Electrical, Computer, and Telecommunication Engineering Department of Rochester Institute of Technology (RIT), first as an Assistant Professor from 2012 and then as Associate Professor from 2018. His research interests are the rendering and perceptual evaluation of spatial audio, digital preservation of aural heritage, and auditory training for hearing rehabilitation. In 2019, Sungyoung received the Japan Society for the Promotion of Science (JSPS) International Fellowships for Research in Japan, 2019. He holds two international patents (granted), US8320590B2 (2012) and US9351074B2 (2016), on spatial sound rendering and wrote a book chapter, "Height Channels," in *Immersive Sound: The Art and Science of Binaural and Multi-Channel Audio*.

Guillaume Le Nost is Executive Director, Creative Technologies, L-Acoustics. With advanced degrees in both acoustics, and electrical and electronics engineering, he has led research projects for some of France's largest corporations, coupled with a start-up entrepreneurial background in audio technology applied to gaming and VR. Guillaume brings a rich cross-disciplinary and future-focused mindset to the L-Acoustics Creative team, using a fresh perspective to resolve technical challenges that would confound most.

Guillaume's keen interest in music extends to the electric bass, and he has played and toured as a professional musician.

Hyunkook Lee is Reader (i.e. Associate Professor) in Music Technology and Director of the Applied Psychoacoustics Laboratory (APL) at the University of Huddersfield, UK. His recent research has advanced understanding about the psychoacoustics of vertical stereophonic localization and spatial impression in 3D sound recording and reproduction. This has provided theoretical bases for the development of several 3D microphone arrays, including Schoeps ORTF-3D. His current research topics include the six-degrees-of-freedom (6DOF) perception and rendering of virtual acoustics, and the creation and evaluation of a multimodal immersive experience for extended reality (XR) applications. From 2006 to 2010, he was a senior research engineer in audio R&D with LG Electronics, South Korea, where he participated in the standardizations of MPEG audio codecs and developed spatial audio algorithms for mobile devices. He received a bachelor's degree in music and sound recording (Tonmeister) from the University of Surrey, Guildford, UK, in 2002 and a PhD in spatial audio psychoacoustics from the Institute of Sound Recording (IoSR) at the University of Surrey in 2006. He is a Fellow of the Audio Engineering Society (AES) and Vice Chair of the AES High Resolution Audio Technical Committee.

Gareth Llewellyn began his sound career in film audio post before moving to Galaxy Studios, Belgium, where he was heavily involved with the pioneering 3D audio work that led to the Auro-3D® sound format. Gareth helped mix the world's first commercial 3D audio feature (*Red Tails*, Lucasfilm) with a team from Skywalker Sound, and went on to remix dozens of Hollywood features – pioneering new creative and technical approaches to deliver audience immersion through channel and object-based sound systems. Gareth went on to co-found production house Mixed Immersion, which developed pioneering location-based XR experiences for international brands, immersive theatre productions, and artists. Gareth is currently working on MR audio technology in his new start-up, MagicBeans and is producing 6DOF audio experiences for artists, musicians, and brands.

Jo Lord is an engineer and educator specializing in live sound and spatial audio. She has 13 years' experience as an industry practitioner across the live music, touring, and event sectors. Jo has been involved with spatial music production for more than half of her career and was a part of the team responsible for the first 3D audio stage at the Glastonbury Festival in 2019. At present, Jo is researching 3D audio technologies for live music and record production, and lectures in audio engineering at various institutions across England, including BIMM London, London College of Music, Academy of Contemporary Music, and Anglia Ruskin University.

Kenneth B. McAlpine is an award-winning composer, academic, and broadcaster who works at the Melbourne Conservatorium of Music, The University of Melbourne as an Enterprise Fellow in Interactive Composition.

After completing a piano diploma with the London College of Music and a bachelor's degree in mathematics, he completed a PhD in algorithmic music composition at the University of Glasgow, bringing together his shared love of music, maths, and technology, and he has continued to work at that interesting point at which different disciplines collide: he has developed interactive soundtracks for live theatre, film, and video games; produced bagpipe music for the Beijing Olympics; created a music streaming app for newborn babies and young children for the Scottish Government and the Royal Scottish National Orchestra; and built a unique digital harpsichord exhibit for the National Trust in London.

Outside of work, Kenny can generally be found rolling around hills on a mountain bike with a GoPro strapped to his handlebars, huffing his way round a marathon course with a GoPro strapped to his chest, or baking bread. Without the GoPro.

Justin Paterson is Professor of Music Production at London College of Music in the University of West London, where he leads the MA Advanced Music Technology course. He has numerous research publications ranging through journal articles, conference presentations, and book chapters, and is the author of the *Drum Programming Handbook*. Justin is Co-chair of the Innovation in Music conference series and co-editor of its books. He is also an active music producer. His current research interests are 3D audio, interactive music, and haptic feedback. Together with Professor Rob Toulson and working with Warner Music Group to release prominent artists from their roster, he developed the variPlay interactive-music format, funded by the UK Arts and Humanities Research Council. His current research – funded by Innovate UK through its Audience of the Future programme with consortium partners including Generic Robotics and the Science Museum Group – is around the development of a novel music- production interface in mixed reality (MR) that utilizes haptic feedback.

Chris Pike is a Lead Research and Development Engineer at the BBC, working in the Immersive and Interactive Content team. He has led the BBC's work on 3D audio in recent years. This has involved 3D audio production on major programmes, such as *Doctor Who*, *Planet Earth*, and the BBC Proms, as well as in award-winning VR experiences, such as *The Turning Forest and Nothing To Be Written*. It has also involved major research collaborations, such as the S3A and Orpheus projects. In 2019, Chris completed his PhD with the Audio Lab at the University of York on binaural technology and its perceptual evaluation.

Diego Quiroz is a recording engineer, musician, and teacher of music production and audio engineering, and immersive audio. He has recorded classical music records in Montreal, Canada with classical orchestras, jazz orchestras, and numerous ensembles of baroque, contemporary, and modern music. He graduated from the Master of Sound Recording at

McGill University in Montreal, Canada, and from the Musical Synthesis major, Magna Cum Laude at Berklee College of Music in Boston, USA. He is currently a PhD candidate in sound recording at McGill University with highly renowned professors such as George Massenburg, Richard King, Wieslaw Woszczyk, and Martha de Francisco.

Diego has been involved in numerous AES conference papers since becoming a PhD candidate on subjects around immersive audio perceptual evaluation, height-channel format perceptual evaluation, gestural control for audio, audio perception in VR/AR and high-resolution audio. Diego's thesis proposal encompasses input devices and gestural control for 3D audio production. His other interests include Ambisonics, binaural audio, and VR/AR in audio.

Pedro Rebelo is a composer, sound artist, and performer. In 2002, he was awarded a PhD by the University of Edinburgh, where he conducted research in both music and architecture.

Pedro has recently led participatory projects involving communities in Belfast, favelas in Maré, Rio de Janeiro, travelling communities in Portugal, and a slum town in Mozambique. This work has resulted in sound art exhibitions at venues such as the Metropolitan Arts Centre, Belfast, Centro Cultural Português Maputo, Espaço Ecco in Brasilia and Parque Lage and Museu da Maré in Rio, Museu Nacional Grão Vasco, and MAC Nitéroi. His music has been presented in venues such as the Melbourne Recital Hall, National Concert Hall Dublin, Queen Elizabeth Hall, Ars Electronica, and Casa da Música, and at events such as Weimarer Frühjahrstage für zeitgenössische Musik, Wien Modern Festival, Cynetart, and Música Viva. His work as a pianist and improvisor has been released by Creative Source Recordings and he has collaborated with musicians such as Chris Brown, Mark Applebaum, Carlos Zingaro, Evan Parker, and Pauline Oliveros, as well as artists such as Suzanne Lacy.

His writings reflect his approach to design and creative practice in a wider understanding of contemporary culture and emerging technologies. Pedro has been Visiting Professor at Stanford University (2007), Senior Visiting Professor at UFRJ, Brazil (2014), and Collaborating Researcher at INEM-md Universidade Nova, Lisboa (2016). He has been Music Chair for international conferences such as ICMC 2008, SMC 2009, and ISMIR 2012, and has been an invited keynote speaker at ANPPOM 2017, ISEA 2017, CCMMR 2016, and EMS 2013. At Queen's University Belfast, he has held posts as Director of Education, Director of Research, and Head of School. In 2012, he was appointed Professor of Sonic Arts at Queen's and awarded the Northern Bank's Building Tomorrow's Belfast prize. He has recently been awarded two major grants from the Arts and Humanities Research Council, including the interdisciplinary project *Sounding Conflict*, investigating relationships between sound, music, and conflict situations. His ongoing research interests include immersive sound design and augmented listening experiences.

Frédéric Roskam is Head of Immersive Audio, L-Acoustics. Together with his Immersive Audio team, Frédéric designs and coordinates the foray of L-Acoustics in multichannel immersive technologies, transforming the way sound is experienced across the globe from stadium concerts down to intimate residential experiences.

Frédéric's extensive experience ranges from real-time music programming, research in consumer electronics, and currently live sound. He holds five international patents in consumer audio and spatial audio and has since contributed to numerous award-winning consumer and professional products.

Born in Paris and an ardent violinist, Frédéric holds a master's degree in electronics engineering from Phelma, and a master's degree in signal processing and acoustics from Paris's Sorbonne University.

Rod Selfridge is a postdoctoral researcher in Sound and Music Computing at KTH Royal Institute of Technology, Stockholm, focusing on sound design, sonic interactions, with an interest in sound-effect synthesis by physical modelling, and musical applications within VR, AR, and MR. During his PhD research at Queen Mary University of London, he was awarded two best paper awards and a silver design award for his work on synthesising aeroacoustic sound effects. He worked as part of a team at FXive developing procedural-audio sound effects for a browser-based library of sound effects. At the University of Edinburgh, he was part of a team that developed a prototype VR experience allowing users to experience historical musical performances.

Kaushik Sunder currently works as the Director of Engineering and leads the Audio and Acoustics Research Team at Embody. Kaushik has spent a great deal of his research career in the field of 3D audio and psychoacoustics. Over the last few years, his research has focussed on understanding the importance of personalized HRTFs, particularly for headphone playback of spatial audio. Prior to working at Embody, Kaushik served as a Research Scientist at Ossic and Postdoctoral Researcher at the Sound Recording Department, Schulich School of Music at McGill University. He is also a visiting research scholar at the Human Factors department at NASA Ames Research Center. Kaushik received his PhD from the Digital Signal Processing Laboratory, Nanyang Technological University, Singapore. He has regularly authored articles that have appeared in the Journal of Acoustical Society of America, IEEE Signal Processing Magazine, Journal of AES, and AES International Conventions.

Preface

Justin Paterson and Hyunkook Lee

Welcome to *3D Audio*, a topic-specific edition in the *Perspectives on Music Production* (POMP) series. 3D audio can loosely be defined as a form of surround listening that extends beyond the horizontal sound field – including an impression of height. As an academic field, 3D audio is diverse. Many people will relate to it in the context of Dolby Atmos® in a contemporary cinema, yet the concept has been exploited in performance since the Renaissance, and it evolved through the latter 20th century as an augmentation of composition, often via the acousmatic school. A seminal moment was the development of Ambisonics by Michael Gerzon in the 1970s, and this format has recently experienced a great surge in research interest and applications, often through its deployment in the zeitgeist that is the (re-)emergence of virtual reality (VR).

There is a huge volume of scientific research – particularly in recent years, often stimulated by recent advances in the technological ecosystem that offer potentially disruptive new modes of listening to the consumer. As mentioned, cinema and VR are in evolution, but binaural playback is poised to pervade mainstream headphone listening via technologies such as Sony 360 Reality Audio, and Apple AirPods Pro earbuds greatly enhance the binaural experience via head tracking. Apple is also expected to shortly enter the extended reality (XR) arena.

The scientific research covers many fields and subfields. Psychoacoustics and perception, Ambisonics to wave field synthesis, binaural and speaker playback, panning algorithms, capture techniques, formats, and much more. In tandem with this, there is pertinent creative-practice research and writing. This ranges from concepts such as engulfment and diffusion in the electroacoustic and acousmatic worlds, from sonic interaction with architecture to music implementation for new media, and from composition to music production. Of course, the link between these two worlds of science and art is technology. This technology is typically born out of the scientific research, yet comes to shape both new music and consumer experiences of now and the future through its deployment in creative practice. As such, it is an important actor when considering the greater arena of 3D audio.

Despite all this, at the time of writing, there are few books that position themselves in 3D audio. Roginska and Geluso's *Immersive Sound* is one that comes to mind, and there are other more specialist texts, for instance Zotter and Mathias' *Ambisonics*. However, both of these excellent books align themselves squarely with the science, and it was clear that the opportunity existed to have a book with a broader perspective, one that drew from science, creative practice, and attendant technologies, presenting not just research, but a perspective of research; this is that book.

The field of 3D audio is far too expansive to be wholly encompassed in a single volume, so the editorial notion of curation comes into play. A large number of potential contributors submitted abstracts for this book. These came from all around the world, from many eminent and established authors – both industrial practitioners and academics – and they identified many exciting topics. The abstracts needed to be carefully scrutinized since many had degrees of overlap and treated their subjects at a variety of academic levels, and it was important that collectively, they should fulfil the predefined scope that this book tacitly aspired to. It became apparent that in order to select the most insightful, necessary, and interesting chapters, it was going to require the creation of something of a smorgasbord. As such, this book makes no effort to wholly encompass 3D audio, but rather offers a range of perspectives over a range of disciplines that attempts to first draw in the interested reader, and then very possibly take them further into specific aspects of 3D audio that they might not yet have fully investigated. Most, although not all, chapters are accessible to those broadly familiar with the field of audio and music technologies. Some chapters provide definitive research – conducted by the authors – and others offer authoritative views of the states of the art. Others embed detailed literature reviews, and others still offer speculative insights into the emergent 3D audio world. The chapters are clustered into two broad thematic areas: *current technologies and research* and then *composition and production techniques*. They can be read in any order.

CURRENT TECHNOLOGIES AND RESEARCH

The treatment of technologies opens with the BBC R&D's Chris Pike charting how broadcasters have employed binaural techniques, and looking at issues around capture and rendering with a historical backdrop. He then develops these areas towards both post and live production, and discusses various pioneering broadcasts and experiments, from legacy to the present day and into the near future. Real-world challenges, such as duplication of production effort, are discussed, and this leads into the potential solution of object-based audio. Its operation and opportunities are explained within the context of "next-generation audio," which leads into a vision of how broadcast audio will be personalized in the future, with 3D audio aspects being just one element of new modes of listening.

From broadcast, we segue into another multi-listener mode; that of live sound. Etienne Corteel, Guillaume Le Nost, and Frédéric Roskam have provided live sound solutions for artists as diverse as Aerosmith, Ennio Morricone, and the Los Angeles Philharmonic. In this chapter, they discuss 3D audio for live sound. They link panning algorithms with speaker placement and stage setup in order to provide 3D sound coverage for the maximum number of audience members in a range of auditoriums while on tour. Covering localization blur and spectral masking, they move on to loudspeaker system design and evaluation, and discuss how their own unique software provides new levels of control for engineers and acts working in emergent 3D audio live presentation.

From multi-listener environments to an egocentric one in which Gareth Llewellyn and Justin Paterson look into a world that has yet to coalesce, that of audio in six degrees of freedom, specifically applied to VR, augmented reality (AR), and mixed reality (MR). In this mode, listeners not only experience surround and elevation, but are at liberty to navigate through the real world and expect whatever audio is superimposed on their ears to respond to it. At the time of writing, there are very few real-world approaches to such audio delivery, and the authors discuss many of the reasons, challenges and trajectories towards this exciting sound world of the future. They do this from both a perspective of current research, and also consideration regarding the practicalities of capture and delivery. Lastly, they look at how existing production workflows need to be rethought in this arena.

While the previous chapter was focused on one particular mode of presenting sound, Diego Quiroz provides an in-depth perspective of gestural control for 3D audio for more general applications. Conventional panning devices have generally been 2D affairs and are fundamentally awkward to apply to 3D space, yet increasingly, the proliferation of mixing 3D content is demanding another axis of control. Quiroz provides a narrative that is densely woven with literature to form a human-computer-interaction-based taxonomy of 3D input devices and their operation. He goes on to discuss ergonomic issues associated with mid-air control and haptic feedback when working with three degrees of freedom.

The research section opens with Hyunkook Lee discussing some of the important psychoacoustic principles of vertical stereophonic perception and their practical implications on the recording and reproduction of 3D audio. Recently developed 3D audio loudspeaker formats employ so-called "height" channels, utilizing loudspeakers in elevated positions, as well as at ear level. However, the perceptual mechanisms of horizontal and vertical stereophonies are fundamentally different. This gives rise to the need for new methods to capture and render a stereophonic sound image for the added height dimension. Lee first describes how humans localize vertically orientated real and virtual sound images from his recent research findings, as well as some of the classic literature. He then explains principles related to the perception and rendering of vertical spatial impression and how they can be exploited in 3D recording and upmixing. This chapter also presents a considerable volume of Lee's original research in an accessible narrative.

The theme of loudspeaker listening is then continued, but from a different perspective as Cal Armstrong and Gavin Kearney provide a comprehensive tutorial on the theories and practice of Ambisonics. They point out that there are misunderstandings about the definition of Ambisonics, even though, today, it is a technology that is widely used for immersive audio recording and reproduction. This chapter clarifies any such misconceptions by providing technical principles, such as spherical-harmonic functions, Ambisonic encoding and decoding principles with various types of decoder matrix weighting, and lastly, binaural rendering of Ambisonic signals, which sets up a forthcoming specific treatment.

Indeed, the preceding two chapters were largely about presentation over loudspeakers, but Kaushik Sunder now provides a thorough overview on headphone-based binaural perception and rendering for 3D audio applications. He looks into the importance of personalized head-related transfer functions (HRTFs) and introduces the state-of-the art techniques of HRTF personalization in great detail. Sunder then describes important cues for distance perception and techniques to model these distance-dependent HRTFs. In augmented reality applications, one of the biggest challenges is to match the reverb of the virtual world with the real world. In order to do this, the early/mid reflections of the real world should be estimated in real time and resynthesized in the virtual world. Sunder also explores various research and techniques in binaural reverberation estimation for AR and MR applications.

Moving away from mechanisms of delivery, Sungyoung Kim discusses how contextual factors can influence the perception of immersive audio. He first provides evidence of contextual biases found in previous subjective studies. Issues such as how the preference rating of multichannel recording techniques can be dependent upon test material, and he then raises a question of whether any contextual factors could modulate subjective judgement on auditory immersion in XR applications. Based upon his own experimental findings, he provides answers for this question in terms of two specific types of contextual factors: socio-cultural background and listening environment. He discusses how the judgement of perceived sound quality of an identical 22.2-channel sound recording can vary, first between two ethnic groups, and subsequently, also in listening rooms with different acoustic characteristics. This concludes the first section of the book.

COMPOSITION AND PRODUCTION TECHNIQUES

The notion of perception is continued, but this second section takes a very different perspective, and now considers issues associated with creativity in 3D audio, starting with music composition. Eminent composer Natasha Barrett explores the opportunities available to composers when working in the spatial domain, with particular reference to Ambisonics. Starting from a context of listening and history, Barrett offers a discerning review of the spatial techniques available to acousmatic composition, ranging from treatment of distance and proximity, through use of recorded sound fields and the composition of images in space. She then covers the notion of either being inside a sound or looking onto a sound scene, and this leads into the concept of composing spatial motion. Although a chapter about composition, this text consciously disregards conventional music theory and instead offers an insightful overview of how any composer working in the spatial domain might fully exploit the medium.

In fact, this notion is founded upon a long-established tradition, and next, Pedro Rebelo takes a conceptual stance, deliberately avoiding a discussion of 3D technologies, but instead looking back at how composers and sound artists have utilized spatialization over the past 500 years. These range from cori spezzati – multiple choirs separated in space, sometimes by height – through Stockhausen and Xenakis, to contemporary examples, including his own work; all of these with an emphasis on "why?" These draw from both 2D and 3D examples, but by investigating concepts and commonalities in a number of case studies, he goes on to form a four-point framework for strategies for working with sound in space, strategies that the contemporary 3D sound designer can employ with the latest technologies – in their own context – whatever that might be.

Continuing the link to architectural space but using cutting-edge technology to form a bridge between "then and now," Kenny McAlpine, James Cook and Rod Selfridge explore how 3D audio might best support historically informed performance. They do this via a case study and model the acoustic properties of two performance spaces in Scotland, not just as they are now, but as they would have been when originally put to use, in one case 600 years ago prior to its current state of ruin. By first combining and comparing techniques such as binaural room impulse response (BRIR) capture, architectural scanning, and ray-trace-based modelling, the authors then record authentic repertoire in an anechoic chamber before superimposing the derived acoustic profiles upon the music in VR. They also explain the relevance of the historical perspective, and define aspects of workflow that might be drawn from this case study and redeployed for related purposes.

Of course, a great deal of music production starts with recording and the final outcomes are dependent upon it. Next, Paul Geluso discusses microphone techniques to capture sound and music performances acoustically in 3D. He classifies these methods into the following three types of systems: binaural, coincident (including Ambisonic and beamforming technologies), and spaced microphone techniques. Starting with an Ambisonic primer for recordists, he builds from theoretical to practical aspects of each of the microphone techniques that are described. Geluso also describes how 3D sound captured using the various methods can be delivered with regard to object-based, scene-based, static binaural, and XR systems.

Jo Lord investigates the musicology and musicality of periphonic record production from an action-based phenomenological viewpoint. She describes a study that she undertook to collate and assess a sound-staging technique that involved aligning musical concepts to sonic schema for a periphonic-binaural 3D audio arrangement. Lord specifically discusses metaphor as a vehicle for enhancing both the immersive musical experience and also creative production practice. Through case studies, this chapter explores ways in which periphony can enhance musical staging beyond that which most current industry record production practice affords, and introduces concepts such as omnimonophonic and polyperiphonic vocal staging. From the sound stage to the film stage…

Kevin Bolen has worked as a re-recording mixer on some of Hollywood's biggest movies. To conclude this book, Bolen interviews a number of leading professionals in the field, and harvests their perspectives on mixing movie sound in 3D. The activities of the re-recording mixer are often considered as something of a dark art, and here, unique insights are presented that illustrate cutting-edge best practice, challenges, and considerations, collectively drawing from decades of experience in Hollywood. Topics discussed include auditoria spaces – their setup and the sweet spot, dialogue panning, placement of music, height/width/depth, and timbral shift. Bolen weaves a range of expert opinions into an engaging narrative that is sure to change your perspective the next time you visit the cinema.

So, it is hoped that this book will provide a useful perspective to a broad range of readers, a range that draws from students to academics, from industrial practitioners to manufacturers/developers, and from producers to composers. For some, certain topics in the scope might well start in the familiar, but the expertise of the contributors is substantial, and most readers will soon benefit from this expertise. Further, the interested reader is strongly encouraged to also explore the reference lists – even for topics that are more familiar to them, since these will yield further fascinating lines of exploration. However, it is likely that the smorgasbord will also lead even the seasoned practitioner into new areas pertinent to their own, perhaps developing a deeper insight into how they are interconnected and developing new interests accordingly.

For others, perhaps only just entering the world of 3D audio, the range of topics should provide some interesting points of entry, and moving through the chapters will provide a solid road map for the 3D audio landscape, with a detailed treatment of many salient aspects. Should one chapter take too much to assimilate, move on and read another (one or two) first, and then come back, and you might be surprised how your perception has grown.

This book contributes to the spirit of the POMP series – to provide all interested parties with a greater perspective of the state of the art in music and audio production. 3D audio is here to stay. Of course, for most humans, it always has been!

This is an ongoing book series, and future additions, together with current calls for chapters, can be viewed at https://research.mottosound. co.uk/POMP.

1

3D audio in broadcasting

Chris Pike

1.1 INTRODUCTION

At its core, broadcasting is about storytelling and the communication of ideas, whether to inform or entertain. The aim of 3D audio is to create a more realistic and immersive auditory impression for the listener. For broadcasters, it is hoped that 3D audio can help to give more compelling and enjoyable audience experiences, providing new creative possibilities for programme makers.

Broadcasting is traditionally a one-to-many radio communication system. Millions of people can receive the same information at the same time, via radio or television (TV), giving it an important role in society. In recent years though, there have been huge changes in the way that people access audio-visual media. Content is efficiently distributed over the internet and mobile data networks using advanced coding technologies, and consumed on internet-connected TVs and mobile devices with significant processing power, in many different environments beyond just the home. This is changing the role of media in peoples' lives and the services that broadcasters should offer.

In 1996, engineers from BBC R&D published a report on production experiments in five-channel (5.0) surround sound [1]. This format was first introduced on free-to-view broadcasting in the UK in 2006, being already established for home cinema uses. Since then, there has been little change in the broadcast audio formats offered to listeners.

3D audio is becoming established in other entertainment areas, such as cinema and gaming. Broadcast audio should keep in step with audience expectations. There are trends towards greater fidelity in TV picture quality with ultra-high definition (UHD). 3D audio is widely seen as the companion to UHD video in new higher-fidelity broadcast services, and broadcasters around the world have been investigating 3D audio technology and content production for years. However, we are at an exciting moment where 3D audio looks set to become a regular part of mainstream services. This chapter reviews the developments that broadcasters have made towards offering 3D audio to their audiences.

1.2 BINAURAL AUDIO IN BROADCASTING

Binaural audio gives headphone listeners a realistic impression of 3D space by precisely recreating how sounds would reach our ears in the real world. The term *binaural* is used because the acoustic signals observed at both ears are directly represented and reproduced, invoking natural spatial hearing processes. Therefore, complex 3D scenes can be represented with a two-channel audio signal. Headphones are most often used to reproduce binaural sound because it is easier to control the sound pressure at the ears independently, although loudspeaker reproduction techniques do exist.

Headphone usage has grown considerably in recent years [2]. With headphones, normal stereo often gives the impression that the sound sources are inside the listener's head [3]. This has motivated broadcasters to improve the headphone listening experience. Binaural audio is appealing to broadcasters because it offers more immersive 3D audio to a large number of listeners without them needing new equipment. Existing stereo infrastructure can also be used for distribution.

1.2.1 Binaural recording

Binaural audio production originates from the binaural recording technique, where real acoustic scenes are captured with microphones placed in the ears of a human listener or an artificial head. The development of binaural microphones began in the early 1930s [4], at the same time as stereo recording techniques. While broadcasters made stereo transmissions as early as 1962 [5], it was not until 1973 that the first binaural broadcast programme emerged. *Demolition* was a binaural audio play produced by the Radio in the American Sector (RIAS) station and showcased at the International Broadcasting Exhibition (IFA) in Berlin that year. The newly released Neumann KU80 was used in production, the first commercial microphone that was purposely designed for binaural recording. The play was broadcast by several German stations and was very well received by both the public and press [6].

In the UK, the first binaural radio programme was broadcast in 1977. *Oil Rig* was a BBC documentary described as "a portrait in sound" of life on a rig in the North Sea. In 1978, the BBC broadcast *The Revenge*, a binaural audio play recorded without dialogue, from a first-person perspective. It was written and performed by Andrew Sachs of *Fawlty Towers* fame and it is still occasionally retransmitted. The earliest binaural productions at the BBC [7] did not use an artificial head microphone, but instead used a matched pair of omnidirectional microphones spaced at approximately ear width with a Perspex® disk as a baffle, as shown in Figure 1.1. In studio tests, this was found to give a similar effect to the KU80 microphone, but it was favoured for reasons of practicality and cost.

Hundreds of binaural radio programmes were produced and broadcast across Europe during the 1970s and early 1980s. However, popularity waned due to incompatible production conventions and listening habits, as well as challenges with spatial and timbral sound quality [6].

Figure 1.1 Early binaural recording at the BBC using a Perspex® disc and a spaced pair of omnidirectional microphones. (a) Lloyd Silverthorne recording for *Oil Rig*, (b) Andrew Sachs during recording of *The Revenge*. Photographs courtesy of Lloyd Silverthorne.

Binaural audio has had a revival in broadcasting in the last decade. Today, there is a range of commercially available binaural microphones, both artificial head and in-ear types. Binaural recordings are still used, but binaural rendering (see Section 1.2.2) is also an important tool now. With binaural microphones, natural sound scenes can be recorded accurately, including environmental acoustics and dynamic source movements. The complexity of the scene has no impact on the complexity of the recording or reproduction process, which is not the case for binaural rendering. However, one of the main limitations of binaural recording is that it can only capture naturally occurring scenes. In broadcasting, it is common to either create sound scenes that do not exist in real life – via sound design and editing, or by making aesthetic changes to real sound scenes – through processing and mixing multiple microphone recordings.

1.2.2 Binaural rendering

It was in the late 1980s that digital signal processing techniques were first applied to achieve accurate binaural rendering – also termed binaural synthesis, which is the processing of an audio signal to give a realistic auditory spatial impression, simulating the binaural signals at the ears [8]. The source signal is processed with filters describing the acoustic transfer function from the desired source position to the ears, commonly called a head-related transfer function (HRTF). A virtual scene can be constructed by simulating multiple sound sources at various directions and distances, as well as environmental acoustic effects. Binaural rendering allows production of auditory scenes that are not based on real acoustic scenes, which is useful for creative applications. Binaural recordings can also be augmented with binaural rendering of additional sound sources.

Binaural rendering can provide listeners with a convincing spatial impression, but its application in broadcast imposes constraints. In the near-term approach, using existing stereo infrastructure, broadcasters must distribute pre-rendered binaural signals. Therefore, rendering cannot be personalized to the listener's own HRTFs, the headphone reproduction cannot be properly equalized, and there can be no adaptation to listener head movements. Despite these limitations, broadcasters believe binaural rendering can give advantages over stereo for its headphone listeners. There is a large body of research into binaural technology and its perceptual evaluation, as reviewed in [9]. In Chapter 7, Sunder offers greater insight into the engineering aspects of binaural audio.

1.2.3 Headphone surround sound

Early applications of binaural rendering in broadcasting focused on repurposing 5.1 surround sound material. This can be called headphone surround sound (HSS) and it works by simulating virtual loudspeakers around the listener. It could potentially improve the headphone listening experience without duplicated production effort. Processing may be applied to create a two-channel binaural signal before distribution, or indeed locally – on the listener's receiver device.

Various studies have evaluated HSS systems and found that HSS offers, at best, only small improvements in the listening experience over a stereo down-mix [10]–[13], with many systems often making it significantly worse. However, there is often variation in performance according to the audio material, as well as varied preferences among listeners. This suggests offering a choice between HSS and stereo versions.

In 2014, the BBC ran an online study to evaluate the HSS of a 5.1 mix of an audio play, *Under Milk Wood* by Dylan Thomas [14]. This programme was distinctive in that the surround channels were used more heavily for foreground scene elements than is common in broadcast production. The 60 survey respondents showed overwhelming preference for HSS over a stereo down-mix. Two HSS versions were used: one used a virtual reproduction room, while the other used anechoic rendering. Listeners were divided in their preferences between the two. Listeners also spent much more time listening to the HSS versions than the stereo down-mix. In this case, the evaluation was not blind, the different versions were labelled in the browser-based playback device as shown in Figure 1.2, but this is representative of the target use case. Of course, these results might be biased by the fact that the audience consciously self-selected the version, and there could also have been a novelty factor that influenced preference.

Radio France has frequently used the HSS approach. In 2013, the broadcaster launched a website called *nouvOson*, with 5.1 and binaural versions of many programmes (see Figure 1.3). A survey of 121 users found that a large majority of listeners (over 85%) perceived and liked the binaural effect [15]. In its first few years, most binaural material on *nouvOson* was created using the HSS approach from 5.1 surround mixes. The broadcaster then began mixing to an equiangular eight-channel horizontal-surround layout, which gave a more effective binaural signal, while also being easy to down-mix to 5.1 for loudspeaker listening. In 2017, the *nouvOson* website space was rebranded as *HyperRadio* [16], and by the start of 2020, it had approximately 400 binaural productions available.

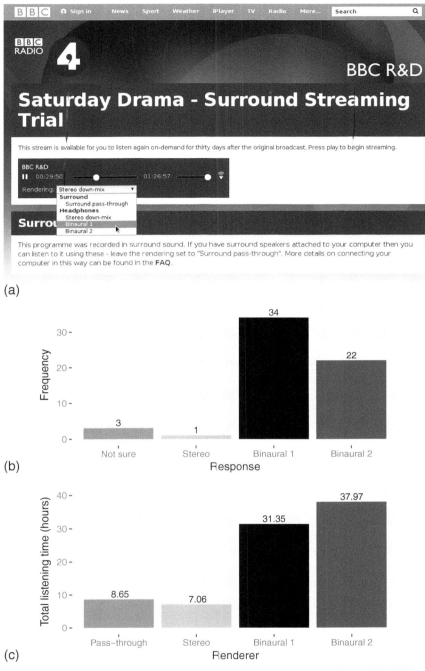

Figure 1.2 Web-based test for *Under Milk Wood* HSS streaming trial. (a) Web player interface, (b) listener preference choices, (c) total listening time.

Figure 1.3 Radio France's *nouvOson* web player, featuring a headphone surround sound option.

Despite disappointing results in blind laboratory studies, it seems clear that HSS is popular for listeners who seek out where audio material has been curated for binaural presentation, often on the websites of broadcasters.

1.2.4 3D binaural production

HSS applications are limited by the input format of 5.1 surround. The loudspeaker layout is horizontal only and loudspeaker placement is sparse outside of the frontal region, so it is not possible to create a convincing impression of a 3D scene. Virtual loudspeaker rendering can instead be used with a 3D layout, but this *channel-based* approach still imposes limits upon the resolution of the sound scene.

For several years, the BBC and Radio France have both utilized an *object-based* rendering approach. Each sound source, or object, is processed independently with its own binaural filter rather than using a virtual loudspeaker layout as an intermediate format. Sources can be rendered to any 3D position, and the most appropriate rendering parameters, including the virtual acoustic environment, can be chosen for each scene and sound source. This approach applies rendering prior to distribution, but is otherwise similar to that used in computer games and virtual reality systems – although these render locally at the moment of listening.

1.2.5 Post-production

Broadcasters have experimented with a wide range of tools and workflows for producing binaural material. Radio France currently uses a tool called

Spat Revolution [17] for object-based binaural rendering. It also features a 3D room-modelling algorithm. The tool is integrated with a digital audio workstation (DAW).

At the BBC, a custom post-production system was developed to allow a range of rendering options for sound objects, including stereo amplitude panning, 3D binaural rendering using a dense array of HRTFs, virtual loud-speaker rendering via a 3D layout with measured binaural room impulse responses (BRIRs), and auxiliary parametric 3D reverb. The producer can select the desired rendering option for each object at any point in time, and often a mix of different techniques is used. A third-party plugin, the IOSONO Spatial Audio Workstation, is used to author the 3D scene in the DAW, which sends open sound control messages to control the rendering software. More detail on this system can be found in [9].

The first production created using this approach occurred in 2014, it was an episode of a radio drama, set during the First World War [18]. In an online listening test with 59 participants, blind comparisons were made between stereo and binaural versions of several excerpts [9]. There was a significant preference for the binaural version for five of the seven excerpts. For the other two, there were no significant preferences found and respondents more often said that they heard no difference for these. Since this work, there have been many different drama productions created using this system, for example, [19], [20].

Other broadcasters have also produced object-based binaural drama productions. In Germany, for example,[1] WDR collaborated with Cologne University of Applied Sciences to use SoundScape Renderer software [21] and BR worked with the IRT using their custom object-based binaural rendering system [22].

In recent years, many different tools that offer 3D binaural panning in post-production have become available. It is not always necessary for broadcasters to develop custom approaches, and at the BBC, a range of different tools are now used.

1.2.6 Live production

There is interest from broadcasters in creating 3D binaural mixes of live music performances, offering a greater sense of presence and immersion. However, the live production scenario imposes increased demands on the efficiency and robustness of tools. Musical source material is also particu-larly sensitive when it comes to the timbral quality of the processing used.

Binaural mixes have been created live at the BBC Proms concert series since 2017. The stereo mix takes priority since it is used for the main broad-cast, so a workflow was developed to allow sound engineers to produce an object-based binaural version efficiently and simultaneously [23]. Each microphone signal used in the stereo mix was fed from the mixing con-sole to the binaural rendering system after gain, delay, equalisation, and dynamics processing had been applied. This ensured that the dynamic level changes made by the mix engineer during a performance were maintained in the binaural mix. Custom rendering software was designed for this scenario (Figure 1.4a) [24]. It allows for spatialization, additional delay and gain to be applied to each object, or to groups, and presents a simple user interface on a touchscreen.

In addition to the microphones used in the stereo production, a microphone array, the ORTF-3D [25], is used to capture an impression of 3D ambience in the mix (Figure 1.4b). Often, the artificial reverberation used in the stereo version is also mixed in to obtain a similar character to the stereo mix.

Microphone arrays like the ORTF-3D offer an alternative to artificial head recording. They can be used to capture a 3D sound scene and then be processed to generate a binaural signal, while also being suitable for playback on 3D loudspeaker arrays. The character and quality of 3D microphone arrays is an important area of investigation for broadcast production

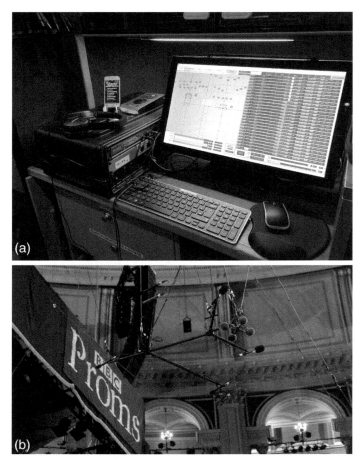

Figure 1.4 Live binaural music production at the BBC Proms. (a) Binaural rendering tool in the outside broadcast vehicle, (b) ORTF-3D microphone array in the Royal Albert Hall.

Figure 1.5 A microphone array test with the BBC Philharmonic Orchestra.

Figure 1.6 Molecule and Hervé Dejardin performing *Acousmatic 360* in 2019. Photos by Goledzinowski©.

research. Figure 1.5 shows a 2019 test session with the BBC Philharmonic Orchestra and researchers from the University of Huddersfield.

In 2019, Hervé Dejardin, the producer behind the *nouvOson* project, embarked on a tour with the artist Molecule called *Acousmatic 360* (Figure 1.6). Electronic music was generated live and the components were spatialized in real time in the venue using the L-ISA system [26]. The same spatial scene information was used to generate a binaural rendering, and a concert video was produced in collaboration with France Télévisions for online viewing [27].

1.2.7 Challenges for binaural audio in broadcasting

Many broadcast programmes have now been produced using binaural audio, but challenges remain for uptake to become mainstream.

Production is more complex, and additional training and experience is required. Broadcasters have developed training courses for staff, but as the volume of binaural production increases, there is a challenge to maintain high-quality output. While facilities often only require additional software tools to enable binaural production, there can be significant cost to doing this across many studios. New recording tools, such as multi-microphone arrays, are also needed.

Whether live or post-produced, one of the challenges with binaural production is that it requires duplication of effort. A stereo broadcast must still be produced for loudspeakers, and this remains the priority since more people will listen to this version. The correct version must then be delivered to each listener according to how the individual is listening. Another challenge is terminology; user research shows that binaural is not a very meaningful term to audiences. Other terms such as immersive or 3D sound have been used, but consistency is needed so that audiences can find the material and understand what it offers to them.

A big question that remains is whether the headphone listening experience is genuinely improved by binaural audio. Blind comparisons between binaural and stereo in listening tests give mixed results, often showing preferences for stereo material, for example, [28]. However, when listeners know they are selecting between 3D and stereo versions, a huge preference for binaural has been found [29]. Recent studies at the BBC showed that listeners clearly appreciate the spatial aspects of the binaural listening experience and the benefits that this can have on the sense of presence and realism. This was the case both for professional sound engineers and general audience members, although professionals more often identify timbral issues with binaural mixes. The static pre-rendered approach clearly has some limitations in terms of realism when compared with the head-tracked interactive rendering used in applications such as VR.

1.3 3D AUDIO OVER LOUDSPEAKERS

Headphone listening offers a more straightforward means of delivering 3D audio to broadcast audiences, but it is clear that loudspeaker reproduction is also needed to enable shared listening. As with surround sound, an array of separate loudspeakers can be used, but with additional speakers placed above ear height, and sometimes below. There are also now advanced soundbars that aim to reproduce 3D audio from a single unit. Loudspeaker 3D audio requires greater technological changes to broadcast production and distribution infrastructure; therefore, it is not yet as widely used as binaural audio.

1.3.1 Use in broadcast services

The Japanese national broadcaster NHK was pioneering in developing a 22.2 multichannel 3D sound system [30] to accompany its 8K UHDTV format Super Hi-Vision (SHV). This was first made public in 2005 at the World Expo in Japan. NHK found that the 22.2 layout, standardized as

System H (9+10+3) in [31], gives a greater sense of presence and realism than 5.1 in its listening tests, and better performance when outside of the central listening position. However, the differences were not as great as between 5.1 and stereo.

A trial of SHV and 22.2 audio was conducted at the London 2012 Olympic Games, in collaboration with the BBC and Olympic host broadcaster OBS [32]. Live event feeds and daily highlights packages were distributed to demonstration theatres in the UK, Japan, and the USA. Figure 1.7 shows the system installed at the BBC Radio Theatre in London.

NHK launched its public SHV service in December 2018, featuring 22.2 multichannel audio. The 2020 Olympic Games in Tokyo[2] would have been the first at which OBS offer 3D audio feeds to rights-holding broadcasters [33], and this would have been a major event for NHK's SHV service. OBS currently use a nine-channel 3D loudspeaker layout that is sometimes termed 5.1.4 since it has four elevated loudspeakers above a typical 5.1 layout, but it is standardized as system D (4+5+0) in [31].

Besides NHK, a number of other broadcasters are now offering multi-channel 3D audio for some programmes, mostly using the 4+5+0 loud-speaker layout. The UK satellite TV service Sky has offered English Premier League football games with 3D audio since 2017 [34] using the Dolby Digital Plus codec with Joint Object Coding (branded as Dolby Atmos®) [35]. It has since broadcast a wide range of other programmes, including more sports events, live music, and drama. In South Korea, SBS began broadcasts of 3D audio with UHD at the 2018 Football World Cup using the MPEG-H 3D Audio Codec and the ATSC 3.0 broadcast standard

Figure 1.7 2012 Olympic showcase of Super Hi-Vision with 22.2 audio at the BBC Radio Theatre in London.

[36]. It plans to steadily increase its UHD output in the coming years. France Télévisions has also run public UHD broadcast trials with MPEG-H Audio during the Roland Garros tennis tournament [37].

1.3.2 Challenges in broadcasting

Production facilities need significant upgrades to support loudspeaker 3D audio production. Studios need 3D loudspeaker arrays and mixing consoles that support multichannel buses and 3D panning. There are still gaps in the 3D audio production toolset, for example, multichannel dynamics and reverberation processors are only just emerging. Sound engineers cannot readily apply their favourite tools from stereo or surround mixing. Also, tools for creative sound design in 3D audio formats are rare. This also comes with the challenge of training staff, as highlighted by OBS's Nuno Duarte in [33].

There is a challenge too around loudspeaker reproduction for audiences, as new receivers and loudspeaker systems are required. A 2018 survey of 1,102 people in the UK found that only 11.5% had a surround sound system installed at home [38], despite many years of widespread availability. By contrast, the survey found that 16.9% owned a soundbar for reproducing surround sound. Multichannel loudspeaker systems are complex to install. While soundbars are easier to set up, it appears that the listening experience is not as good as with discrete surround sound, or even stereo loudspeakers [39]. The survey did not ask specifically about 3D audio reproduction capabilities. Discrete 3D loudspeaker systems are available to consumers, but the setup complexity only increases with elevated loudspeakers, although soundbars that can support reproduction of 3D audio are now becoming available. Controlled studies of the sound quality of the 3D loudspeaker reproduction options that are available to broadcast audiences are needed in future.

Codecs exist for the delivery of multichannel 3D audio, but these should be widely supported in consumer equipment before the roll-out of services. Availability is steadily increasing, with many new TVs and mobile devices supporting at least one such codec, and this makes such services viable in the near future.

With multiple different 3D loudspeaker layouts and codec technologies, there is a need for open formats in production and archiving. The *audio definition model* (ADM) [40] was developed to provide metadata describing audio formats, including the many variants of 3D audio, and this can be stored in broadcast wave files [41]. Such files are now supported in several popular DAWs, but wider support throughout the broadcast chain is needed in the future.

As with binaural audio, duplication of production effort is a concern since this will increase costs and time requirements. For example, rather than use a down-mix process, Sky currently produces separate mixes for 3D and 5.1 versions of its sports programmes to maintain the quality of the existing 5.1 service [34].

1.4 OBJECT-BASED AUDIO

Object-based audio (OBA) systems aim to solve the problem of duplicated production effort by allowing a single mix to be created and then adapted on the receiver to suit the available reproduction system. This can include 3D audio, both for headphones and loudspeakers, as well as legacy stereo and surround formats. With OBA, components of the audio mix are delivered separately with metadata describing them and how they should be reproduced. Rendering algorithms on the receiver device then adapt and combine these objects to achieve the best experience with the available setup. This also introduces the potential for the personalization of binaural rendering and use of listener head tracking.

1.4.1 Object-based audio in TV and radio

The broadcast industry uses the term *next-generation audio* (NGA) to describe codec systems that support the OBA approach. Three standardized NGA codecs are available: MPEG-H 3D Audio [42], Dolby AC-4 [43], and DTS-UHD Audio [44]. These systems are supported in recent standards for broadcast and internet delivery of media [45], [46].

The ADM supports OBA representations. An online ADM tutorial from the European Broadcasting Union (EBU) demonstrates this [47], and a renderer specification for converting ADM formats to loudspeaker signals has recently been standardized [48]. This was based upon input from EBU broadcasters and NGA codec providers. It is planned to be applied in production systems in the near future.

OBA also enables personalization of the audio signals for listeners and their listening context. This can improve the accessibility and enjoyment of broadcasts. The balance between objects can be adjusted to give improved dialogue intelligibility and understanding of narrative, for instance, for hearing-impaired listeners [49] or for people listening in loud environments [50]. Objects can also be added, for example to provide audio description for the visually impaired [51], or switched within the mix, for example allowing efficient selection between different languages, or choice of different perspectives in sports commentary. EBU broadcasters conducted a test of NGA production at the European Athletics Championships in 2018, producing a 3D audio mix, while also delivering multiple audio objects [52]. This allowed a choice between French and English commentary with dialogue-level adjustment, as well as the availability of audio description within the same stream.

Personalization features – particularly for improving accessibility – are more important to broadcasters than 3D audio. The BBC recently ran a public trial of OBA personalization for improving the understanding of narrative using an episode of the TV drama *Casualty* [53]. This received a great deal of press attention in the UK, highlighting the importance of intelligibility issues.

The Orpheus research project investigated an end-to-end pipeline for production of OBA for radio and developed a mobile application that features a range of test content from the broadcasters involved. Extensive studies were carried out to evaluate the user experience in this context and the novel features were highly appealing to users, particularly dialogue-level control and 3D audio playback [29].

Results from both these, and future trials and user studies will inform how OBA systems should be deployed in broadcast services in the future. While 3D audio is not seen as the main benefit of OBA for broadcasters, the roll-out of NGA systems will nonetheless enable the delivery of 3D audio services to mass audiences.

1.4.2 Advanced uses of object-based audio

To tackle the challenge of audiences installing 3D loudspeaker systems, the BBC and the S3A research project developed a concept called *device orchestration* [54]. This makes use of the many existing loudspeakers often found in the home, which are in devices connected to the internet or local network, such as mobile phones and smart speakers. Services are run to synchronize playback between these devices and to distribute objects across them to form an ad hoc spatial audio array, thereby orchestrating the devices to provide an immersive listening experience. User research has found that this can give a quality of listening experience equivalent to a 5.1 system, or even better when outside the central listening position [55]. A short science fiction audio drama, *The Vostok-K Incident*, was created for orchestrated devices and made available on the web. Initial user evaluation showed that it was positively received, though users often only connected a small number of additional devices [56]. BBC R&D is now working to create tools that allow anyone to experiment with device orchestration for immersive audio using this web framework.

Objects can be discrete in time, as well as representing layers within the mix. With suitable metadata models and rendering algorithms [57], this enables experiences that respond in more sophisticated ways to the needs and preferences of the user. For example, variable-length programmes [58] can respond to how much time the listener has available to listen, or personalized narratives can be offered through manual user interaction, or even automatically based on user profile data [59]. By incorporating the multitude of sensors available on modern devices, *perceptive media* could automatically adapt the media presentation based on the characteristics and activity of the user and the individual's environment [60].

1.5 SUMMARY

Broadcasters are working with industry partners to develop the technology and skills for 3D audio production. Some are already regularly providing 3D audio material to audiences. Headphone delivery using binaural audio is more readily achieved since it requires less technological change. The installation of 3D loudspeaker setups in homes is potentially challenging, but 3D soundbars make loudspeaker delivery more viable.

3D audio material can be delivered efficiently using NGA codecs, which use an object-based representation. These systems allow automatic adaptation to the listener's chosen reproduction format. They also offer personalization of the audio presentation, which has perhaps greater importance for broadcasters, particularly to improve accessibility. However, the roll-out of object-based audio for this purpose opens the door to widespread availability of 3D audio. There are still challenges ahead, but it seems likely that 3D audio will be a key part of more immersive and personalized broadcast services in the near future.

NOTES

1   WDR: Westdeutscher Rundfunk, BR: Bayerischer Rundfunk, IRT: Institut für Rundfunktechnik
2   The event was deferred due to the COVID-19 crisis, which is ongoing at the time of writing.

REFERENCES

[1]   D. G. Kirby, N. A. F. Cutmore, and J. A. Fletcher, "Program origination of five-channel surround sound," *Journal of the Audio Engineering Society (AES)*, vol. 46, no. 4, pp. 323–330, 1996.
[2]   Futuresource, "Worldwide headphone market outlook April 2018." 2018.
[3]   G. Plenge, "On the differences between localization and lateralization," *The Journal of the Acoustical Society of America*, vol. 56, no. 3, pp. 944–951, Sep. 1974.
[4]   S. Paul, "Binaural recording technology: a historical review and possible future developments," *Acta Acustica united with Acustica*, vol. 95, pp. 767–788, 2009.
[5]   BBC, "Start of experimental stereo broadcasting, 28 August 1962," *History of the BBC*. www.bbc.com/historyofthebbc/anniversaries/august/experimental-stereo-broadcasting.
[6]   S. Krebs, "The failure of binaural stereo: German sound engineers and the introduction of artificial head microphones," *International Journal of Constitutional Law*, vol. 23, pp. 113–143, 2017.
[7]   Interview with Lloyd Silverthorne, Nov. 27, 2019.
[8]   E. M. Wenzel, F. L. Wightman, and S. H. Foster, "A virtual display system for conveying three-dimensional acoustic information," *Proceedings of the Human Factors and Ergonomics Society Annual Meeting*, vol. 32, no. 2, pp. 86–90, Oct. 1988.
[9]   C. Pike, "Evaluating the perceived quality of binaural technology," University of York, 2019.
[10] C. Pike and F. Melchior, "An assessment of virtual surround sound systems for headphone listening of 5.1 multichannel audio," in *Proceedings of 126th AES Convention*, 2013.
[11] A. Silzle, B. Neugebauer, S. George, and J. Plogsties, "Binaural processing algorithms: importance of clustering analysis for preference tests," in *Proceedings of 126th AES Convention*, 2009.

[12] G. Lorho and N. Zacharov, "Subjective evaluation of virtual home theatre sound systems for loudspeakers and headphones," in *Proceedings of 116th AES Convention*, 2004.

[13] J. Kiene and M. Meier, "Binaural surround sound for streaming (Binaurale Präsentation von Mehrkanalton für Streaming)," in *29th Tonmeistertagung – VDT International Convention*, pp. 97–105, November 2016.

[14] C. Pike, T. Nixon, and D. Evans, "*Under Milk Wood* in headphone surround sound – BBC R&D," *BBC R&D Blog*, 2014. www.bbc.co.uk/rd/blog/2014-10-under-milk-wood-in-headphone-surround-sound (accessed Dec. 15, 2018).

[15] R. Nicol, M. Emerit, E. Roncière, and H. Déjardin, "How to make immersive audio available for mass-market listening," *EBU Technical Review*, July 2016.

[16] Radio France, "HyperRadio." hyperradio.radiofrance.com/son-3d/.

[17] Flux::, "Spat Revolution." www.flux.audio/project/spat-revolution/.

[18] C. Pike and F. Melchior, "*Tommies* in 3D – binaural headphone mix," *BBC R&D Blog*, 2014. www.bbc.co.uk/rd/blog/2014-10-tommies-in-3d (accessed Dec. 15, 2018).

[19] C. Pike, "Fright night," *BBC R&D Blog*, 2015. www.bbc.co.uk/rd/blog/2015-10-fright-night-binaural-horror.

[20] C. Pike, "Sounding special: *Doctor Who* in binaural sound," *BBC R&D Blog*, 2017. www.bbc.co.uk/rd/blog/2017-05-doctor-who-in-binaural-sound.

[21] M. Hassler, "The future of illustrated sound in programme making," in *ITU-R Workshop on Topics on the Future of Audio in Broadcasting*, 2015.

[22] IRT, "Bringing object-based audio to the production," 2017. https://lab.irt.de/bringing-object-based-audio-to-the-production/.

[23] T. Parnell and C. Pike, "An efficient method for producing binaural mixes of classical music from a primary stereo mix," in *Proceedings of 144th AES Convention*, 2018.

[24] M. Firth, "Developing tools for live binaural production at the BBC Proms," *BBC R&D Blog*, 2019. www.bbc.co.uk/rd/blog/2019-07-proms-binaural.

[25] H. Wittek and G. Theile, "Development and application of a stereophonic multichannel recording technique for 3D audio and VR," in *Proceedings of 143rd AES Convention*, 2017.

[26] L-Acoustics, "L-ISA – immersive sound art." http://l-isa-immersive.com/.

[27] France Télévisions, "Molécule 'Acousmatic 360°' en live à La Cigale," 2019. www.fip.fr/emissions/fip-360/molecule-acousmatic-360-en-son-binaural.

[28] T. Walton, "The overall listening experience of binaural audio," in *Proceedings of International Conference on Spatial Audio*, 2017.

[29] A. Silzle, R. Schmidt, W. Bleisteiner, N. Epain, and M. Ragot, "Quality of experience tests of an object-based radio reproduction app on a mobile device," *Journal of the AES*, vol. 67, no. 7, pp. 568–583, 2019.

[30] K. Hamasaki, T. Nishiguchi, R. Okumura, Y. Nakayama, and A. Ando, "A 22.2 multichannel sound system for ultrahigh-definition TV (UHDTV)," *SMPTE Motion Imaging Journal*, vol. 117, no. 3, 2008.

[31] ITU-R, "Recommendation BS.2051-2 – advanced sound system for programme production," vol. 2. 2018.

[32] B. M. Sugawara *et al.*, "Super Hi-Vision at the London 2012 Olympics," *SMPTE Motion Imaging Journal.* vol. 122, no. 1, pp. 29–38, 2013.

[33] S. Harvey, "Immersive audio at the Tokyo 2020 Olympics," *Pro Sound News*, 2019. www.prosoundnetwork.com/post-and-broadcast/immersive-audio-at-the-tokyo-2020-olympics.

[34] F. Ringrose, "Next generation audio summit: inside Sky's journey through Dolby Atmos production," *SVG Europe*, 2018. www.svgeurope.org/blog/headlines/next-generation-audio-summit-inside-skys-journey-through-dolby-atmos-production/.

[35] H. Purnhagen, T. Hirvonen, L. Villemoes, J. Samuelsson, and J. Klejsa, "Immersive audio delivery using joint object coding," in *140th AES Convention*, pp. 1–9, 2016.

[36] S. Meltzer and A. Murtaza, "First experiences with the MPEG-H TV audio system in broadcast," *SET International Journal of Broadcast Engineering*, pp. 47–52, 2018.

[37] O. Jouinot, "Roland Garros 2019: an ultra HD event channel with MPEG-H audio," *idfrancetv.fr blog*, 2019. http://idfrancetv.fr/roland-garros-2019-an-ultra-hd-event-channel-with-mpeg-h-audio/.

[38] C. Cieciura, R. Mason, P. Coleman, and M. Paradis, "Survey of media device ownership, media service usage, and group media consumption in UK households," in *Proceedings of AES 145th Convention*, Oct. 2018.

[39] T. Walton, M. J. Evans, F. Melchior, and D. Kirk, "A subjective comparison of discrete surround sound and soundbar technology by using mixed methods," *BBC Research and Development White Paper*, vol. 320, 2016.

[40] ITU-R, "Recommendation ITU-R BS.2076-2 – audio definition model," 2019.

[41] ITU-R, "Recommendation BS.2088 – long-form file format for the international exchange of audio programme materials with metadata," 2015.

[42] ISO/IEC 23008-3:2015, "Information technology – high efficiency coding and media delivery in heterogeneous environments – part 3: 3D audio," 2015.

[43] ETSI TS 103 190-2, "Digital audio compression (AC-4) standard – part 2: immersive and personalized audio," V1.2.1. ETSI, 2018.

[44] ETSI TS 103 491, "DTS-UHD audio format; delivery of channels, objects and ambisonic sound fields," V1.1.1. ETSI, 2017.

[45] ETSI TS 101 154, "Digital video broadcasting (DVB); specification for the use of video and audio coding in broadcast and broadband applications," V2.4.1. ETSI, 2018.

[46] HbbTV Association, "HbbTV 2.0.2 specification," V2.0.2. 2018.

[47] EBU, "ADM guidelines," 2019. https://adm.ebu.io/.

[48] ITU-R, "Recommendation ITU-R BS.2125 – audio definition model renderer for advanced sound systems," 2019.

[49] L. A. Ward and B. G. Shirley, "Personalization in object-based audio for accessibility: A review of advancements for hearing impaired listeners," *Journal of the AES*, vol. 67, no. 7–8, pp. 584–597, 2019.

[50] T. Walton, M. Evans, D. Kirk, and F. Melchior, "Exploring object-based content adaptation for mobile audio," *Personal and Ubiquitous Computing*, vol. 22, no. 4, pp. 707–720, Aug. 2018.

[51] J. Paton, "Audio description – embracing the next generation of audio," *AbilityNet*, Nov. 15, 2019. https://abilitynet.org.uk/news-blogs/audio-description-embracing-next-generation-audio (accessed Sep. 23, 2020).

[52] F. De Jong, D. Driesnack, A. Mason, M. Parmentier, P. Sunna, and S. Thompson, "European Athletics Championships: lessons from a live, HDR, HFR, UHD, and next-generation audio sports event," *SMPTE Motion Imaging Journal*, vol. 128, no. 3, pp. 1–10, 2019.

[53] L. Ward, "*Casualty*, loud and clear – our accessible and enhanced audio trial," *BBC R&D Blog*, 2019. www.bbc.co.uk/rd/blog/2019-08-casualty-tv-drama-audio-mix-speech-hearing.

[54] J. Francombe *et al.*, "Qualitative evaluation of media device orchestration for immersive spatial audio reproduction," *Journal of the AES*, vol. 66, no. 6, pp. 414–429, Jun. 2018.

[55] J. Woodcock, J. Francombe, R. J. Hughes, R. Mason, W. J. Davies, and T. J. Cox, "A quantitative evaluation of media device orchestration for immersive spatial audio reproduction," in *Proceedings of AES International Conference on Spatial Reproduction – Aesthetics and Science*, 2018.

[56] J. Francombe and K. Hentschel, "Evaluation of an immersive audio experience using questionnaire and interaction data," in *23rd International Congress on Acoustics*, pp. 6137–6144, 2019.

[57] J. Cox, M. Brooks, I. Forrester, and M. Armstrong, "Moving object-based media production from one-off examples to scalable workflows," *SMPTE Motion Imaging Journal*, vol. 127, no. 4, pp. 32–37, May 2018.

[58] M. Armstrong, M. Brooks, A. Churnside, M. Evans, F. Melchior, and M. Shotton, "Object-based broadcasting – curation, responsiveness and user experience," in *International Broadcasting Convention (IBC) 2014 Conference*, p. 12, Aug. 2014.

[59] M. Armstrong *et al.*, "Taking Object-Based Media from the Research Environment Into Mainstream Production," *SMPTE Motion Imaging Journal*, vol. 129, no. 5, pp. 30–38, 2020.

[60] BBC R&D, "Perceptive radio." www.bbc.co.uk/rd/projects/perceptive-radio.

2

3D audio for live sound

Etienne Corteel, Guillaume Le Nost, and
Frédéric Roskam

2.1 INTRODUCTION

A paradigm shift is transforming the live sound reinforcement industry,
moving away from traditional left/right channel-based systems and
towards the adoption of object-based audio workflows. Different technical
solutions have appeared on the market, all of which rely on a minimum
of five full-range sources (point-source or line-source [1] loudspeakers)
above the stage, and these are sometimes supplemented by surround or
height loudspeakers.

For the audience, these new systems bring localization accuracy,
allowing for improved audio-visual consistency across a very large part
of the venue. The ability to differentiate instruments in space – spatial
unmasking – is also greatly improved. For live sound engineers, for which
left/right systems are more often dual-mono systems, these new technologies open the doors to new creative possibilities, and they often refer to
them as "game-changing."

One can therefore wonder why it took so long for these technologies
to reach the live sound market when they were already deployed in other
entertainment sectors, for example, cinema and gaming. Some specifics
of large-scale touring productions made the adoption difficult. First, these
events are fast-paced, with a show going from build-up to break-down in
less than 12 hours. Second, these productions require scalable and portable
technologies since they might move from small rehearsal spaces to a series
of different arenas, each with more than 10,000 fans.

In order to ensure a successful deployment of these technologies in a live
context, rather than just a simple shift from channel-based to object-based
audio, a full-system approach is necessary. This requires:

- Spatial audio algorithms adapted to the specific constraints of live sound
- A loudspeaker system design methodology and evaluation metrics that
  can define and optimize the loudspeaker system required for a specific venue
- Adapted mixing techniques and creative tools, such as 3D reverberation, adapted to large-scale reproduction systems and precedence
  constraints

• Signal processors and software tools capable of handling large input and output counts at low latency, typically below 5 ms

Many solutions are available at the time of writing, either from loud-speaker manufacturers, such as L-ISA from L-Acoustics [2], Soundscape from d&b audiotechnik [3], and Spacemap LIVE from Meyer Sound Laboratories [4], or from independent manufacturers or software editors, for example, Astro Spatial Audio [5], TiMax from Outboard [6], and Spat Revolution from Flux and IRCAM [7]. All these solutions use an object-based approach, but differ by the available panning algorithm, 3D reverberation, loudspeaker system design approach, and software and hardware implementation.

Section 2.2 presents panning approaches and discusses their applicability for live sound. Section 2.3 proposes a unified evaluation framework that allows comparison of the performance of these algorithms in terms of both audio-visual consistency and auditory separation, when considering a five-speaker layout. Section 2.4 details an evaluation toolkit for the design of loudspeaker systems. Section 2.5 presents 3D reverberation algorithms that have been developed to fulfill the scalability requirements of live sound. In Section 2.6, recent use-cases encompassing different music genres will be discussed, highlighting the perceptive benefits for both an audience and live productions.

2.2 PANNING ALGORITHMS FOR FRONTAL SYSTEMS

2.2.1 Scene-based: higher-order Ambisonics

Higher-order Ambisonics (HOA) describes a sound field as a combination of spherical harmonics [8]. The higher the order, the more harmonics are required and the more accurate the description of the sound field becomes. The reproduction of HOA content over loudspeakers involves a decoding step that aims to reproduce the harmonics with the available loudspeaker system. This step is critical and does not perform well with a frontal-only system or irregular loudspeaker layout.

Moreover, the reproduction of an object at a precise location in space usually involves all loudspeakers, including surround loudspeakers, if available, even for a frontal source. Depending on the order of the description and the decoding scheme, the remaining spatial crosstalk between the intended localization and the loudspeakers either to the sides or at opposite directions may be insufficient. This may cause improper localization for off-centre listening positions due to the precedence effect [9], [10], with the direct sound (coming from the intended direction) becoming perceived as an echo.

HOA is therefore not suitable for large-scale sound reinforcement and is not discussed any further in this chapter. Most of the available commercial products have instead adopted one of the two following panning-algorithm categories.

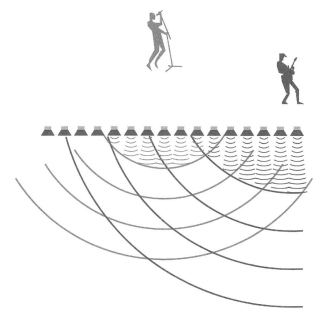

Figure 2.1 Wave field synthesis: the perceived localization and distance of the musicians is synthesized by the interference patterns arising from many individually driven loudspeakers.

2.2.2 Delay-based: Wave Field Synthesis

Wave Field Synthesis (WFS) is a delay-based panning technique that theoretically reproduces the physical sound field emitted by an audio object across the entire audience, accounting not only for the direction of an audio object but also its distance [11]. This, however, requires an infinite number of loudspeakers that can be individually controlled and amplified (see Figure 2.1).

Reducing the number of loudspeakers restricts the positioning of audio objects, most often to two dimensions, and limits the accuracy of the reproduction to only low frequencies. Large loudspeaker spacing restricts physically accurate reproduction below 100 Hz for typical large-scale systems (loudspeaker spacing > 4 m) [12]. In this case, there is no physical wavefront reconstruction, and the localization is driven by the first loudspeaker contribution that reaches the listener (precedence-based localization). WFS with such a small number of loudspeakers therefore presents some analogies with delta stereophony [13].

2.2.3 Amplitude-based: vector-based amplitude panning

Vector-based amplitude panning (VBAP) is an extension of two-channel amplitude panning techniques to an arbitrary loudspeaker layout that considers only loudspeaker directions [14]. VBAP considers triplets of

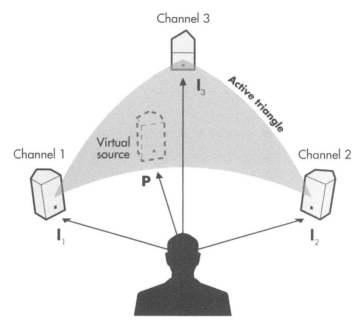

Figure 2.2 A single VBAP base.

loudspeakers (a base) to create a virtual source for audio object positioning in an "active triangle," a portion of a spherical cap (see Figure 2.2), which could resume to a pair of loudspeakers in a horizontal-only layout. The active triangles of multiple bases can be conjoined in a non-intersecting manner to allow the audio object to be panned across these bases, thus increasing the range of potential placement according to the extent of the loudspeaker layout, frontal only or full 3D sphere depending on the application. Multiple-direction amplitude panning (MDAP) was proposed as an extension of VBAP to offer additional control of object width [15].

The L-ISA algorithm is based on VBAP and MDAP, bringing two major enhancements to the original approaches:

- A low-frequency-build-up compensation algorithm for objects positioned between loudspeakers
- An original decorrelation algorithm for the control of the width of objects that minimizes tonal colouration and temporal artefacts when combining multiple loudspeakers

2.3 PANNING ALGORITHM EVALUATION

2.3.1 Evaluation framework

The proposed evaluation framework considers the following conditions:

1. Five loudspeakers regularly spaced (4 m) along a 16 m wide stage, located 6 m above the stage

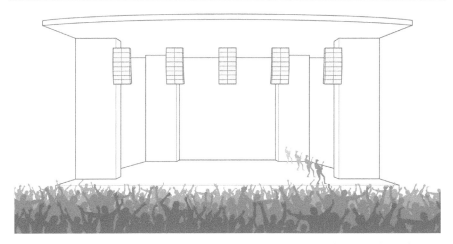

Figure 2.3 Reference stage (1) – left-right, and object (2) – front-back positions.

2. Four object positions at 6 m house-right: 1, 2, 5, and 10 m behind the loudspeaker system
3. A total of 81 positions in the audience (16 × 16 m, every 2 m)

A representation of how 1 and 2 align is shown in Figure 2.3.
The evaluation is conducted along two perceptual dimensions:

• Location error for audio-visual consistency
• Spatial unmasking, evaluating the benefit of spatializing audio objects compared to a typical mono situation in live sound

Instead of conducting perceptual experiments, auditory models from the Auditory Modeling Toolbox [16] are used here. Auditory models have proven their accuracy for localization estimation of loudspeaker-based spatialization systems [17]. Here, they allow the testing of many conditions and a number of listening positions, which would not be possible with a classical listening experiment. Also tested were more stage widths (12 and 20 m), object lateral positions (centred, and off-centred), and audience sizes, but these are not reproduced here since results were very similar and showed the same tendency.

Auditory models are fed with binaural signals corresponding to sound waves arriving at both ears of a listener in a concert situation. Head-related transfer functions measured in an anechoic environment are used to simulate a free-field situation [18], not considering the reverberation of the environment, but concentrating on the direct sound only.

2.3.2 Audio-visual consistency

Relative to a frontal stage, performers are positioned along the width and depth of the stage, on or above the stage, depending on the scenography of the show or the venue.

Audio-visual consistency should therefore be envisaged in all three dimensions.

In the vertical dimension, audio-visual consistency is entirely related to the loudspeakers' height. The height should be set to create less than 30° of vertical separation between the performers and loudspeaker system to allow for audio-visual consistency in the vertical dimension [19].

In the horizontal dimension, an error of 7.5° between the target and actual auditory positioning is considered as negligible to guarantee audio-visual consistency [19]. Angular localization depends on the position of the performer along the width, but also along the depth of the stage, particularly for offside listeners. The arrangement can be seen in Figure 2.4.

Audio-visual consistency in the horizontal dimension therefore depends on:

1. Performer position (with respect to the listener) along the width and depth of the stage (with respect to the venue)
2. The loudspeaker system resolution (i.e. number of loudspeakers)
3. The panning algorithm (amplitude-based or delay-based)

The Auditory Modeling Toolbox function wierstorf2013_estimateazimuth is used here to provide an estimate of the horizontal localization and the associated standard deviation [16]. When assessing the localization error, the spread of participant responses is simulated so that the standard deviation of all responses corresponds to the estimated localization blur. The localization error is calculated as the absolute difference between simulated participant-responses and the target localization for each test-listening position.

The values in Figure 2.5 represent the median localization error (diamond) and the 25th and 75th percentile (lower and upper end of the vertical bar). The results show that the localization error is less than 7.5° in the

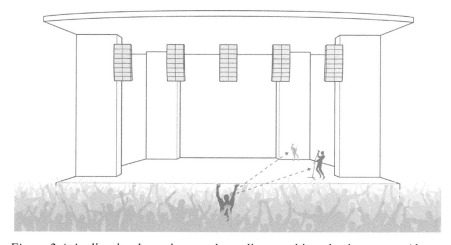

Figure 2.4 Audio-visual consistency depending on object depth on stage (downstage: singer, or upstage: guitar player).

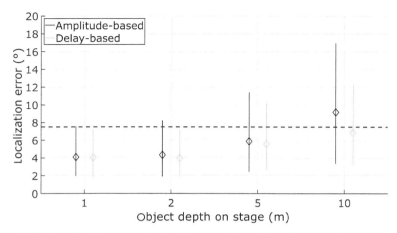

Figure 2.5 Localization error for the house-right audio object position.

horizontal dimension for most performer and listener locations. Amplitude-based and delay-based algorithms perform similarly in terms of localization error, even for objects located upstage. There is no clear benefit to the delay-based algorithm in terms of audio-visual consistency, although it accounts for the perceived on-stage depth of the object.

2.3.3 Localization accuracy: precision vs. blur

When assessing localization precision, participants are asked to perform a localization task, indicating where they perceive a given sound stimulus. The results of localization tests are typically analyzed according to two dimensions, both being available through the auditory model:

- Accuracy: average response among participants
- Blur: uncertainty in localization, corresponding to the spread of participants' answers

The amplitude-based algorithm only makes use of the right outer-two loudspeakers with approximately equal levels to create the house-right audio object, creating a localization blur of about 5°. A delay-based algorithm induces a localization blur that increases when the object gets further upstage and can typically go up to 10° (median, diamond in graph at 10 m upstage). Indeed, in delay-based algorithms, the further the source moves upstage, the more loudspeakers are active with significant level, thus increasing localization blur (see Figure 2.6).

The localization blur can be minimized by placing the source directly on-axis to a loudspeaker, for example, at the centre of the stage. In amplitude-based algorithms, this reduces the blur to 2°, which is the lowest threshold of the auditory system [20] (see Figure 2.7). In delay-based algorithms, the blur increases gradually with object depth, reaching 8° at 10 m distance, in the same range as that shown in Figure 2.6 for the house-right position.

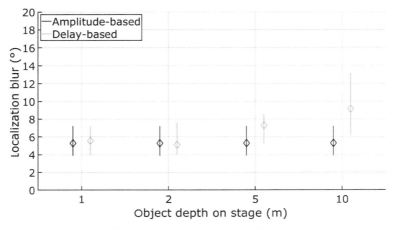

Figure 2.6 Localization blur for the house-right audio object position.

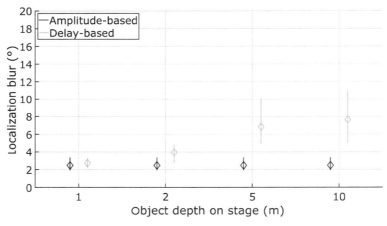

Figure 2.7 Localization blur for the centred audio object position.

2.3.4 Spatial unmasking

Spatial unmasking is the ability to identify what a specific performer plays, among concurrent performers (e.g. lead singer against background vocal or keyboard) [21]. This typically reduces the need of compression or equalization on the background objects to maintain the intelligibility of the lead vocal. Spatial unmasking is calculated here in dB using the jelfs2011 function of the Auditory Modeling Toolbox. This model estimates the increase in speech intelligibility of the target when the target and interferer are spatially separated [22]. It accounts for the localization accuracy and is typically affected negatively by the increase of localization blur of the interferer.

In the example in Figure 2.8, the lead vocal is centred, and the background object is located at 6 m house-right. Both objects are tested at the same depth onstage (1, 2, 5, and 10 m behind the loudspeaker system).

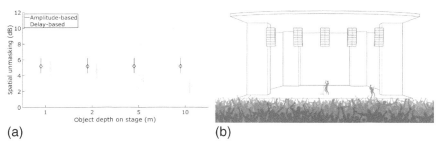

(a)   (b)

Figure 2.8 Spatial unmasking of singer against guitar player depending on object's depth on stage.

The results show that spatial unmasking is larger for amplitude-based than delay-based algorithms. In delay-based algorithms, the apparent angular separation of performers decreases, and the localization blur increases when performers move upstage, which in turn reduces the spatial unmasking by half. Amplitude-based algorithms maintain a large spatial separation of performers and minimize the localization blur.

2.4 LOUDSPEAKER SYSTEM DESIGN

Loudspeaker system design is the process of planning the physical deployment and the electronic settings of loudspeaker enclosures. The principal objective of the design process is that the loudspeaker system's output covers the entire audience. The coverage is the area over which the loudspeaker system provides a direct sound within an acceptable range of frequency response. In an immersive system, this should be achieved for every possible audio object location, reproduced by one or several full-range sources.

The loudspeaker system design presented in Figure 2.9 describes loudspeaker system components and evaluation metrics that have been defined by L-Acoustics in the context of L-ISA. However, the established principles could be extended to other approaches. The system design can be described according to two components:

1. A principal system that is fed with discrete spatialized outputs of the spatial-sound processing unit
2. Complementary or secondary systems that are used to extend coverage to a specific audience area or object directions

2.4.1 Principal loudspeaker system description

As stated, the minimum loudspeaker system for spatial sound reinforcement consists of five full-range sources located above the stage and spanning its full width. This defines the *scene* system. Full-range sources should be configured to maximize their shared coverage area, which defines

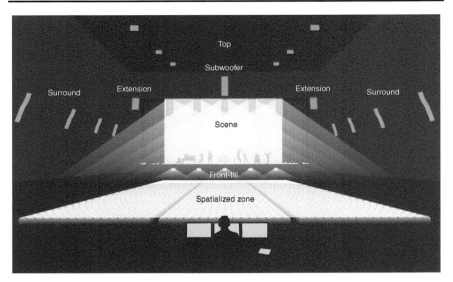

Figure 2.9 Loudspeaker system layout.

the area of the audience where sound sources can be precisely localized (spatialized zone).

An optional *extension* system can be used to extend the panorama – the perceived total width of the sound scene. This consists of one or multiple full-range sources located on either side of the scene system, configured so that their coverage maximally overlaps with the spatialized zone.

The scene and extension system form the essential component of the principal loudspeaker system that receives the main contributions in a live front-facing act. This is where the main performers are (spatially) located and where most of the acoustical energy needs to be created.

Whenever possible or appropriate, additional full-range sources can be added to complement the panorama of the principal loudspeaker system (surround, height, and bottom). These additional full-range sources open creative possibilities, but are complex to include in a touring scenario where the show moves from venue to (physically different) venue. They are more easily implemented in long-term residencies or permanent installations.

All loudspeakers in the principal system should be fed with discrete signals from the spatial sound processing unit.

2.4.2 Principal loudspeaker system evaluation

The loudspeaker system is designed to respect six constraints or evaluation criteria:

1. *Maximum sound pressure level* (SPL) capability of the system: this is often evaluated in dBA and averaged over the spatialized zone. The target value is driven by the music program, typically 10 to 12 dBA more for hip hop than classical music. The higher the value, the more versatile the system.

2. *SPL distribution* over the spatialized zone: it can be calculated as the interval length between the minimum and maximum SPL values in dB (across a bandwidth of 1–10 kHz) over 95% of the covered zone. The target is to achieve a variance of less than 6 dB. This ensures that all audio objects can be heard at a similar level within the audience, regardless of their spatial location. This is an essential aspect when scaling 3D audio applications to large audiences.

   The optimization of the SPL distribution is primarily related to the design of individual full-range sources. This is particularly challenging for full-range sources that should be located at or about the ear height of the audience – where the variation of distance between the source and audience members varies the most, hence also the level for point sources. In such cases, line sources should be considered. Typical modern line sources provide an attenuation in the range of 1 dB for every 10 m from the closest point to the audience outwards, with limited variations in the frequency response (+/– 3 dB in the operating bandwidth) [23]. In some cases, a constant level over distance can be obtained, but often at the expense of a degraded frequency response. Point sources should be reserved to elevated loudspeaker locations or short-throw applications where the distance from the source does not vary significantly over the audience.

3. The *spatial resolution* of the loudspeaker system, which is simply the number of full-range sources in the frontal system: this criterion closely relates to the separation of objects with spatial unmasking. As mentioned previously, the separation is maximized with minimized localization blur, which in turns is realized when audio objects are located on a single full-range source. The more full-range sources, the better the separation of audio objects providing more possibilities of precisely positioned audio objects in the panorama that is formed.

4., 5.   The fourth and fifth criteria are horizontal and vertical localization accuracies, respectively, which are related to the ability to realize audio-visual consistency. These are derived directly from geometrical considerations (full-range sources positions, position in the spatialized zone, and position of the performer on stage). The goal is to maximize the portion of the spatialized zone where the *localization* error is less than 7.5° in the *horizontal* dimension and less than 30° in the *vertical* dimension. The criteria are separated to help the optimization of the loudspeaker system-design task. The horizontal localization accuracy is guaranteed for most of the audience with enough full-range sources for the scene system only, with five being enough, in most circumstances. Vertical localization accuracy is directly related to the height of the loudspeaker system.

6. The *panorama* that a loudspeaker system can offer within the spatialized zone: it is calculated as the horizontal angular difference between the extreme full-range sources of the system at each listening position. The target is to guarantee a panorama of at least 40° throughout the audience, corresponding to what can be experienced by the mixing engineer on a typical left-right system. This can be challenging for long venues,

like sports arenas, where the audience can be located at more than 100 m away from the stage. A wide enough extension system can solve this problem without the need for a full surround configuration.

2.4.3 Complementary loudspeaker system description

Full-range sources are often complemented by a *subwoofer* system to extend the operating bandwidth of the loudspeaker system. The subwoofer system is preferably configured in a central position above the stage. This is the configuration that maximizes efficiency and the homogeneity in the low-frequency range. It is typically fed by a mono down-mix of all objects located within a certain angular pan range.

There are often areas of the audience that need so-called "fill" systems to complement the coverage of the principal system. The fill systems are designated according to both (1) the portion of the audience that they need to cover, and also (2) where they are placed:

* Front-fill (AKA lip-fill): (1) directly in front of the stage, (2) located along the stage
* In-fill: (1) directly in front of the stage, (2) located on either side of stage
* Out-fill: (1) on either side of the stage, (2) located on either side of stage
* Under-balcony fill: (1) below a balcony where the frontal system is physically masked, (2) located along the edge of the balcony
* Delay-fill: time-delayed fills, used over large distances where the principal system cannot provide enough intelligibility or SPL

In-fill, out-fill, and delay-fill systems often comprise individual full-range sources addressing a specific area with little or no overlap between themselves and the principal system. They are therefore fed by a mono down-mix of all objects located within a certain object angular range.

When multiple full-range sources are fed with correlated signals, the timing offset due to time propagation from different placement can create comb-filtering in their shared coverage, creating artefacts such as sound colouration or, in the worst case, the perception of an echo [24]. Careful placement, orientation, and horizontal-directivity control of full-range sources can reduce or even eliminate the area where such artefacts could be audible. Delays and gain adjustments can also be applied to individual full-range sources to limit the artefacts; this process is referred to as time alignment.

In this context, time alignment can be assessed with a seventh criterion related to *time*. It can be evaluated as the portion of the audience inside a time window (typically below 10 ms) within their shared coverage or covered by only one full-range source (6 dB louder than any other full-range source within the 1–10 kHz bandwidth). The target is to minimize the disruption of the principal system by fill systems in the spatialized zone. As shown in Figure 2.10, this can be visualized in Soundvision with a specific colour code within the audience as:

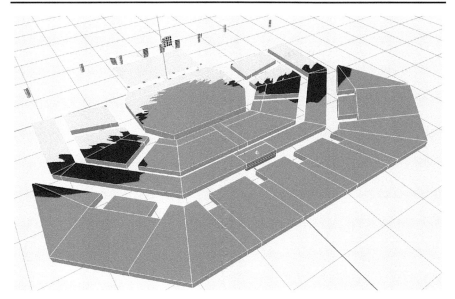

Figure 2.10 Audience qualification in Soundvision.

Shared, aligned: time-aligned portion of the shared coverage of the scene system (dark grey)

Shared, unaligned: misaligned portion of the shared coverage of the scene system (black)

Unshared: coverage by fill system fed with mono downmix (light grey)

Not covered: should be minimized

Secondary systems may also include spatial-sound reproduction. This is the case for under-balcony or front-fill systems that may be used to replicate the frontal system in their respective area. The surround system may also be replicated at different levels in rooms with multiple balconies.

In the case in which there is enough cross-coverage possible between the individual loudspeakers of a secondary system, its full-range sources use individual feeds from the spatial processing. In the case of limited cross-coverage of the secondary system, a virtual replica of the principal loudspeaker system could be created with delay-based algorithms. This virtual loudspeaker technique has been previously used in WFS for the playback of multichannel content (stereo, or 5.1 sources) [25]. It allows for expanding the coverage of single full-range sources, and creating localization while limiting the necessary realignment of the loudspeaker system for each audio location.

2.4.4 Summary

All the proposed metrics can be grouped in a radar graph that aims to provide a compact view of loudspeaker-system performance. This view can help both loudspeaker system designers to create and optimize projects,

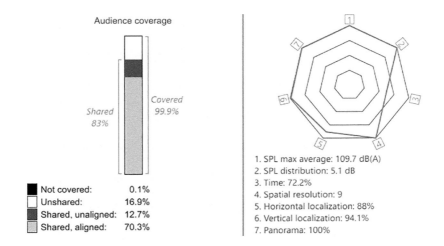

Figure 2.11 Performance metrics summary in Soundvision.

and also production teams to compare different solutions and evaluate the impact of potential trade-offs with easy-to-understand performance metrics. See the performance metrics summary in Figure 2.11.

2.5 ROOM ENGINE FOR LIVE 3D RENDERING

In the context of live sound, the concept of reverberation has specific requirements and constraints for the creation of high-quality virtual spaces. The goal here is not to use reverberation as an effect that blends with the signal of an audio object (as does a spring reverb), but rather to create a configurable virtual space (for all objects) that can further enhance the listener immersion. The creation of this space also allows control of the depth perception of audio objects, adjusting the balance between direct sound and the associated room information, such as reflection and reverberation.

The live sound constraints are presented in Section 2.6.1. Algorithms and control parameters are then described in Sections 2.6.2 and 2.6.3.

2.5.1 Live sound constraints and requirements

The constraints of live sound are the following:

- Scalability, from studio to large-scale venues
- Adaptability and portability from venue to venue

The main requirement is that from pre-production to live shows, the spatial impression and the associated mixing decisions remain valid, as there is often very little time available to adapt to a new environment. As for direct sound processing, complying with these constraints is best possible

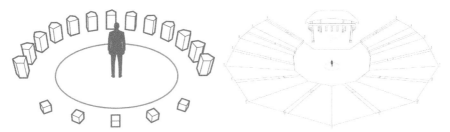

Figure 2.12 Studio environment (left) and large-scale venue (right), in each venue, the circle around the listener represents a propagation time of about 5 ms.

through a combination of optimum loudspeaker system design and well-adapted algorithms with an optimized processing cost to account for many audio objects at low latency (< 5 ms).

One important aspect is that no listener in the venue detects single full-range sources playing reverberation, but rather feels immersed in a natural reverberation field. According to Sonke *et al.* [26], this requires a minimum of eight decorrelated channels surrounding the listener with "similar enough" levels. The latter requires proper loudspeaker system design, with each full-range source being specified to minimize level differences over the audience. This ensures that direct sound is perceived at a similar level throughout the audience, and is also beneficial for creating a perceptually valid diffused field.

Live shows also need to adapt to many different loudspeaker layouts, from frontal-only systems to surround systems to full 3D, depending on the constraints of the production and the venue. The room engine should therefore be capable of adapting seamlessly from both a channel count perspective, and also the shape and size of the loudspeaker system and listening area (see Figure 2.12).

Associated with this is the ability of the room engine to guarantee the precedence of direct sound against reverberation at any listening position. In a studio setup or a small room, all loudspeakers are time aligned against the listener or present only small propagation time differences. This is not the case in large environments where there can be significant propagation time differences between the front and rear loudspeakers. Rear listeners may therefore hear the reverberation signals much earlier (up to hundreds of milliseconds) than the direct sound. In the worst case, again, the direct sound could even be perceived as an echo. It is therefore mandatory that the room engine preserves the precedence of direct sound for every listening position and audio object position, and adjusts automatically for any loudspeaker setup.

2.5.2 Algorithms

The signal flow in Figure 2.13 illustrates how the different constraints can be implemented.

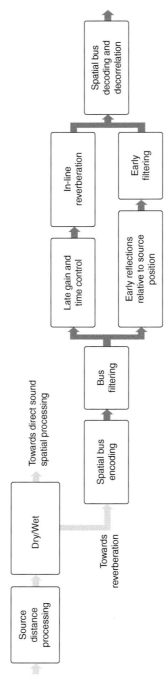

Figure 2.13 Room engine block diagram.

The input source object is initially processed to simulate the effect of distance – with overall level reduction and further high-frequency attenuation due to air absorption [27]. This processed sound is then distributed between two parts, one that will be the direct sound spatial processing, and the other that will go into reverberation processing. This is linked to the dry/wet control familiar to sound engineers for reverberation.

The spatial bus is a crucial asset to deliver control and scalability: it is defined by a fixed number of directional vectors [28]. Their orientation optimally follows the loudspeaker layout, scaling from a 2D frontal system to a 3D half-dome immersive configuration. The source pan and elevation information are used to encode the signal using VBAP. This minimizes the crosstalk between the spatial bus input signals, for instance, compared to an HOA encoder. Any bus filtering can then be applied efficiently, for example, with high-pass filtering, which prevents sending low-frequency reverberation down to the subwoofers.

The signal is then split for the independent control of early reflection generation and late reverberation. Late reverberation includes the first stage of both gain shaping (distribution of energy in space) and spatial time control. The latter defines the late reverberation onset time relative to a source position, for both creative purposes and to ensure the precedence of the direct sound. Precedence preservation is realized if any loudspeaker does not start playing reverberation before it has been "reached" by the direct sound. The pre-delay to be applied on the reverberation channels therefore depends on both the object and loudspeaker positions. Object-by-object processing at the outputs would require deploying as many reverberation engines as objects.

Instead, each spatial bus direction can be processed individually, being redistributed over all channels of the spatial bus with appropriate delays to guarantee precedence, and anticipating the decoding step of the spatial bus over the loudspeaker setup [28]. The processed spatial bus then feeds as many independent in-line reverberation processing engines [29] as there are channels in the spatial bus to create strong decorrelation between channels, and to guarantee precedence for any audio object and listener position.

The early reflections are generated in parallel, replicating the different specular wall reflections relative to a spatial bus direction. Some care can be taken to match an actual physical space's delay times, or to give more creative control to the sound engineer in order to obtain the desired listening experience. Doing this at the spatial bus level avoids the cost of processing audio at source level.

Finally, the spatial bus channels need to be decoded over the loudspeaker setup, both accounting for their direction and using a similar panning algorithm as for the direct sound. Width may be applied to feed multiple loudspeakers with one channel – using additional decorrelation – especially in the case where there are more loudspeakers than spatial bus signals. In this case, spatial bus decoding requires special attention; additional decorrelation principles can be considered (e.g. the width algorithm described earlier).

2.5.3 Controls

Standard controls like (frequency-dependent) reverberation time (RT) and pre-delay must be available to the sound engineer, as well as the independent control of early reflections from late diffuse reverberation. The target is to offer the sound engineer a comprehensive set of parameters to create virtual spaces that can be transported from the studio to large-scale venues. An ensemble of factory presets should be available to create baseline settings that can be fine-tuned to the needs of the show or venue.

With the chosen room-engine approach, more innovative controls can be proposed, linked to the notion of object distance. In an object editor, such as the L-ISA Controller software shown in Figure 2.14, the distance parameter can almost be used as a fader, with the possibility to focus on small or larger distances. The sound engineer can also choose to activate some of the processing elements (e.g. low-pass filtering or level attenuation) for each audio object separately.

A distance-shaping parameter allows the adjustment of the amount of reverberation depending upon the distance of the source. In a dry studio environment, the desired level of reverberation may marginally vary in order to emulate a larger space. However, in larger and more reverberant venues, it can be important to reduce the amount of reverberation for audio objects positioned at short distances in order to maximize intelligibility. In the L-ISA controller software, the reverberation-level variation with distance can be adjusted at minimum and maximum distances, as well as the linearity of the variation in between. These global parameters, therefore, enable adaption to the acoustics of the venue with minimal changes to the show programming.

The temporal shape of the diffuse reverberation tail can be modified with a "linearity" parameter, the effect of which is shown in Figure 2.15. When this parameter is positive, the diffuse energy decays faster at the end of the tail than at the beginning. When negative, the opposite happens. This parameter allows adjustment of the early decay time (EDT) against the overall RT. The EDT is closely linked to the running reverberance – the amount of reverberation that is perceived during actual playing of an instrument – whereas the RT is more related to the "stop chord" reverberance that is perceived when an instrument stops [30].

Other parameters that can be labelled as "spatial response" (see Figure 2.16) allow better shaping of the spatial distribution of the reverberation. Gain shaping allows the concentration of the reverberation energy in the direction of the audio object, or in the opposite direction to favour intelligibility of the objects. Delay shaping adjusts the pre-delay to increase the dimensions of the room, either in the direction of the audio object, or in the opposite direction to increase the perceived size of the room at rear positions. Thanks to the proposed algorithm architecture, these spatial behaviors are maintained for a given object, even if its position (azimuth/elevation) changes, which makes them a very powerful tool for spatial audio mixing.

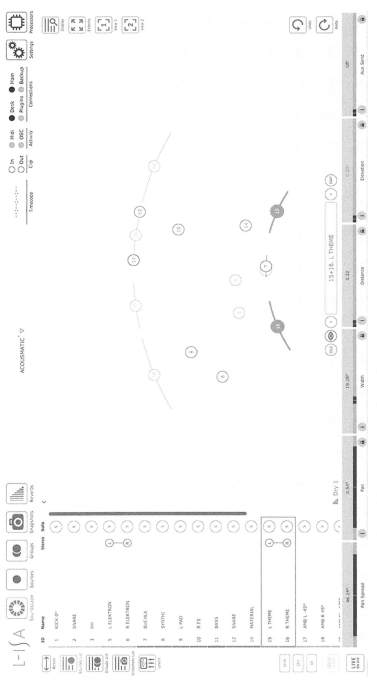

Figure 2.14 Soundscape view of the L-ISA Controller software.

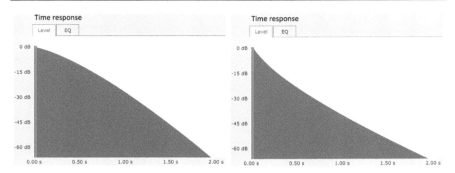

Figure 2.15 Reverberation-decay linearity adjustment (left: positive, right: negative).

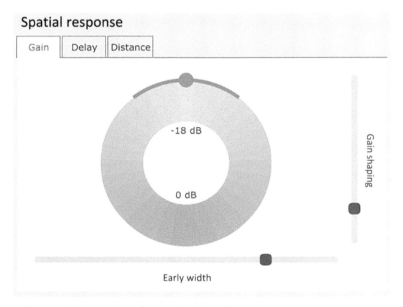

Figure 2.16 L-ISA room engine spatial shaping.

2.6 KEY APPLICATIONS

At the time of writing, large-scale live events increasingly use multiple full-range sources, from frontal configurations to 3D layouts. For venues such as amphitheatres or arenas (up to an audience capacity of 15,000), such frontal configurations can dramatically improve intelligibility and localization.

2.6.1 Frontal loudspeaker systems

A first natural application is to deploy a frontal configuration of five or more full-range sources above classical orchestras. Numerous productions have been deploying such configurations, such as Ennio Morricone's *60*

Years of Music European tour, BBC Proms 2018 and 2019 at the Royal Albert Hall, and the Los Angeles Philharmonic in their 2019 Korea tour. As all the audio inputs are acoustical instruments, the benefits provided by an accurate match of sound and visuals are huge, to a point where the listener forgets that the sound is reinforced.

For pop music, such as Mark Knopfler's *Down the Road Wherever* tour in 2019, which entailed 56 dates in Europe and 28 dates in North America, frontal configurations brought benefits in terms of spatial unmasking and "freedom to pan." This was indeed the decision factor for this tour, which involved a very large number of electric and acoustic instruments. The system provided improved voice intelligibility, even in large venues, such as the O2 Arena in London. In this venue, delay-fill systems are usually deployed to reinforce the direct sound for the most distant audience. With a multi-array frontal configuration, no additional fill systems were needed, and voice intelligibility was still very high, even at the furthest tribune.

For moving audio sources, such as actors in a musical, or chief executive officers in a corporate presentation, the capacity to adjust the location of the reinforced sound for every member of the audience in real time gives a similar perceptive effect to the classical orchestra example: the listener quickly forgets that the voice is being reinforced, providing a more natural, engaging experience.

2.6.2 To surround and full 3D

Expanding the "frontal concert" experience, some productions use surround systems to further immerse listeners into the show. As an example, alt-J used a complete 360 system for two shows in 2018 in the Forest Hills stadium in New York (audience size 14,000) and the Royal Albert Hall in London (audience size 5,000). These types of configurations are well suited to music with electronic components. Such audio objects have no physical anchor and can therefore be positioned freely in space – if the timing consistency between different parts of the music is maintained. In the case of alt-J, reverberation and electronic layers were sometimes present in the surround systems, and this created some highly immersive moments in the show [2].

Beyond the concept of stage, an increasing number of large-scale events are using surround (circular 2D) or dome-shaped (spherical) speaker configurations. Indeed, the challenge of this type of deployment comes with their scale: within a 50 m-diameter circular deployment, timing discrepancies between speaker signals can go past 150 ms. Care must be taken to think the space into sections where temporal aspects of the composition are coherent for the entire audience. Each section should maintain coherence, but may evolve independently from the others, for example, rhythmic sections, effects, and background. Artists like Molecule (France) and festivals like Wonderfruit (Thailand), Coachella (USA), and Tomorrowland (Belgium) have deployed such systems for audiences up to 5,000. For the artist, it is a brand new dimension to play with, and we will surely see many creative uses of such systems in the next decade [2].

2.7 CONCLUSION

Algorithms, technology, and system-design processes are now mature enough to gain a rapidly growing market share for live 3D audio solutions. From a creative standpoint, new possibilities are opening, with the possibility to localize sounds to where the visual action takes place – and from an audience standpoint, this further enhances the feeling of immersion. From a venue-owner perspective, solutions can now combine 3D direct sound positioning with room acoustic enhancement solutions in a single system, which opens the range of shows that can be programmed in the same space.

Some challenges remain: first, with the mechanical positioning constraints of loudspeakers, which manufacturers are tackling via optimized product designs. Second, manufacturers still need to work on the interoperability of various solutions, to ensure that 3D audio will become as standard as the stereo format. In that regard, standardization initiatives such as the Audio Definition Model [31] should be deployed in the live sound industry. Finally, the tipping point for spatial audio in live sound will be when it can benefit from greater access to the creative tools in a mobile or home studio environment. When the spatial dimension is at the heart of the creation process, it quickly becomes as important as the time and spectral dimensions.

REFERENCES

[1] P. Bauman, M. Urban, and C. Heil, "Wavefront sculpture technology," in *Proceedings of 111th Audio Engineering Society (AES) Convention*, New York City, USA, Nov. 2001. www.aes.org/e-lib/browse.cfm?elib=9813.

[2] L-ISA, "L-ISA – shaping the future of sound," *L-ISA*, 2020. www.l-isa-immersive.com/ (accessed Nov. 06, 2020).

[3] d&b audiotechnik, "The d&b soundscape – More art. Less noise," *d&b audiotechnik*, 2020. www.dbsoundscape.com/global/en/ (accessed Nov. 06, 2020).

[4] Meyer Sound, "Meyer Sound," 2020. https://meyersound.com (accessed Nov. 06, 2020).

[5] Astro Spatial Audio, "Astro spatial audio | creativity through simplicity," *Astro Spatial Audio*, 2020. www.astroaudio.eu/ (accessed Nov. 06, 2020).

[6] OutBoard, "OutBoard–TiMax 3D Spatial Audio Showcontrol, Rigging, PAT," *OutBoard TiMax*, 2020. http://outboard.co.uk/ (accessed Nov. 06, 2020).

[7] IRCAM, "Spat Revolution," *Spat Revolution*, 2020. www.flux.audio/project/spat-revolution/ (accessed Nov. 06, 2020).

[8] J. Daniel and S. Moreau, "Further study of sound field coding with higher order Ambisonics," in *Proceedings of 116th AES Convention*, Paper 6017, Berlin, Germany, May 2004. www.aes.org/e-lib/browse.cfm?elib=12789.

[9] R. Y. Litovsky, H. S. Colburn, W. A. Yost, and S. J. Guzman, "The precedence effect," vol. 106, no. 4, pp. 1633–1654, 1999.

[10] H. Wierstorf, A. Raake, and S. Spors, "Localization in wave field synthesis and higher order Ambisonics at different positions within the listening

area," in *Proceedings of German Annual Conference on Acoustics (DAGA)*, Mar. 2013.

[11] A. J. Berkhout, D. de Vries, and P. Vogel, "Acoustic control by wave field synthesis," vol. 93, no. 5, pp. 2764–2778, 1993.

[12] É. Corteel, "On the use of irregularly spaced loudspeaker arrays for wave field synthesis, potential impact on spatial aliasing frequency," in *Proceedings of the 9th International on Digital Audio Effects (DAFx'06)*, Montreal, Canada, pp. 209–214, Sept. 2006.

[13] G. Steinke, "Delta stereophony – a sound system with true direction and distance perception for large multipurpose halls," *Journal of the AES*, vol. 31, no. 7/8, pp. 500–511, 1983.

[14] V. Pulkki, "Virtual sound source positioning using vector base amplitude panning," *Journal of the AES*, vol. 45, no. 6, pp. 456–466, 1997.

[15] V. Pulkki, "Uniform spreading of amplitude panned virtual sources," in *Proceedings of the 1999 IEEE Workshop on Applications of Signal Processing to Audio and Acoustics. WASPAA'99 (Cat. No. 99TH8452)*, pp. 187–190, 1999.

[16] P. Søndergaard and C. Majdak, "The Auditory Modeling Toolbox," in *The Technology of Binaural Listening*, J. Blauert, Ed. Springer Science & Business Media, Berlin, Heidelberg, Germany, 2013, pp. 33–56.

[17] H. Wierstorf, A. Raake, and S. Spors, "Binaural assessment of multichannel reproduction," in *The Technology of Binaural Listening*, J. Blauert, Ed. Springer Science & Business Media, Berlin, Heidelberg, Germany, 2013, pp. 255–278.

[18] H. Wierstorf, M. Geier, and S. Spors, "A free database of head related impulse response measurements in the horizontal plane with multiple distances," in *Proceedings of 130th AES Convention*, London, UK, May 2011, www.aes. org/e-lib/browse.cfm?elib=16564.

[19] E. Hendrickx, M. Paquier, V. Koehl, and J. Palacino, "Ventriloquism effect with sound stimuli varying in both azimuth and elevation," vol. 138, no. 6, pp. 3686–3697, 2015.

[20] J. Blauert, *Spatial Hearing: The Psychophysics of Human Sound Localization*. The MIT Press, 1997.

[21] M. L. Hawley, R. Y. Litovsky, and J. F. Culling, "The benefit of binaural hearing in a cocktail party: effect of location and type of interferer," vol. 115, no. 2, pp. 833–843, 2004.

[22] S. Jelfs, J. F. Culling, and M. Lavandier, "Revision and validation of a binaural model for speech intelligibility in noise," *Hearing Research*, vol. 275, no. 1–2, pp. 96–104, 2011.

[23] L-Acoustics, "L-Acoustics training module," *L-Acoustics Training*, 2019. www.sseaudio.com/Training/L-Acoustics-Training (accessed Nov. 07, 2020).

[24] F. Toole, *Sound Reproduction: The Acoustics and Psychoacoustics of Loudspeakers and Rooms*. CRC Press, 2009.

[25] G. Theile, H. Wittek, and M. Reisinger, "Potential wavefield synthesis applications in the multichannel stereophonic world," in *Proceedings of AES Conference: 24th International Conference: Multichannel Audio, The New Reality*, Jun. 2003. www.aes.org/e-lib/browse.cfm?elib=12280.

[26] J.-J. Sonke, J. Labeeuw, and D. de Vries, "Variable acoustics by wavefield synthesis: a closer look at amplitude effects," in *Proceedings of 104th AES*

Convention, Amsterdam, The Netherlands, May 1998. www.aes.org/e-lib/browse.cfm?elib=8468.

[27] International Organization for Standardization, "ISO 9613-1:1993," *ISO*, Jun. 1993. www.iso.org/cms/render/live/en/sites/isoorg/contents/data/standard/01/74/17426.html (accessed Nov. 07, 2020).

[28] L. N. Guillaume, R. Frederic, C. Etienne, and H. Christian, "Method and system for applying time-based effects in a multi-channel audio reproduction system," EP3518556A1, Jul., 2019.

[29] V. Välimäki, J. Parker, L. Savioja, J. O. Smith, and J. Abel, "More than 50 years of artificial reverberation," in *AES conference: 60th international conference: dreams (dereverberation and reverberation of audio, music, and speech)*, Leuven, Belgium, Jan. 2016. www.aes.org/e-lib/browse cfm?elib=18061.

[30] T. Lokki, H. Vertanen, A. Kuusinen, J. Pätynen, and S. Tervo, "Auditorium acoustics assessment with sensory evaluation methods," in *Proceedings ISRA*, pp. 29–31, 2010.

[31] S. Füg, D. Marston, and S. Norcross, "The audio definition model – A flexible standardized representation for next generation audio content in broadcasting and beyond," in *Proceedings of 141st AES Convention*, Los Angeles, USA, Sept. 2016.

3

Towards 6DOF: 3D audio for virtual, augmented, and mixed realities

Gareth Llewellyn and Justin Paterson

3.1 INTRODUCTION

Questions soon arise for any engineer looking to create audio for virtual reality (VR), augmented reality (AR), and mixed reality (MR) applications. The questions are the same as those that have been with us since the dawn of recorded sound. What is the purpose of this sound reproduction? What is the audience experience supposed to be? However, to inform these questions in this context, perhaps the route to understanding a major demand that modern spatial audio presents is found in the answer to the following question:

> *Where* is the audience in relation to this sound?

Until very recently, the major goal of sound reproduction has been to place the listener "in front of" (and latterly, with surround formats, in the middle of) the sound being presented, or otherwise, in some abstract space that conforms to "nowhere in particular." The listening experience has been linear in that it is played back the same way each time – and the listener has had a fixed relationship to the spatiality of the sound reproduction, regardless of the listener's orientation to the speakers. This is no longer the case in the experiential landscape that is unfolding before us.

Technology company Qualcomm coined the term extended reality (XR) [1], and this will be used in this chapter to represent the combined set of VR, AR, and MR applications [2]. Many extant sonic conventions need to be rethought to best serve the XR listener experience. XR applications have particular "procedural-audio" requirements – "sound as a process rather than sound as data" [3, p. 1] – and such extant conventions are currently being upended by emergent technological and creative advances. Such changes are set to eventually replace some significant portion of the ubiquitous linear listening experiences that we are so accustomed to across all media, and they will come to revolutionize sound reproduction in both form and purpose.

This chapter aims to give an overview of procedural and spatial audio technology and practice as they relate to XR applications, and it will consider

some of the ways they are deployed in the creation of XR experiences – a
field that is undergoing rapid evolution. It should be noted that, at the time
of writing, this field is young and very much still formative, and so rather
than attempt to define the state of the art, the principal thrust of this chapter
is to offer the reader insight into the various salient aspects that together
will shape the evolution of this new sound world. To proceed with this discussion, it is worth highlighting the fundamental differences that XR audio
applications engender in comparison to what has gone before. The fundamental differentiator between XR experiences and what has gone before is
that an XR audience requires a first-person experience coupled with six-
degree-of-freedom (6DOF) movement, which requires a procedural-audio
environment to facilitate it.

"Degrees of freedom" is an engineering term used to describe the kinds
of movement that are possible for a rigid body in a given space. Having
zero degrees of freedom implies that this rigid body cannot move in any direction in this environment. One degree of freedom implies the possibility
of straight line movement along a single axis (think of a train confined to a
straight track – it can only move forwards or backwards).

Three degrees of freedom (3DOF) can refer either to movement along
each of the three axes – x, y, and z – that make up a 3D space, or to rotation
around any of these three axes. In this way, having six degrees of freedom
means that our rigid body is free to move along the x, y, and z axes of a 3D
space and that it is also able to rotate itself around any of these three axes,
as shown in Figure 3.1. Confusion can arise using degrees-of-freedom
terms where it remains undisclosed if the degrees in question are rotational
or axial. Thus, we have 3DOF experiences (e.g. 360° videos) that offer
rotational movement around an origin point – only. These are distinct from
the 6DOF experiences that are of primary concern in this chapter.

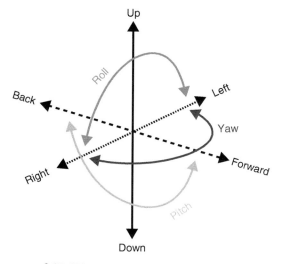

Figure 3.1 The axes of 6DOF.

In this discussion of 6DOF audio, the "rigid body" in question is the actual listener and the space is the 3D environment that the listener can move through as defined or permitted by the XR application. 6DOF audio implies that the listener is free (within the constraints of the experience) to move around a sonic space in all six degrees – and that the audio played back will conform in some meaningful way to the changing position and orientation of the listener's head (specifically the ears) in relation to that space. Of games, Goodwin offers a helpful metaphorical definition; "a listener works like an idealised microphone placed in the game world" [4, p. 125]. True 6DOF listening is, in fact, exactly what one's everyday experience of listening to sound in the real world is. In theory, one has the freedom to move freely around a space and orientate the body and ears relative to sound sources to both better access sonic information, and to better understand the spatial relationships between sounds in the environment. Of course, such actions are often subconscious.

Further, Smalley [5] attested that localization was the most significant "type" of space that might be represented in a recording. One could therefore argue that the recreation of realistic 6DOF audio should be the ultimate limit of any sound reproduction system – since presumably any and all other playback systems are merely a subset of this overarching 6DOF system. How far away we are from this end goal is unclear at this present time, but suffce to say, we are at the bottom of a technical and creative mountain that will be progressed for many years to come, if not indefinitely.

The movement of a character in 6DOF has long been part of the mainstream of video games of "first-person shooter" (FPS) style. The first example is likely to be Exidy's *Starfire* in 1979 [6], with *Quake* being a more contemporary title. Until very recently, much of what can be described as procedural spatial audio has been at the service of these FPS games, but as will be discussed in more depth shortly, there is a complete layer of abstraction between the character's 6DOF movement and the audio experience that is delivered to the player at any given moment. In the case of 6DOF audio for XR applications, it is the first-person perspective of listeners themselves that is the key differentiator and which makes these new applications worthy of separate analysis.

3.2 PERSPECTIVE

There are two key aspects that differentiate the various kinds of listening experience that are reproducible today. The first is *linearity*. Traditional recorded sound plays "rendered" audio from a storage medium to the listener in "a straight line." One may jump around tracks, fast forward, and rewind, but regardless, the playback of material from any given moment in a recording is converted to sound waves for an audience in a fixed and predictable manner, a mode that Gracyk [7] termed autographic. In contrast to this, "procedural" audio is created at runtime – that is, it is generated algorithmically in real time when an application is running, according to a predefined set of rules. This can be regarded as akin to object-based audio [8] as discussed by Pike in Chapter 1. While procedural audio may contain linear elements (samples of sound, or playback of complete sections of audio), it also encompasses synthetically generated real-time audio and audio effects that are all created and brought together only at runtime. Audio playback becomes a real-time expression of the rules encoded in the program itself. This kind of audio can, in principle, be "different every time" – although there are constraints to this, of course. It makes sense to think of the degree of linearity as a continuum as opposed to a dichotomy – there may be many varied degrees of linearity/nonlinearity between the two opposite ends of the "manner of playback" spectrum.

The second key aspect is *perspective*. In this case, we mean the physical perspective of the audience in relation to the "scene" that the audio describes. Sometimes, the intended perspective of a piece of audio is relatively clear – an example might be an audiophile reproduction of a piece of orchestral music. An engineer might set up an A-B stereo pair of cardioid microphones at a certain position in front of the orchestra that gives a pleasing balance of direct and reverberant information. The engineer's intention is to create a convincing stereo image alongside a good impression of the room acoustics – from an idealized listener position – and the hope is that playback over a good pair of stereo speakers will give a rough analogue of this original experience. By contrast, a piece of abstract electronica may have no obvious "perspective" on the sound, yet it is still likely to have an interesting stereo image and be somewhat spatially pleasing on the same stereo speakers. Both scenarios represent different "landscapes," which Wishart and Emmerson define as expressing "the source from which we *imagine* the sounds to come" [9, p. 136], but it is useful to extend this concept when moving to the far more complex world of 6DOF, hence specifically adopting the term "perspective" here.

Adding surround channels to these examples (which might include both horizontal surround and height channels) can increase the realism of a listener's sensation of being in front of or (depending on the engineer's intent) in the middle of any piece of action. The same material may, of course, also be played back on headphones, and some of the original sense of perspective might even survive. However, the perspective on many elements of the original material is likely to collapse into some abstract point in the middle of the listener's head, although there may also be stereo or binaural effects present, that to a greater or lesser degree can externalize the sound in some semi-realistic manner. Such collapse can also be a deliberate consequence of hybrid mixing [10], [11], where suitable elements are binauralized, but other elements of a track are staged with conventional panning and might be expected to be in-head. This approach can be used to give speaker compatibility for a mix with binaural elements, and an audio example from the latter reference can be found at [12]. The interested reader might also explore a comparative study of externalization across various binaural renderers by Reardon *et al.* [13].

It is sometimes hard to step back from any such audio experiences to understand the full extent to which they offer misleading or "canned" versions (either by design or consequence) of the kinds of experiences they ostensibly seek to represent. This concept might be readily understood in the visual realm, where cinematography can more easily be seen as the art of making a representation of reality look more interesting or aesthetically pleasing than our everyday reality. Zagorski-Thomas explored this concept via his "sonic cartoons" [14], where recorded sound is used to create schematic representations of musical activity. In practice, this means that specific features of a music production might be exaggerated or diminished for dramatic/aesthetic effect. The same goes here for spatialization in 3D sound, although it is notable that words such "realism" and "immersion" are often marketing misnomers for the creation of interesting and aesthetically pleasing sound experiences in some form of surround mode. Although only concerned with monaural and stereo sound, rather appropriately, Camilleri [15, p. 202] coined the term "sonic space" to be "a three-dimensional space divided into: localised space, spectral space and morphological space," and went on to discuss how they interact to form a sonic profile of a piece. These "dimensions" are not *xyz* axes, but rather, the last two are the sonic qualities that might be exhibited within the first, which in the context here will instead be 6DOF.

As has been hinted above, perspective is dependent on both the material itself and the playback system that reproduces the sound. Take, for instance, our original stereo piece of orchestral music. This could be enjoyed on a pair of cabinet speakers, a monaural radio placed high upon a shelf somewhere, or via a pair of headphones – but what is the real difference? Well, clearly, the underlying material itself does not conform to these different playback systems, and the original intentions of the audio engineer remain intact – so what has changed? The "landscapes" certainly have. One aesthetic approach to such an intention is to adopt a coordinate system and frame of reference [16]. An egocentric frame of reference represents or encodes the position of an audio object relative to the listener – as might be expected when listening on headphones. The opposite is an allocentric frame of reference, which encodes the object relative to the environment, for instance, a sound scene that is independent of a single observer's position, such as when mixing for cinema.

While object-based audio [17] might seek to redress playback differences with point-of-delivery rendering attuned to a given device, a fundamental difference in these instances comes down to the expression of spatial audio cues that the listener is able to decode; those that will allow the listener to construct a mental spatial "image" (or not) of the audio that the listener is hearing. These spatial cues are both a function of elements that are present in the recording itself, as well as from cues that are generated from the listener's physical relationship to each of these playback systems. In the case of the speaker-based stereo playback system, there are timing and intensity differences between the original left and right audio signals that are sufficiently intact upon reproduction that a phantom stereo image is created by the two speakers. This effect is further complicated for stereo speakers where there is acoustic excitement of the room, as well as by the head shadowing and pinnae effects that the brain deciphers as the sound follows complex trajectories passing from each speaker to either ear.

So, perspective is really an expression of the extent to which spatial cues are reproduced and/or synthesized for the listener at the point of playback. Perspective might be limited by the particular physicality of the sound reproduction system and the acoustic environment in which it sits, as well as in the engineer's intention and ability to encode reproducible spatial information into the original recordings. Sound reproduction system evolution has focused on greater numbers of channels and more speakers, and latterly with object-based rendering, to an arbitrary number of speakers at the moment of playback, and even here, perhaps the most requested optimization "is a change in the position of individual objects" [18]. However, all these approaches still leave the listener stuck in a bubble of more or less spatial sound – with a sweet spot in the middle that tends to collapse the spatial image as one moves outside of it. So how does one get to 6DOF from here, and why would one want to do this anyway?

3.3 TOWARDS 6DOF

Clearly, two things need to be present for us to have the ability to leave the fixed confines of our traditional listening experiences – the audio would need to be able to conform in real time to the listener's physical movement through an environment, either real or simulated. Work is currently being conducted to evaluate the sensation of motion via binaural cross fading, for instance, by Boerum *et al.* [19]. Further, enough of the sounds that we typically experience, and acoustic phenomena that we have evolved to interpret from real-world environments would both need to be present or be synthesized in suffcient detail and quality, that a meaningful reconstruction of the spatial relationships between sounds could be interpreted by the listener. The browser-based FXive is one such system that offers a library of "synthesis models, audio effects, post-processing tools, temporal, and spatial placement" [20, p. 1]. There are only really four ways that the above can be achieved.

The first approach would be to place speakers everywhere in the environment that one needs sound in a 6DOF experience. Somewhat expersive and impractical, this approach allows us to present audio in a physical space and leave the generation of acoustic cues to the environment itself, as well as through stereophonic images created between speakers. As well as multiple individual speakers to emanate point sources, this approach can also incorporate pairs of speakers or 3D speaker arrays to create a navigable 6DOF audio environment with sweet spots and sound-emitting spaces spread around the area. There are many problematic issues – technical, commercial, and aesthetic – to this kind of approach, although it can lead to highly pleasing spatial results. The listener's ears after all provide perfect spatialization for the sounds in such an environment, notwithstanding the issues that speakers and the acoustic spaces themselves introduce. Theme parks and haunted houses frequently offer a primitive version of this kind of experience, and there are good precedents in immersive theatre and art-music experiences for this approach.

A second approach might be to attempt to recreate the required local-ization using wave field synthesis (WFS). This uses the Huygens–Fresnel principle, which means that it is possible to synthesize wavefronts from numerous elementary spherical waves coming from a large array of speakers, thus attempting to recreate the sounds of individual sound sources in an acoustic space [21]. It is further facilitated by the application of the Kirchhoff–Helmholtz integral, which expresses the wave field in an acoustic volume – a space – inside (or outside) its perimeter surface in terms of the field's sound pressure and velocity on the surface [22]. Localization is not dependent upon the listener position, and so 6DOF is possible. Because audio can appear to be *within* the listening area, it nullifies some of the "trapped in a bubble" issues mentioned above. The practicalities of a WFS speaker array mean that, in addition to the large number of speakers, room reflections returning into the listening area become serious issues and further, WFS is both computationally and financially expensive. However, WFS can create good listening environments, and it has begun to see a much more widespread adoption form in TV soundbars and sometimes automotive playback [23]. Indeed, although WFS reproduction performs well in refined conditions – as do Ambisonic speaker arrays – both tend to suffer excessively from issues with calibration, phase colouration, and room acoustics. There are good reasons why the larger cinema formats instead employed a mix of standard channels to recreate their sound stereo-phonically alongside simple amplitude panning of audio objects.

The third is compensated amplitude panning (CAP), a system that adjusts individual speaker gains in response to head tracking, thus modulating interaural time difference cues and hence offering dynamic localization [24]. Using two speakers, this method can produce stable images – including above or behind the listener – giving 6DOF perception (below 1 kHz). It is worth noting that, unlike binaural methods, CAP is agnostic towards head biometrics.

The fourth approach is to employ binaural headphone reproduction, and to seek to recreate soundscapes in real time using procedural audio coupled with head-tracked headphone audio. This is the approach that will be discussed in more depth below, as it is the most prevalent and likely to be the mainstream way of accessing 6DOF audio experiences in a world of spatial computing [25] – that is, computer interaction that references objects and spaces in the real world.

3.4 PRECEDENTS

As touched on earlier, video games have been the commercial test bed for much of procedural-audio development, and as such, some discussion offers a context of, and trajectory towards, the current 6DOF state of the art. A plethora of publishers and developers create thousands of new games each year, using in-house code or otherwise utilizing customizable ready-made game engines, such as Unity3D and Unreal Engine as their devel-opment environments. Audio middleware also exists (Wwise and FMOD being current industry leaders) that can provide useful extensions to the built-in audio systems of these game engines and (perhaps more important-antly) can offer more uniform audio integration and working environments for sound designers – allowing sound development to progress with less demands on the time and attention of core (visual-orientated) coding teams.

Game engines have long worked with the concept of audio objects to build sonic experiences that are able to adapt in real time to the many changing states the game can find itself in at runtime. Audio objects are a class of game object that feed into the physical modelling, geometry, and audio digital-signal-processing systems built into the game engine – they provide instructions for how to render sound at every point in the game. At its most basic level, an audio object is simply a piece of audio with some accompanying metadata that describes the way it should be rendered at runtime. This information often involves data like the sound's position and trajectory in game space, a description of how the sound might change over time, and information that will help to determine the ways in which to render it in the physics engine – for example, what material it is made out of, and how sonically reflective/absorbent it is. An audio object may be comprised of many different components that each define different aspects of the object's functionality, and this audio object in turn may be part of a larger object or set of objects that together make up parts of the game.

The arrival of audio object-based rendering in feature films using formats such as Dolby Atmos®, DTS:X®, and AuroMax® is a clear sign that this approach is slowly but surely set to revolutionize the mainstream – outside of gaming. Audio objects in feature films at present contain only panning and volume automation. In fact, many of the current marketing claims of their benefits remain somewhat overblown, but it seems that a likely strategic path might be towards the confluence of various playback tropes into a single highly descriptive format utilizing procedural playback methods. Fraunhofer's "container" formats MPEG-H and MPEG-I are cases in point. These seek to create licensable container technologies that can reconstruct multifarious configurations of channels, objects, and higher-order Ambisonic (HOA) elements at the point of playback, the latter container including video and with the potential to be optimized for 6DOF operation. Such powerful formats facilitate any number of emergent applications, but as was the case with MP3, market dominance by a lossy format might also be viewed with scepticism by an industry aspiring towards development and creativity.

There is currently tremendous growth in both interest and investment in the new creative and commercial realms being opened up by the contemporary crop of XR applications. Most, if not all, of these applications ideally require some level of 6DOF audio to complete the experience. While phone-based AR audio has traditionally suffered the same abstraction away from the first-person perspective (you listen through the perspective of the phone as opposed to the perspective of one's own ears), MR wearables such as HoloLens 2, Magic Leap, and nreal all seek to correct this perspective issue by placing the tracking device directly upon the user's head. At the time of writing, AirPods Pro earbuds have just been released with head-tracked spatial audio (which also calculates the difference between the orientation of the phone and the user's head – a related patent can be explored at [26]), and especially along with the long-anticipated Apple entry into the XR world, such approaches are likely to become truly mainstream. We are seeing the beginnings of audio-centric experiences that utilize 6DOF sound to develop new forms of audiovisual expression that can exist in the real world, and a new generation of wearables will ultimately allow these experiences to become as popular and pervasive as mobile applications are for audiences today.

The 2019 launch and cross-company acceptance of both MPEG-H and object-based Sony 360 Reality Audio [27] demonstrate that the record and streaming industries are keen to explore immersive audio formats. If coupled with compatible Sony headphones, the system can already estimate head-related transfer functions (HRTFs) (see below) using self-taken photographs of the user's pinnae, and playback can only become significantly better if head-tracked with (say) the soon-to-be-released Dolby Atmos®-enabled Apple AirPods Pro. At the time of writing in 2020, the versatile video coding (VVC) of MPEG-I – which is optimized for 3DOF+ (pitch, yaw, and roll with limited translational movement along the three axes, which can, for instance, facilitate parallax) is due to be standardized this year; however, further work is expected later to reduce the huge amounts of data (typically a couple of hundred Mbps) required for true 6DOF operation with high-quality video [28], and just as when the internet struggled to convey pulse-code modulation (PCM) audio, a lossy solution might be the only pragmatic route to proliferation.

Procedural playback requires both central processing unit (CPU) time and rules; these can now be detailed in the container program itself, and decoding is not hardware dependent, as was always required in the past. The branding and appropriation of basic procedural-audio functionality ought to be resisted where possible, and agreement on open formats in the mould of MIDI would undoubtedly be the best outcome for the industry at large.

A small number of scientific studies have been conducted that illustrate approaches to 6DOF, and some of these now follow. In 2009, Southern *et al.* [29, p. 715] demonstrated a method that facilitated "real-time dynamic auralization for wave-based approaches." This paper outlined how acoustical "walk-through" auralization could be achieved, although room impulse responses (RIRs) needed to be precomputed (to save computational power) at defined points in the listener's path. In 2011, Kearney *et al.* [30] built upon the well-established process of data-based auralization; convolving anechoic recordings of music with measurements of acoustic spaces to emulate performance within those spaces. They demonstrated how real-world acoustic and architectural measurements of Christ Church Cathedral, Dublin could be combined with computational-based auralization to realize a real-time walk-through with an associated computer-graphic visualization. "Hybrid reverberation was utilized, where early reflection synthesis was achieved using image source modelling, combined with the real-world diffuse field measurement," and a recording of a choir was made, both close mic'ed as a source, and ambiently as a reference for subsequent evaluation.

In 2018, Rivas Méndez *et al.* [31] conducted a case study around the recording of a jazz ensemble to compare potential 6DOF workflows: "object-based using spot microphones and diffuse field capture microphone arrays, Ambisonics with multiple-placed sound-field microphones and hybrid approaches that utilize the prior two methods." This employed the Google Resonance software development kit (SDK) to deploy virtual listeners, sources, and sound fields in a game engine model of Abbey Road Studio 3, and users could traverse a model of this environment in VR. Also in 2018, Plinge *et al.* [32] proposed a binaural 6DOF method using parametric modelling of first-order Ambisonics (recorded at a single point in space), and then rotating the field and applying distance-dependent gain, relative to dynamic vectors for both the sound source and listener. Lastly, and also in 2018, Bates *et al.* [33] took Lee's PCMA [34] approach to synthesizing virtual microphones for horizontal surround and replaced his five coincident pairs with four first-order Ambisonic microphones – the Perspective Control Ambisonic Microphone Array (PCAMA). The researchers then generated four virtual microphone response patterns encoded using fifth-order Ambisonics. Wearing a head-mounted display to track motion, a listener could navigate in space and the virtual patterns could be morphed from (say) a rear-facing cardioid to an omnidirectional and then a forward-facing cardioid, thus changing the aural perspective. This perspective was further enhanced by changing distance cues, and the virtual microphone's position could also be moved. Although described as a "preliminary investigation," this approach shows promise.

In traditional FPS games, there has tended to be a combination of fixed-channel-based assets (e.g. stereo music, 5.1 ambience) that get triggered according to game requirements, combined with some real-time audio objects that need to move around the listener in some way and are amplitude panned around the circle of speakers (usually standard home cinema formats, such as 5.1 or 7.1) to match the character's orientation. Headphone versions of the FPS game experience would often binauralize these panned elements or otherwise just fold them down to standard stereo. A sense of appropriate perspective on the sound effect in question would often be generated very simply through the roll-off of high frequencies according to distance (air absorption modelling) and/or the selection of a sound effect with appropriate "distance" on it (e.g. a real gunshot may be recorded at 1 m, 20 m, 100 m, and 1000 m, and the appropriate version selected at runtime). More sophisticated FPSs might include slap-back delays and occlusions timed to give approximations of the distance and direction of nearby buildings and walls. The UNITY version of dearVR offers real-time auralization – reflections – based upon room geometry, and the Steam Audio plugin similarly features an integrated toolset to control propagation, reflections, and occlusions. When a sound source passes behind a (virtual) object that in some way shadows the direct sound reaching the listener, some form of occlusion-emulation processing must be undertaken. Such principles have evolved from ray-tracing lighting emulations, which are based around mesh models of the virtual world, and audio can respond to this world via dedicated sound area meshes for "sounds" and effect area meshes for reverberation. An excellent and in-depth discussion of all such issues can be found in Goodwin [4], and the concept will be further amplified for 6DOF, below.

For a more evolved first-person 6DOF audio experience, further acoustic modelling is required to provide the information that is either not needed in the above FPS example or would otherwise be delivered by the physical position of virtual sound sources – or perhaps speakers – around the listener. The sound must continuously appear to be in the correct position in relation to the user's ears as the individual moves through the experience, and enough of the acoustic information that we would normally use to interpret position and geometry of sound sources (and environments) must be generated in real time and delivered in versions that match the left and right ear's different perspectives. To facilitate 3DOF perception of localization, a virtual 3D analogue of the (actual or an average) listener's head – an HRTF – must now also be modelled that will modify the sound in several ways before it gets rendered to the left and right headphone outputs. The HRTF supplies appropriate spectral filtering of sounds and timing differences that would normally occur as sound waves pass across and over the physical structures of the user's head and pinnae to each ear independently. A generalized expansion of this is offered in [35].

In order to achieve 6DOF, such well-established modelling needs to be augmented with the parameters associated with traditional FPS movement, as described above. Because of this, game engines make a good environment for undertaking 6DOF work. It is important to make the distinction that for games – or indeed the completely synthetically generated environments of VR – the geometric meshes are preordained. However, things are much more complicated for an XR system where audio might be generated in the application, yet requires treatment to sound as if it is emanating from the real-world environment surrounding the listener – at an arbitrary distance. Further, some degree of real-time volumetric capture is necessary to model not just the world, but also the reflectivity and absorption characteristics of its various components – hence its acoustic behaviour. Thus, such a hybrid process could deliver real-time rendering of the timing and level cues our brain requires to make sense of a 3D sound scene.

3.5 PRACTICALITIES

So, for a 6DOF scene, an algorithmic process takes a raw description of the scene and turns it into a real-time 6DOF experience for the listener.

Four interconnected parts can be thought of as follows:

1. The "dry" audio elements (sound sources) making up the 6DOF scene: the audio elements of the scene could be any kind of audio signal – pre-recorded, live input, or synthesized, and each of these might be monophonic, stereophonic, or Ambisonic.
2. The acoustic modelling of the surrounding world: while volumetric capture is steadily evolving via devices such as HoloLens 2, ascertaining the acoustic properties of environmental objects is nascent. Artificial

       intelligence offers opportunities in this regard since distinctive shapes like a table or sofa have relatively predictable acoustic properties.

3. The creation of listener position-dependent spatial cues for these audio elements: the most important spatial cues are the first few early reflections (which should arrive at each ear with both appropriate timing and spectral filtering), with the subsequent diffuse reverb being much less important for directionality, although still highly important for realism.

4. The rendering of 1, 2, and 3 through an appropriate head model: once these various elements are determined, they can then be binauralized through appropriate HRTF processing to create the real-time binaural feed that the user will ultimately hear. In order to offer a convincing and smooth effect as the user traverses the environment, this will need to be updated frequently, and interpolated between each scene transition.

Point 1 has been implicit in the preceding text; however, with regard to point 2, one aspect of early reflections and sound sources for 6DOF in general is in their ability to aid in the rendering of the occlusions mentioned above. The complexity involved in the real-time calculation of the various complex paths, reflections, and spectral changes of even a simple scene is immense and, in fact, gross simplifications of the actual physical processes are undertaken instead – namely low-pass filtering of the dry sources, as well as changing the paths of early reflections. Recognition and estimations of scene, lighting, material, and the geometry of environments through machine vision is seeing great R&D activity at present. Modern XR SDKs now create models by performing estimates of lighting and light sources as collected through a phone camera (especially multiple aperture versions), and they use this data to attempt to match the models' lighting and shadows. The "inside out" style of XR head-mounted display constantly scans its environment, and the latest generation of iPad features light detection and ranging (LiDAR) scanning, doubtless an integral feature of future iThings. The same also needs to happen for *audio* objects being placed in real-world environments – where conflicting reflection and reverberation characteristics in the audio (i.e. applying the "wrong" reverb to a virtual audio object in a real-world environment) can entirely break immersion and make things appear artificial. Again, these kinds of calculations are set to have greater available resources available to them in the coming years and we can expect to see considerable improvements, even on traditionally more processor-limited mobile applications, perhaps more so as edge computing proliferates.

    To elaborate on point 3, an important factor concerning the use of monophonic audio objects in game engines for 6DOF use cases is in the ability to add real-time early reflections and reverb to the sound sources so that they appear to sit realistically in the scene. The importance of this should not be underestimated as it is the main route (other than the dry sound level) by which a complex audio scene can be deciphered by the human hearing system. Early reflections are the primary way in which we make sense of a sound object in an echoic environment. Most modern spatializers designed for game engines offer at least first-order real-time early reflections – which give acoustic information about an object's directivity and position in relation to the listener and the space that the individual is in. It is computationally intensive to generate complex scenes with early reflections for many objects, especially when higher orders of reflection are calculated. Progress in this area – perhaps more so than in other popular research areas such as personalized HRTFs – is most likely to dramatically improve 6DOF audio in the future, and improvements in available computing power will facilitate this.

In the case of point 4 – if going from Ambisonic source material to a binaural headphone feed – a two-stage process has to occur. First, a virtual speaker array is set up around the listener's head and the Ambisonic material is then decoded – just as would be decoded to speakers in the real world, at a polling rate satisfactorily close to real time. The polling update ensures that spatial information is updated as the listener moves through the environment. The "live" outputs of this virtual speaker array are then sent through the multiple filters that comprise the HRTF system to achieve a binaural render to headphones. The resolution of the Ambisonic decode is thus dependent both on the resolution (or "order") of the Ambisonic recording and also upon the number of virtual speakers that are rendered. Resolution is also greatly dependent upon listener-HRTF matching. To achieve higher orders of rendering – and therefore greater angular resolution – very large numbers of channels need to be held in memory, processed, and rendered in real time (e.g. 36 channels for a single fifth-order source). Comprehensive technical discussions can be found at [36] and are also discussed in Chapter 6 by Armstrong and Kearney.

3.6 CAPTURE CONSIDERATIONS

The fundamental and unresolvable issues of Ambisonic microphone recordings are not mitigated even in the "perfect" conditions that can be modelled in a virtual environment. Ambisonics as a mathematical construct works well, but when transposed to the real world is founded upon the concept that the listener is at a single point in space – which clearly one (and one's two ears) are not; an Ambisonic decode essentially decodes for a point of origin somewhere in the middle of one's head. Inversely, Ambisonically recorded material can capture intensity differences from various different directions, but not timing differences, which are the precise form of information that the brain uses to decode spatial information via the ears. In this case, the timing differences that are delivered to one's ears are a function of the HRTF decode through the virtual speaker array and not from the material itself – these spatial cues are not present in the original material and cannot be extracted from it, nor superimposed upon it except by such artificial methods. There is something analogous here to the depth and parallax issues that 3D cinema has – where a fixed relationship between depth elements is set across both eyes so that the normal decoding of parallax by the visual system is impossible. The strain and unnatural nature of 3D glasses tends to be more evident than with sound, but is certainly there with Ambisonic recordings and playback, and these issues are inherently unresolvable.

The countervailing argument is that only lower frequencies (below around 800 Hz) are affected by this; therefore, the importance of this short-coming is ostensibly minimal. However, this makes several assumptions about the subtleties of audio cognition. There are a number of factors that remain opaque, but it is possible that lower-frequency directionality is important in determining environmental geometry (e.g. making sense of axial room modes, particularly in the horizontal plane, to which our local-ization is most sensitized), as well as making complete sense of the overall directionality of wideband noise (which includes many natural sounds we hear in our environment). To have directional information from the higher frequencies (level and spectral differences), but not corresponding timing differences from the lower frequencies creates a perceptual conflict that the brain must work hard to ignore or to integrate into the 3D image in some other way. This unresolvable cognitive conflict is a major reason why Ambisonic recordings can sound less spacious and psychologically "relaxing" than spaced microphone arrays – since the latter (when appro-priately designed) resolve many of the timing issues for the listener, who subsequently has to do far less subconscious mental gymnastics to compen-sate for the conflicting and artificial information being processed. Having said this, there is research that argues for HOA (with MPEG-H) as being the optimal delivery format for audio in XR [37].

One obvious approach following this is to deploy the virtual speaker array concept into the game engine itself. Since sound is being fairly accur-ately modelled in these systems anyway, in principle, what works in real environments ought to also work to some degree "in the box" – and this is somewhat true. By placing virtual speakers (which are in fact, just ordinary sound sources – perhaps with some directionality) in a procedural-audio scene, the effects of a speaker array can be simulated and the timing differences between elements can remain roughly intact. To expand upon this notion, one may move between arrays of speakers and construct large navigable spaces that can be covered by virtual speakers (or other audio objects that have suitable directionality) – much as was discussed in the real-world "haunted house" example earlier. This is only akin to the standard placement of monophonic sound objects in game engines and utilizes improvements to available spatialization of objects and their HRTF processing to attain something comparable to real-world results. Mach1 is one commercial solution for this type of post-production-orientated work-flow and is optimized for 360° video. This system uses the Shell Processing Support (SPS) format, a coincident alternative to Ambisonics [38].

One can worry about phase and timing issues, but when virtual speakers that mimic microphone positions are kept close to original layouts, these issues are little different from similar utilization of speakers in real-world environments, where we tend to have reasonable expectations of what will and won't work. The major issue then becomes auto-correcting levels and directionality – where necessary – for each sound source as the user traverses the scene. At the time of writing, modelling microphones such as the Antelope Edge Duo are evolving that can dynamically modify their polar patterns – the directionality and width of various sources – to accommodate the relative positions of a listener in a 6DOF environment, and this can even be done after actual recording via software. These are likely to become ubiquitous tools for managing complex and diverse microphone setups for 6DOF reconstruction.

Ambisonic and HOA microphones (e.g. the 64-capsule fourth-order em64 Eigenmike® by mh acoustics) can also be utilized for this kind of spaced microphone array. The combination of information from these multiple coincident directional microphone capsules becomes a perhaps even more useful setup for recreating 6DOF scenes than their use as single microphones – for the reasons detailed above. In this instance, virtual microphones of various widths and directions can be reconstituted from the information gathered by these multiple capsules, and most conceivable kinds of real-time virtual microphone reconstruction is possible when reconstituting a 6DOF scene using these materials. Zylia is a company that recommends that its third-order microphones can be set up in such a room-size grid. They use a bespoke renderer to create an HOA field for any listener position by interpolating multiple HOA signals from different ZM-1 microphones, thus offering a 6DOF scene [39], navigable within VR.

However, in general, the difficulties are twofold: one is the impractical number of tracks required and the enormous streaming and CPU load that this entails, and the second can be issues around timbre and microphone placement for these reconstructions. With regard to the first, even creating a modest "quad" spaced array with em32s would involve recording and manipulating 128 (pre-Ambisonic-encoded) channels, which is challenging for any commercial system at the time of writing, although 122 were successfully recorded simultaneously in the Abbey Road Studio 3 test recordings [40]. The current Zylia demonstration uses 30 microphones, hence 480 channels. Facebook Reality Labs is currently developing a 6DOF research dome equipped with 420 Core Sound TetraMics as part of its "Codec Avatars" project to render virtual humans [41]. The dome captures audio-visual data at a rate of 180 GB/s.

The second issue is timbre and the aesthetic qualities of the capture. What can get excluded in purely academic approaches to 6DOF reconstruction is that there is an art and aesthetic to achieving commercial and tasteful results, and that pure veracity of reconstruction is neither the aim nor the value of such recordings. Microphones with higher directivity patterns are less able to capture and recreate lower frequencies. Different capsule designs and manufacturing standards also impact upon the sound and "feel" of different microphones. Even without understanding all the principles involved, these two points are self-evident to anyone who has experience of using different microphones for their work. While multiple Ambisonic microphone capture seems like the obvious approach to capturing 6DOF scenes, practical commercial reality is such that they may not be ideal approaches, unless in those instances where one *is* looking for a highly realistic acoustic simulation of a space. However, artistic control over the results of a recording is enormously important to commercial engineers and should not be discounted when designing 6DOF capture and reconstruction.

For this reason, being able to choose a microphone and its placement with regards to sound sources in the scene are just as important as they ever were with more traditional stereophonic recordings – and one should not assume that placement of higher-order microphones or spatial arrays is the end of their work – it is more likely just the beginning. Moore asserts that "authenticity is ascribed to, rather than inscribed in a performance" [42, p. 220], but this does not necessarily hold for sonic capture. The engineer should not be swayed too much by a desire to capture everything with utter truth, rather to represent an artistic version of the scene at hand, and inscription might be necessary. It is by far the best approach to capture the absolute best quality material one can with actors and/or musicians at hand. This involves the traditional skills of microphone placement and using the ears to get suitable balances between direct and reverberant sound, and thus achieving pleasing timbre and spaciousness. One can always create something spatial with well-recorded sound from individual microphones, but the inverse is not necessarily the case – however "nice" a given room is – and often, that room is not ideal for every specific purpose.

This leads to an important point regarding 6DOF production in general – why bother with spatial capture (meaning multiple microphones being deployed to capture everything at once, with reproducible correlation between microphone information) of sound in the first place? There are good arguments for creating 6DOF sound scenes from scratch just using standard monophonic and stereophonic sound sources and utilizing synthetic acoustic simulation to create all the cues that are necessary to place these sounds believably in 6DOF environments. The advantages of this are many, the most important being complete control over the 6DOF scene and its elements. In fact, spatial capture should only be considered one kind of approach for specific kinds of materials with a specific audience experience in mind. Whereas a spatial capture is often the best approach for capturing a "realistic" snapshot of a scene (e.g. an acoustic musical ensemble), as above, pure reality is often the last thing people want in an artistic endeavour such as making music or cinematic scenes. "Reality" and "pleasing or believable spaciousness" are often conflated concepts, but they are, in fact, distinct – although they may overlap to a greater or lesser degree [43].

There are numerous (typically first-order) Ambisonic sound effect libraries currently available. These offer pre-recorded ambiences, which might include horizontal surround and elevation, and they can have individual audio sources superimposed upon them to create an immersive audio environment. Since XR budgets remain tighter than those in games, film, and (sometimes), music, it is understandable that practitioners might wish to utilize these to expedite the deployment of "non-focal" content, such as ambiences. However, there is also an argument to suggest that many of the kinds of scene for which these ambiences are intended could be better created using more controlled and controllable capture. These might be designed and mixed into bespoke HOA beds and then integrated with the discrete sources as above, to create fully tailored 6DOF scenes.

3.7 PROCESSING

Professional engineers adopt such bespoke approaches through a requirement for complete creative and technical control over the end results. Accordingly, audio in most high-end productions is entirely built upon the manipulation and processing of such simple sources, even though "surround," and now "surround with height (and objects)," have been the main output formats for many years. Of course, music – especially cinematic orchestral scores – is the exception to this scenario. However, powerful increments over the pre-recorded ambience libraries are emerging. Lee and Milln's microphone array impulse response (MAIR) library of [44] is an open-access library of RIRs created with 39 microphone arrays that include many with height, and it comes with a digital audio workstation (DAW)-compatible renderer. The renderer convolves up to 13 signals with the user-selected MAIR, thus being a powerful tool to simulate high-quality 3D recording with discrete sources. This was extended into a set of binaural and Ambisonic RIRs at multiple receiver positions and rotation angles [45] – specifically designed for 6DOF work. This library offers 13 positions in a reverberant concert hall with 100 dummy head rotations at each.

There are other production workflow considerations for 6DOF. The first and most major shift in working practice is the subversion of the concept of "post" production. Since the moment of rendering is now at the point of playback (not when recording down masters in a studio), this means that what has previously been deemed post-production is now, in some ways, a *pre-production* of elements. Post-production becomes the manipulation of that amalgam of rules that will determine rendering conditions at runtime – and is done in the game engine, middleware, or another bespoke real-time environment. This is rather self-evident for game-audio stalwarts, but can be something of a leap for engineers coming from other fields.

Nonlinear processing such as compression and reverb operate fundamentally differently under 6DOF conditions, and cannot be utilized in quite the same ways. Since these limitations are largely a CPU/resource issue, once again, things are likely to improve over time as computer processing architecture and capacity grows. The control of dynamics is fundamentally different in 6DOF environments since compression is only feasible on mono or stereophonic sources, and not on HOA beds or 6DOF scenes at large. In fact, compressing a "scene" makes very little sense when one has freedom of movement through it – and what in fact must instead be deployed is intelligent dynamic control of sound source levels in real time. Part of this dynamic level control is inherent to the built-in physics engines (e.g. frequency and level roll-off curves over distance), but further bespoke level and frequency rules may need to be created when these standard physical rules do not fully serve the *artistic* requirements of a given scene.

3.8 CONCLUSION

So, it can be seen that while 6DOF is in the ascendancy and is sure to shape future audio experiences in the rapidly pervading world of XR, as yet – as a technological paradigm – it is very much in its infancy. It is clear that procedural object-based audio is a key driver, and that much can be drawn from the world of games, where 6DOF character movement is long established. Game audio has also pioneered many of the mechanisms that 6DOF acoustics and perception must engage with, but for XR – which must map the physical world in real time, these mechanisms must be developed considerably. Ambisonics presents many opportunities, but it also brings limitations, and while it is undoubtedly a powerful 6DOF tool, it does not necessarily represent a complete solution in itself. Traditional music production and even the concept of post-production must evolve to facilitate 6DOF audio. Scientific research in the field seems to be in its nascence, yet the potential of virtual and extended worlds, and the large-scale investments currently seen in attendant technologies are sure to amplify this.

In conclusion, while at present it is only possible to offer a broad insight into 6DOF audio, the reader should feel confident that the sound engineer's traditional skills of listening, tasteful collation, and design of sound and music are as important in this world as they ever were. This new arena is too important to leave to game developers alone, and people with traditional audio skills are very much needed to push these boundaries forward – as they always have been. There is certainly a need for more intuitive and XR-friendly audio workflows – and these will come. These and all of the other challenges are being tackled by many people on many fronts, and we can expect to see and hear continuous improvements to the new sound world of 6DOF audio that will provide experiences to entertain and captivate us for many years to come.

REFERENCES

[1]   Qualcomm, "Extended reality XR | immersive VR," *Qualcomm*, 2017. www.qualcomm.com/invention/extended-reality (accessed Jul. 02, 2020).

[2]   R. Shumaker and S. Lackey, *Virtual, Augmented and Mixed Reality: 7th International Conference, VAMR 2015, Held as Part of HCI International 2015, Los Angeles, CA, USA, August 2-7, 2015, Proceedings*. Springer, 2015.

[3]   A. Farnell, *Designing Sound*. The MIT Press, 2010.

[4]   S. N. Goodwin, *Beep to Boom: The Development of Advanced Runtime Sound Systems for Games and Extended Reality*, 1st edition. Routledge, 2019.

[5]   D. Smalley, "Space-form and the acousmatic image," *Organised Sound*, vol. 12, no. 1, pp. 35–58, 2007.

[6]   N. R. Tringham, *Science Fiction Video Games*. CRC Press, 2014.

[7]   T. Gracyk, *Rhythm and Noise: An Aesthetics of Rock*. London, UK: IB Tauris and Co Ltd, 1996.

[8]   C. Pike, R. Taylor, T. Parnell, and F. Melchior, "Object-Based 3D Audio Production for Virtual Reality Using the Audio Definition Model," in *Proceedings of Audio Engineering Society (AES) International Conference*

on Audio for Virtual and Augmented Reality, Los Angeles, USA, Sep. 2016, Accessed: Mar. 21, 2017. [Online]. Available: www.aes.org/e-lib/browse. cfm?elib=18498.

[9]   T. Wishart and S. Emmerson, *On Sonic Art*. Psychology Press, 1996.

[10] S. Fontana, A. Farina, and Y. Grenier, "Binaural for popular music: a case of study," Jun. 2007. https://smartech.gatech.edu/handle/1853/49982 (accessed Jul. 08, 2020).

[11] P. White, "Binaural panning in logic pro," *Sound On Sound*, Oct. 2019. www.soundonsound.com/techniques/binaural-panning-logic-pro (accessed Jul. 08, 2020).

[12] Cydonia Collective, "Because it's Los Angeles," Apr. 2019. https://cydoniacollective.bandcamp.com/track/because-its-los-angeles (accessed Jul. 08, 2020).

[13] G. Reardon, G. Zalles, A. Genovese, P. Flanagan, and A. Roginska, "Evaluation of Binaural Renderers: Externalization," in *Proceedings of 144th AES Convention*, Milan, Italy, May 2018, [Online]. Available: www.aes.org/e-lib/browse.cfm?elib=19506.

[14] S. Zagorski-Thomas, "The musicology of record production," *Twentieth-Century Music*, vol. 4, no. 02, pp 189–207, 2007.

[15] L. Camilleri, "Shaping sounds, shaping spaces," *Popular Music*, vol. 29, no. 2, pp. 199–211, May 2010.

[16] J. Riedmiller, S. Mehta, N. Tsingos, and P. Boon, "Immersive and personalized audio: a practical system for enabling interchange, distribution, and delivery of next-generation audio experiences," *SMPTE Motion Imaging Journal*, vol. 124, no. 5, pp. 1–23, Jul. 2015.

[17] M. Armstrong, M. Brooks, A. Churnside, M. Evans, F. Melchior, and M. Shotton, "Object-based broadcasting – curation, responsiveness and user experience," presented at *IBC*, Amsterdam, The Netherlands, Jan. 2014, pp. 12–2, doi: 10.1049/ib.2014.0038.

[18] J. Woodcock, W. J. Davies, F. Melchior, and T. J. Cox, "Elicitation of expert knowledge to inform object-based audio rendering to different systems," *Journal of the AES*, vol. 66, no. 1/2, pp. 44–59, 2018.

[19] M. Boerum, B. Martin, R. King, and G. Massenburg, "Lateral Listener Movement on the Horizontal Plane (Part 2): Sensing Motion through Binaural Simulation in a Reverberant Environment," in *Proceedings of AES Conf.: 2016 AES International Conference on Audio for Virtual and Augmented Reality*, Los Angeles, USA, Sep. 2016, [Online]. Available: www.aes.org/e-lib/browse.cfm?elib=18507.

[20] P. Bahadoran, A. Benito, T. Vassallo, and J. D. Reiss, "FXive: A Web Platform for Procedural Sound Synthesis," in *Proceedings of 144th AES Convention*, Milan, Italy, May 2018, [Online]. Available: www.aes.org/e-lib/browse.cfm?elib=19529.

[21] A. J. Berkhout, D. de Vries, and P. Vogel, "Acoustic control by wave field synthesis," *The Journal of the Acoustical Society of America*, vol. 93, no. 5, pp. 2764–2778, 1993.

[22] O. A. Godin, "The Kirchhoff–Helmholtz integral theorem and related identities for waves in an inhomogeneous moving fluid," *The Journal of the Acoustical Society of America*, vol. 99, no. 4, pp. 2468–2500, Apr. 1996.

[23] T. Ziemer, *Psychoacoustic Music Sound Field Synthesis: Creating Spaciousness for Composition, Performance, Acoustics and Perception*. Springer, 2019.

[24] D. Menzies, M. F. S. Galvez, and F. M. Fazi, "A low-frequency panning method with compensation for head rotation," *IEEE/ACM Transactions on Audio, Speech, and Language Processing*, vol. 26, no. 2, pp. 304–317, 2017.

[25] T. Caelli and H. Bunke, *Spatial Computing: Issues in Vision, Multimedia and Visualization Technologies*. World Scientific, 1997.

[26] D. A. Satongar, A. Family, S. J. Choisel, and J. O. Merimaa, "Coordinated tracking for binaural audio rendering," US10278003B2, Apr. 30, 2019.

[27] Fraunhofer IIS, "MPEG-H Audio," *Fraunhofer Institute for Integrated Circuits IIS*, 2020. www.iis.fraunhofer.de/en/ff/amm/broadcast-streaming/mpegh.html (accessed Jul. 10, 2020).

[28] G. Lafruit, A. Schenkel, C. Tulvan, M. Preda, and L. Yu, "MPEG-I coding performance in Immersive VR/AR applications," presented at the IBC, 2018.

[29] A. Southern, J. Wells, and D. Murphy, "Rendering walk-through auralisations using wave-based acoustical models," in *Proceedings of 17th European Signal Processing Conference*, Glasgow, UK, Aug. 2009, pp. 715–719.

[30] G. Kearney, M. Gorzel, and F. Boland, "Real-time walkthrough auralization of the acoustics of Christ Church Cathedral Dublin," in *Proceedings of 8th International Conference on Auditorium Acoustics*, Dublin, Ireland, May 2011.

[31] D. Rivas Méndez, C. Armstrong, J. Stubbs, M. Stiles, and G. Kearney, "Practical Recording Techniques for Music Production with Six-Degrees of Freedom Virtual Reality," in *Proceedings of 145th AES Convention*, New York City, USA, Oct. 2018, [Online]. Available: www.aes.org/e-lib/browse.cfm?elib=19729.

[32] A. Plinge, S. Schlecht, O. Rummukainen, O. Thiergart, T. Robotham, and E. A. P. Habets, "Evaluating Binaural Reproduction Systems from Behavioral Patterns in a Virtual Reality—A Case Study with Impaired Binaural Cues and Tracking Latency," in *Proceedings of 143rd AES Convention*, New York City, USA, Oct. 2017, [Online]. Available: www.aes.org/e-lib/browse.cfm?elib=19292.

[33] E. Bates, H. O'Dwyer, K.-P. Flachsbarth, and F. M. Boland, "A Recording Technique for 6 Degrees of Freedom VR," in *Proceedings of 144th AES Convention*, Milan, Italy, May 2018, [Online]. Available: www.aes.org/e-lib/browse.cfm?elib=19418.

[34] H. Lee, "A New Multichannel Microphone Technique for Effective Perspective Control," in *Proceedings of 130th AES Convention*, London, UK, May 2011, [Online]. Available: www.aes.org/e-lib/browse.cfm?elib=15804.

[35] J. L. Paterson and G. Llewellyn, "Producing 3-D audio," in *Producing Music*, 1st edition, R. Hepworth-Sawyer, J. Hodgson, and M. Marrington, Eds. Routledge, New York, NY, 2019.

[36] F. Zotter and M. Frank, *Ambisonics: A Practical 3D Audio Theory for Recording, Studio Production, Sound Reinforcement, and Virtual Reality*. Cham: Springer Nature, 2019.

[37] S. Shivappa, M. Morrell, D. Sen, N. Peters, and S. M. A. Salehin, "Efficient, compelling and immersive VR audio experience using scene based audio/higher order Ambisonics," presented at the AES Conference on Audio for

Virtual and Augmented Reality, Los Angeles, USA, 2016. www.aes.org/e-lib/browse.cfm?elib=18493.

[38] A. Farina, "SPS and Mach1 spatial audio formats," 2018. http://pcfarina.eng.unipr.it/SPS-conversion.htm (accessed Jul. 14, 2020).

[39] "ZYLIA 6DoF VR demo," *Zylia Portable Recording Studio. Multi-Track Music Recording With One Mic*, 2018. www.zylia.co/zylia-6dof-vr-demo.html (accessed Jul. 14, 2020).

[40] H. Riaz, M. Stiles, C. Armstrong, A. Chadwick, H. Lee, and G. Kearney, "Multichannel Microphone Array Recording for Popular Music Production in Virtual Reality," in *Proceedings of 143rd AES Convention*, New York City, USA, Oct. 2017, [Online]. Available: www.aes.org/e-lib/browse.cfm?elib=19333.

[41] FRL, "Facebook is building the future of connection with lifelike avatars," *Facebook Technology*, Mar. 13, 2019. https://tech.fb.com/codec-avatars-facebook-reality-labs/ (accessed Dec. 11, 2020).

[42] A. Moore, "Authenticity as authentication," *Popular Music*, vol. 21, no. 2, pp. 209–223, 2002.

[43] A. Bourbon and S. Zagorski-Thomas, "Sonic cartoons and semantic audio processing: using invariant properties to create schematic representations of acoustic phenomena," presented at the *2nd AES Workshop on Intelligent Music Production*, Sep. 2016.

[44] H. Lee and C. Millns, "Microphone Array Impulse Response (MAIR) Library for Spatial Audio Research," in *Proceedings of 143rd AES Convention*, New York City, USA, Oct. 2017, [Online]. Available: www.aes.org/e-lib/browse.cfm?elib=19307.

[45] B. I. Bacila and H. Lee, "360o; Binaural Room Impulse Response (BRIR) Database for 6DOF Spatial Perception Research," in *Proceedings of 146th AES Convention*, Dublin, Ireland, Mar. 2019, doi: www.aes.org/e-lib/browse.cfm?elib=20371.

4

Gestural control for 3D audio

Diego Quiroz

4.1 INTRODUCTION

Ever since stereophonic recording and playback systems came into existence, methods and input devices designed to pan a sound source have been evolving alongside. State-of-the-art audio formats can now present sound around an entire spherical acoustic space, and the sound engineer can exert control in a 3D manner (left-right, front-back, up-down). Input devices that operate in three degrees of freedom (3DOF) or more have been implemented in digital musical instruments (DMI) and bespoke interfaces, and these are also able to control sound-design and spatialization parameters in the studio, live performance scenarios, or virtual reality (VR). The implementation of sophisticated software with data output from more than 3DOF control devices can enable gesture recognition methods for spatial manipulation. However, research on the interaction between gestural controllers and mixing engineers undertaking 3D audio tasks in a production environment is still in its infancy. The intention of this chapter is to cover important aspects related to the subject of gestural control for 3D audio, as well as to provide some insight into research in related subjects.

4.1.1 Evolution of spatial audio

A great deal of R&D has coalesced into the field of 3D or so-called "immersive audio," which can completely envelop listeners with sound – potentially coming from any direction, either via loudspeaker or headphone reproduction. Immersive audio content has also grown in terms of popularity and is becoming more accessible for consumers via cinematic formats, 360° video, and VR. In 2004, Hamasaki *et al.* [1] developed the "integrated surround sound panning system," and since then, tools for a mixing engineer to work in 3D have evolved. The panning and spatialization of monophonic and multitrack audio sources have become commonplace in the workflow of post-production processes in these immersive audio formats (examples can be seen in [2] and [3]). The production of immersive audio is in its infancy, but bespoke workflows are now being defined and formalized [4]–[6]. There are now numerous 3D audio plugins that can manage the spatial parameters of a production session, render the audio signals into the designated channels, and send them to the main system's outputs, according to various formats such as horizontal surround, and formats with height and binaural renders – and custom tools can be built, for instance, with the Spat~ object in Max [7].

The level of control required for the fluent manipulation of these extended spatial characteristics often surpasses the conventional control mechanisms that are currently embedded in typical audio systems. The dynamic panning of sound sources aims to provide the listener with the perception of movement when reproduced with a discrete set of loudspeakers or binaural headphone reproduction systems. Many methods have been developed to create the most seamless movement of an audio source along a trajectory from one position to another. Vector-based amplitude panning (VBAP), distance-based amplitude panning (DBAP), and wave field synthesis (WFS) are some examples of boundary panning (bounded by discrete loudspeakers), and each of them has strengths and weaknesses when implemented in practical applications. An interesting comparison of perceived localization can be found in [8]. Object-based audio has also garnered attention and interest from the academic community, as well as commercially in broadcasting and interactive audio applications [9], [10], and more detail can be seen in Chapter 1 by Pike. Ambisonics and virtual-source panning for headphone reproduction have also been developing in the field of immersive audio for VR and other "extended realities" – an argument for higher-order Ambisonics in VR can be found in [11] and a detailed discussion of the principles can be found in Chapter 6 by Armstrong and Kearney. Dolby Atmos® is an object-based audio format that originated in cinema, but is now extending to television and video games. Atmos can reproduce height, either via speaker systems or headphones [12].

4.1.2 Evolution of panners/controllers

While substantial research progress has been made for 3D audio recording and playback systems, the evolution of related control devices for audio engineers has been much slower [13]. Most of the traditional input devices that audio engineers and industry professionals have available for manipulating audio objects can be categorized into the principal areas of faders, rotary knobs, touchscreens, and mouse/keyboard systems [14]. Rotary pots have been employed extensively in stereophonic systems to pan between left and right channels, and have become the standard for stereo mixing consoles in the audio industry. In the 1990s, the 2D formats 5.1 surround sound and then later, 7.1 became commonplace in audio for video for both cinema and home theatres via MPEG-2 [15], and from the outset, mix engineers relied upon traditional mixing interfaces, such as faders and knobs – drawn from stereophonic production, as well as using them for new modes of signal processing, such as source size, divergence, and panning in two dimensions (left-right and front-back). To support this, mixing consoles came to be enhanced with the implementation of joysticks (sometimes motorized [16]) and then later, touchpads, alongside the conventional

Figure 4.1 3Dconnexion SpaceMouse controller.

mouse/keyboard methods of control for in-the-box mixing environments –
digital audio workstations (DAW). Joysticks, trackpads, and computer mice
can be very efficient in manipulating 2D-pan-control parameters, but
become less effective when working with 3D audio mixing environments
[17]. Recently, some DAWs and professional mixing consoles for film
have been paired with a 3D input device or gestural control to provide the
user with an enhanced control scheme in order to access these novel added
dimensions, for instance, the Fairlight *EVO* console with the Leap Motion
controller [18], and Merging Technologies' *Pyramix* with 3Dconnexion's
3D Mouse series [19]; the controller can be seen in Figure 4.1.

4.1.3 Evolution of gestural controllers

The study of human gestures in a musical context has been around for
a long time, and gesture-acquisition technology and methods of control
for sound creation and manipulation has advanced along with it [20]–[22].
The earliest example of a mid-air (non-contact) gestural controller was
the Theremin, built around 1920 by Lev Termen (later known as Leon
Theremin) [23]. The Theremin consists of two antennae that sense the
proximity of the user's hands and control frequency (pitch) and amplitude
(loudness) via heterodyning. More recently, *The Hands* by Michel Waisvisz
(1984) propelled the development of gestural controllers towards sound
manipulation and music composition. *The Hands* contraption combined a
number of different types of sensors, such as pressure sensors and mercury
switches, into a pair of "data gloves" that enabled gestural control of the
different elements of a musical performance in real time with the user's

hands [24]. Mid-air gestural devices like Don Buchla's 1991 *Lightning* and the Interactive Light (later Roland) *D-Beam* in 1996 followed, employing infrared technology for motion detection [25], [26].

The definitions of gestural controllers and DMIs were tacitly established soon after, and some developers continued to follow approaches that projects like *The Hands* had developed [27]. The simultaneous control of multiple parameters became a highly researched subject in gestural control of computer-generated sound [28],[29]. The technology of devices that can capture human gestures and movements can be regarded as having reached an advanced stage, including both mid-air movement devices and those that function through manipulation [30]. Most gestural controllers have been designed for controlling the parameters of live musical performance and composition [31], [32].

4.2 3D INPUT DEVICES AND CLASSIFICATION

All input devices have their own strengths and weaknesses. Similarly, each task in an immersive audio environment has its own unique demands. As a designer, a challenge resides in acquiring a match between the application and input technology, which involves identifying the parameters that distinguish the application's demands. Devices must be chosen to give the best response to the demands of their range of tasks, and the breadth of this response is a crucial characteristic in every type of application [33]. Some of the criteria that may be influential in such interaction with immersive audio tasks are described below:

4.2.1 Isotonic vs. elastic vs. isometric

Isometric devices, also known as pressure/force devices, sense force but do not (perceptibly) move, and form a connection between the human limb and machine through force or torque, while an isotonic device achieves this through movement. Thus, isotonic devices, also known as displacement or free-moving devices, should have zero or constant resistance. Elastic devices fall in between these two classes – with variable resistance [34].

4.2.2 Direct vs. indirect input

Direct input is when the position of an on-screen pointer is the same as the position of the physical input device (e.g. stylus on a tablet, or finger on a touchscreen). An indirect input device doesn't have the same physical position as the pointer, but the movement translates from the user interface [35]. Direct input for the 3D position of an audio object is currently available for VR 3D mixing environments in projects like DearVR *Spatial Connect* [36], and Ultraleap's *Project North Star* [37] will surely be employed in due course for mixed reality applications. However, indirect input has been also used very effectively with certain input devices and panning system combinations.

4.2.3 Absolute vs. relative

Whether a device perceives values in an absolute mode or relative mode strongly affects the level of articulacy that any given system can allow between the input device and user [38]. A tablet is operated by the user in absolute terms, where the positioning of the pointing device (the user's fingertip) determines the position of the pointer device in the tablet. Conversely, a computer mouse works in relative terms, where the pointer's position is relative to the mouse's position and translation, which is scaled traditionally by a range control, also called the sensitivity parameter. Some studies point towards the relative mode being more effective when accurate operation is paramount.

4.2.4 Integral vs. separable

"Attributes that combine perceptually are said to be *integral*; those that remain distinct are *separable*. For example value (lightness) and croma (saturation) of a color are perceived integrally, while size and lightness of an object are perceived separably" [39]. An input device that operates with more than one degree of freedom can be categorized as being integral or separable if it is possible to move diagonally across dimensions. Performance improves when the structure of the perceptual space of the task reflects that of the control space of the input device. It is easier to draw a square with an "Etch A Sketch®" than freehand with a pencil, but conversely more difficult in the case of a diagonal line, which requires the modification of both horizontal and vertical axes simultaneously.

4.2.5 Position vs. rate control

Position control is where the transfer function from human operator to object movement is a constant and has a one-to-one correspondence between the input and output. *Rate control* maps human input to the velocity of the object movement. *Acceleration control* is said to be a hybrid of position and rate control [40].

4.2.6 Translation vs. rotation – six degrees of freedom

Some devices can only move in three dimensions (left-right, front-back, up-down) but not rotate; such movements are termed translations. Other devices can also rotate around those axes (roll, tilt, and yaw). Devices that can translate and rotate in all six degrees of freedom simultaneously are termed 6DOF [41]; Llewellyn and Paterson discuss this concept further in Chapter 3. When machine learning technologies are used to augment 6DOF gesture-recognition devices, highly sophisticated forms of interaction can be provided to the user [42].

4.3 CONSIDERATIONS AND ISSUES FOR GESTURAL CONTROL OF 3D AUDIO

There are a number of further concerns that are relevant when applying gestural control to 3D audio. These will now be considered.

4.3.1 Frame of reference

A strategic "frame of reference" is required when panning an audio source in a mixing environment. An *allocentric* frame of reference represents an audio object's location with reference to both the locations and directions of other objects in the environment. This method is commonly used in speaker-based production and playback. Conversely, an *egocentric* frame of reference represents an audio object's position relative to the location of the listener [43]. This type of reference is mostly used in binaural production and playback, and also in sound installations where the observer's experience is the main focus in the production.

4.3.2 Panning algorithm

The panning method that is employed in a given mixing environment is crucial to the perception of audio objects, particularly when the location of a source is inside the boundaries of a speaker array. Some disparity exists about the deployment of Cartesian (x, y, and z) versus polar coordinates (azimuth, elevation, and distance) for panning audio objects in 3D audio. For example, considerations might differ depending on whether VBAP or DBAP is utilized in the mixing environment. In the VBAP scenario, audio source localization relies upon the location of the speakers relative to a referenced "sweet spot" (where the listener sits) [44], and panning controls for azimuth and elevation parameters would give the engineer stronger localization cues when moving a sound source around such a *dome*-shaped environment. In this case, input devices and gestural controllers that are suited for separable tasks will outperform those who are not designed as such. This can also be the case for devices that can provide haptic feedback to aid movement constriction to the "virtual walls" of the environment [45]. The DBAP panning method makes no assumptions about the speaker layout or location of the listener, which means it is not sweet-spot dependent, and the localization perception of a sound source when its position moves *through* the mixing environment can improve dramatically [41], [46]. Isotonic devices that can operate in 3DOF or more are particularly appropriate for employing Cartesian coordinate-based methods of panning.

4.3.3 Mapping

Most of the research around the implementation of mapping techniques between input devices and audio resides in the areas of music performance and live sound spatialization (the interested reader might consult [40], [47]–[49]), but little exists around the mapping of existing spatial audio

DAWs and spatialization software. One-to-many, many-to-one, multiple layers, metaphorical gestural control, and machine learning methods have been applied for many years in performance and musical gesture acquisition: research on gestural control for 3D audio could benefit greatly from this knowledge base if the implementation of these mapping schemes was applied to 3D post-production tasks and scenarios. One-to-one mapping relationships have generally been employed when approaching gestural control using devices with 3DOF (or more) for 3D audio object positioning and spatialization. There is significant knowledge to be gained by investigating innovative and efficient methods of panning in multiple dimensions with 6DOF devices [17].

4.3.4 Range control/sensitivity

Absolute methods of mapping may have some advantages when the low-level control of a sound source is required; however, they may produce poor results when accurate simultaneous positioning in three dimensions is required. The relative control method relies on a sensitivity setting to scale the user's movement to an appropriate displacement of the audio object being panned and a "clutch" to indicate the start and end points of such a displacement (analogous to how a mouse button works for click-and-drag operations). Proper sensitivity settings can alleviate operational fatigue and induce improved stability overall [50].

4.3.5 Localization performance

Localization performance spans two aspects: *localization error*, which is the accuracy of the auditory system in estimating the spatial position of a sound source, and *localization blur*, which is a measure of the perceptual resolution of the smallest change in the position of a sound source. Previous research into localization indicates that the following have a significant effect when controlling sound sources in a 3D audio format:

1. Localization accuracy is best for sound sources in front of the listener
2. Localization accuracy is worst – by a wide margin – for sounds above and slightly behind the listener
3. In the front hemisphere, optimal performance extends over polar angles from -45° to +30°
4. In the front hemisphere, responses are biased towards locations just above the horizontal plane
5. In the rear hemisphere, responses are biased forward and towards the left and right poles

4.3.6 The sense of proprioception and ego-location

Although not strictly a "sense," *proprioception* is the human awareness of the position and movement of the body and its component anatomy. Proprioception provides feedback for executing motor control tasks, and together with vision and vestibular inputs, generates feedforward signals to the motor system. Human operators make use of proprioceptive cues provided by input devices to perform tasks, and the type of interaction with the devices corresponds to whether those devices are isometric, isotonic, or elastic. Another important feedback closely coupled with proprioception is *egolocation,* which is the awareness of one's overall position within the defined operation space, or relative to objects in that space [32].

4.3.7 Haptic feedback

Substantial human-computer interaction (HCI) research has been performed on the human haptic channel (the sensation of touch and kinesthesia, other channels being, for instance, visual or auditory) and its interaction with input devices [38], [51]. Numerous investigations have been specifically conducted in the field of haptic feedback and gesture interactions with music [32], [52]. Some studies have investigated the use of haptic feedback devices and 3D audio, for example, [53], and they produced evidence that suggests that devices enhanced with haptic feedback are strongly preferred over those that don't, and in some cases, subjects even prefer such a control scheme without accompanying visualization [18], [54]. The controller used in some of these studies can be seen in Figure 4.2.

4.3.8 Ergonomics and fatigue

Since 3D audio object positioning and spatialization can be a slow and deliberate process, ergonomic qualities in devices are important. Mid-air

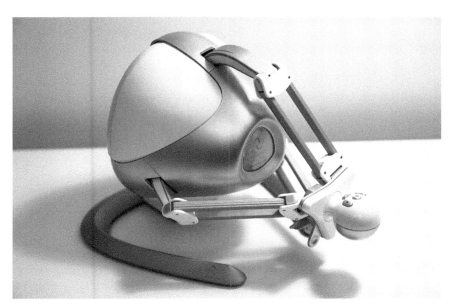

Figure 4.2 Novint Falcon haptic controller.

controllers and devices that are not operated by direct manipulation tend to create user fatigue issues, but proper training and generating familiarity with such devices may mitigate such ergonomic flaws. Proper sensitivity settings and visual feedback might also aid in relieving some of these issues [55].

4.3.9 Dominant vs. non-dominant hand

Applications of 3D audio gestural control could improve if the design process drew from investigations conducted into the interaction of 3D devices and asymmetrical bimanual tasks. An asymmetrical bimanual task is one that employs both arms (or hands) in a continuous and coordinated manner, where one hand is assistive and the other dominates the manipulation, such as holding a nail to be hammered or a needle to be threaded. Buxton stated that in order to design interfaces that make use of existing everyday skills in asymmetrical bimanual tasks, we need to adhere to some observations [51]:

- The non-dominant hand regulates the frame of action of the dominant hand; consider the nail and needle examples
- The order of action is the non-dominant hand, then the dominant; this is also seen in the examples
- The action of the non-dominant hand is of course relative to the action of the dominant hand; the positioning of a needle is not as demanding as the accuracy required to insert the thread

4.3.10 Visual feedback and other channels of perception

Visual aid and graphic representation of manipulated objects have been pervasive in HCI experiments with input devices, while in the realm of digital music performance, gestural controllers for DMIs rely mostly on direct audio feedback from the instruments and their output setup themselves [56], [57]. Some studies have attempted to understand the interaction between input devices and 3D audio, comparing scenarios with and without a computer screen; these might shed some light on the future design of intuitive visualization methods [18], [58].

4.4 GESTURAL CONTROL FOR 3D AUDIO

Input devices that function in 3DOF and 6DOF have been evolving over the past several decades, and extensive research has been conducted in the area of the gestural capture of music instrument performers in order to remap movement to DMIs, but the same cannot be said in the area of 3D audio production [59], [22]. The practice of 3D audio mixing and spatialization is no different from a musical performance in this respect. The gestures used by engineers convey clear expressions and have complex metaphorical significance, yet insufficient research has been conducted to understand how they can be adapted to a 3D audio workflow where multiple dimensions of control could be exploited [60]. Advanced recording techniques, rendering methods, and mixing interfaces for 3D audio are constantly evolving, and the integration of available 3D input devices could be advantageous in improving workflow design for the future generation of DAWs [60]. Investigations into effective functions, primary working circumstances, and workflow ergonomics are required to define the design and practicability of gestural user interfaces in 3D audio mixing [13]. Gestural controllers like the Leap Motion, and custom-built data gloves have been manufactured to capture gestures from the hands of the user in 6DOF or more. Alongside, technologies in gesture recognition have grown considerably, and together with powerful machine learning algorithms, new ideas around the interaction of these devices and standard audio environments have emerged [60], [61].

4.4.1 Gestural control for 1D/2D audio

Once gesture recognition came to be available to user interface designers, hand gestures were investigated for implementation in the control of various parameters involved in traditional stereo or surround audio production sessions. Some of the aforementioned devices even employ a library of gestures mapped to control parameters common in a typical mixing session – pan position, volume and reverberation, compression and other effects, and, in some cases, even transport controls, such as play and stop [62]–[64]. In these studies, the use of metaphoric gesture acquisition taken from traditional mixing workflows can be observed, such as sliding a hand "in and out" for volume, like an engineer would perform on a mixing console, or the use of a "stage metaphor" rather than a "channel-strip" metaphor for panning the audio channels in the stereophonic space [63]. The subjects in these experiments were left with impressions of creativity and intuitiveness when operating with these methods, and observed a decrease in need for visual feedback while operating the system. This allowed the auditory channel to predominate, ultimately resulting in more focused mixing sessions [65], [66]. However, Wakefield *et al.* also indicated that there was a limit to audio objects for which one can have appropriate control at any time as the visual interface began to appear cluttered [67].

Other studies implemented gesture recognition algorithms for some innovative ways of control. In [65], unconventional functions like *solo* and *mute* were controlled by making hand gestures, such as pointing or making a fist, while leaving the controller's 3DOF to control the volume (using the mixing console metaphoric gesture), pan, and reverb send level. Another study implemented the Leap Motion controller as the input device, and the results demonstrated that a lack of haptic and tactile feedback in this device – intrinsic characteristics of a mid-air controller – affected the controller's performance dramatically [50]. Furthermore, when working in VR environments, control schemes generally use the physical controllers included in the VR set, and a recent study provided evidence that suggests physical controllers are preferred over the Leap Motion controller [55].

Figure 4.3 The Leap Motion controller.

These results highlighted the observation that there was a clear match in the perception of physical controllers to that of the VR mixing environment.

4.4.2 Panning control for 3D audio with 3DOF devices

As discussed above, one of the most fundamental aspects of spatialization is panning, and object-based 3D audio has become widespread in cinema [68], [12] and is also pervading broadcasting. Despite this, R&D in mixing environments, control surfaces, and devices that can manipulate audio sources within 3D playback formats are still in their early stages. This contrast between industry and academia advancement conditions creates a shortage of appropriate interfaces and methods of control for accurate and intuitive manipulation.

Input devices utilized in standard systems, such as faders, rotary knobs, trackpads, and computer mice, only provide 1D or 2D simultaneous control. To attain separable control of height parameters, current DAWs have had to deploy a modifier key (e.g. shift) or an additional dial in order to enable 3D input in their mixing environments and panner windows [69]. Pro Tools, for example, utilizes a "steering" metaphor to gesturally control trajectories that include access to the height speakers [70], [56]. The trajectory is still manipulated in a 2D manner, but the graphical user interface (GUI) object rises as it approaches the listener position (centre of the room), effectively transforming distance coordinates into elevation and azimuth changes, conforming to a dome-shaped stroke.

Such methods allow any 2D input device – such as a standard joystick (traditionally used for surround panning) or even a computer mouse – to be employed for the control of two axes simultaneously. Although it is not without limitations, the steering method can produce fast and practical trajectories that are suitable for what is becoming a standard approach to object-based audio mixing and automation. However, the steering method might render poorly when not only all three axes (*XYZ*), but also tilt, yaw, and roll (the other three degrees of freedom) require simultaneous control. The traditional 2DOF horizontal surround panner approach of mapping two parameters to control spatial placement over two axes can be extended to 3D – if using an appropriate 3DOF device. Almost every generic input device or system that allows the capture of 3D gestures, from motion tracking to infrared technologies, has been used in one way or another in an attempt to control audio objects in a 3D audio space [17]. Merging Technologies *Pyramix* is one of the only DAWs to offer out-of-the-box support of a 3D input device with the 3Dconnexion Space Mouse [71], and has a 6DOF isometric sensor. Such reliance on pressure can make both the footprint and operational space smaller than isotonic devices, which rely on movement.

Recent studies show that operating the *Space Mouse* for 3D audio panning is intuitive and has a low-level entry barrier for first-time users of a 6DOF isometric device [65]. Isometric devices such as the *Space Mouse*, have no mechanical travel time, so they are quick to control, but are disadvantageous when accurate positioning is required since they do not provide appropriate displacement cues proportional to the user's output. Conversely, Quiroz and Martin [72] provided some evidence that indicates that isotonic devices such as the Leap Motion, the Novint Falcon, and even a traditional computer mouse, produce faster results in some tasks that involved the accurate positioning of a sound source in a 3D space than their isometric counterparts. Contrastingly, the study also pointed out that the Space Mouse was significantly faster than all others when the localization task involved panning towards speakers located in the "corners" of a room [65]. In the same study, the Leap Motion was one of the most preferred devices and rated most appropriate for a 3D panning task, a result that was mostly attributed to intuitiveness and ease of control over 3DOF. However, being the only mid-air controller of the ones applied in the experiment, the Leap Motion provided no haptic feedback, forcing users to rely on their own proprioception cues for sensing the dimensions of the panning space in relation with their hand position. This also made the hand less steady, thus resulting in less accuracy, and such an operation also induced fatigue in some participants. Additionally, the Novint Falcon was highly ranked in appropriateness and preference, most likely due to the haptic feedback it provided and the fact that perceptually, its control range mapped to the acoustic space in a very intuitive manner. This appeared to corroborate Gelineck and Overholt [18], who showed that not only did the haptic cues provided by the input device (Falcon) significantly influence the general perception of control, but also that the compatibility of the control scheme with the task was crucial in deeming it appropriate and preferred over others.

It appears that, throughout these studies, devices that can operate in 3DOF or more are advantageous for 3D panning tasks since they are generally embraced with a feeling of intuitiveness and freedom of control, although no statistically significant difference in performance to the old 2D counterparts was found by Melchior *et al.* [50]. Investigations into intuitive and powerful control schemes employing these devices are most important for understanding the perception of the interaction between the user and controller, the underlying principles at work, and minimization of the cognitive load imposed upon the operator.

4.4.3 Gestural control for 3D audio and spatialization

The evolution of contemporary software offers further spatialization to a high level of sophistication and complexity [71], [3]. Radiation, spread, and barycentric rotation are some of the multi-parametric functions that are available for manipulating a sound source's energy. Beyond just spatialization, when working with multichannel stems in a 3D audio project, other manipulation opportunities arise, and all of these might utilize a 3DOF level of control and thus also benefit from intuitive approaches to that control. Additionally, perceptual factors around "a source in a room," such as the presence, warmth, and spatial qualities of the room and its reverberation, are also parameters that could benefit from being mapped to a multi-dimensional gestural control method that could handle more than 6DOFs. In a recent project, the author successfully programmed the Leap Motion to control various aspects of the panner windows of different types of tracks (mono, stereo, and multitrack) within a 22.2 NHK format production session [42]. With this application, which was built on the Max platform, the user was able to manipulate up to seven dimensions of control (x, y, z or azimuth/elevation/distance, tilt, yaw, roll, and divergence/source size) mapped from the Leap Motion controller. While manipulating a multichannel track with the 6DOF of the Leap Motion, the *grab* gesture of the operating hand (opening and closing the hand) was mapped to one more dimension of control, the options being source size, divergence, and a low-frequency effect (LFE) feed. This control method requires further evaluation to be pursued in order to fully validate its potential. Furthermore, this application can enable the use of both hands as a gestural controller in a similar way to the single-hand method, producing some enhanced intuitiveness in some users. Since this method relies on the relative values of control parameters, a foot controller was added to operate as the "clutch" to permit takeover. Lastly, the foot controller can also be enabled as the clutch for the single-hand option, relieving the non-dominant hand of its role, and allowing it to perform other low-level duties, like track selection and transport controls (play, stop, back, etc.).

4.5 THE FUTURE OF GESTURAL CONTROL FOR 3D AUDIO

The number of parameters that need to be control-enabled in 3D audio workstations and production workflows has increased dramatically, yet adding control-surface strips with dozens of knobs is not appropriate as an intuitive method of control. Mastering for media in 3D audio formats, automation of 3D trajectories of sound effects for film, and panning and spatialization of multiple sources for live audio events are just some of the scenarios that are beginning to require both appropriate methods of control, and proper implementation of appropriate gestural controllers and input devices. Research that focuses on the interaction of these controllers and immersive audio tasks is paramount for laying the foundation of the design of intuitive and robust methods of gestural control – for 3D audio that will be the norm in our future.

REFERENCES

[1] K. Hamasaki, W. Hatano, K. Hiyama, S. Komiyama, and H. Okubo, "5.1 and 22.2 multichannel sound productions using an integrated surround sound panning system," in *Proceedings of 117th Audio Engineering Society (AES) Convention*, San Francisco, USA, Oct. 2004, www.aes.org/e-lib/browse.cfm?elib=12883.

[2] B. Martin and R. King, "Three dimensional spatial techniques in 22.2 multi-channel surround sound for popular music mixing," in *Proceedings of 139th AES Convention*, New York City, USA, Oct. 2015. www.aes.org/e-lib/browse.cfm?elib=17988.

[3] Flux Software Engineering, "Spat Revolution." www.flux.audio/project/spat-revolution/ (accessed Jul. 27, 2020).

[4] C. Sexton, "Immersive audio: optimizing creative impact without increasing production costs," in *Proceedings of 143rd AES Convention*, New York City, USA, Oct. 2017. www.aes.org/e-lib/browse.cfm?elib=19321.

[5] J. L. Paterson and O. Kadel, "Immersive audio post-production for 360°; video: workflow case studies," in *Proceedings of AES International Conference on Immersive and Interactive Audio*, York, UK, Mar. 2019. www.aes.org/e-lib/browse.cfm?elib=20433.

[6] R. J. Ellis-Geiger, "Music production for Dolby Atmos and Auro 3D," in *Proceedings of 141st AES Convention*, New York City, USA, Sep. 2016. www.aes.org/e-lib/browse.cfm?elib=18479.

[7] T. Carpentier, M. Noisternig, and O. Warusfel, "Twenty years of Ircam Spat: looking back, looking forward," in *41st International Computer Music Conference (ICMC)*, Denton, TX, United States, Sep. 2015, pp. 270–277. https://hal.archives-ouvertes.fr/hal-01247594.

[8] M. R. Thomas and C. Q. Robinson, "Amplitude panning and the interior pan," in *Proceedings of 143rd AES Convention*, New York City, USA, Oct. 2017. www.aes.org/e-lib/browse.cfm?elib=19250.

[9] D. Jang, J. Seo, K. Kang, and H.-K. Jung, "Object-based 3D audio scene representation," in *Proceedings of 115th AES Convention*, New York City, USA, Oct. 2003. www.aes.org/e-lib/browse.cfm?elib= 12385.

[10] R. Oldfield, B. Shirley, and J. Spille, "An object-based audio system for interactive broadcasting," in *Proceedings of 137th AES Convention*, Los Angeles, USA, Oct. 2014. www.aes.org/e-lib/browse.cfm?elib=17471.

[11] S. Shivappa, M. Morrell, D. Sen, N. Peters, and S. M. A. Salehin, "Efficient, compelling and immersive VR audio experience using scene based audio/higher order Ambisonics," in *Proceedings of AES Conference on Audio for*

Virtual and Augmented Reality, Los Angeles, USA, 2016. www.aes.org/e-lib/browse.cfm?elib=18493.

[12] L. Inc. Dolby, "Dolby Atmos cinema sound," 2018. www.dolby.com/us/en/technologies/cinema/dolby-atmos.html.

[13] J. D. Mathew, S. Huot, and B. F. G. Katz, "Survey and implications for the design of new 3D audio production and authoring tools," *J. of Multimodal User Interfaces*, vol. 11, no. 3, pp. 277–287, Sep. 2017.

[14] J. W. East and P. A. Frindle, "The ergonomic challenge of dynamically automating all controls on a digital audio mixer without clutter," presented at the *101st AES Convention*, Nov. 1996. www.aes.org/e-lib/browse.cfm?elib=7385.

[15] F. de Bont, P. Dillen, L. van der Kerkhof, and W. ten Kate, "The high density DVD," presented at the de Bont, Frans, et al. "The high density DVD." presented at *11th AES Conference Audio for New Media (ANM)*, London, UK, Mar. 1996. www.aes.org/e-lib/browse.cfm?elib=7098.

[16] D. Ives and S. Demonbreun, "Audio console with motorized joystick panning system," US6813530B1, Nov. 02, 2004.

[17] A. Churnside, C. Pike, and M. Leonard, "Musical movements – gesture based audio interfaces," in *Proceedings of 131st AES Convention*, New York City, USA, Oct. 2011. www.aes.org/e-lib/browse.cfm?elib=16022.

[18] S. Gelineck and D. Overholt, "Haptic and Visual feedback in 3D audio mixing interfaces," in *Proc. of the Audio Mostly 2015 on Interaction With Sound – AM '15*, Thessaloniki, Greece, 2015, pp. 1–6.

[19] Merging Technologies, *3Dconnexion Mouse – Configuration Guide*, 2016. https://confluence.merging.com/pages/viewpage.action?pageId=14615125 (accessed Jul. 28, 2020).

[20] C. Cadoz and M. M. Wanderley, "Gesture – music," in *Trends in Gestural Control of Music*, IRCAM – Centre Pompidou, 2000, pp. 71–94.

[21] T. M. Nakra, "Searching for meaning in gestural data," in *Trends in Gestural Control of Music*, IRCAM – Centre Pompidou, 2000, pp. 269–300.

[22] B. Bongers, "Physical interfaces in the electronic arts," in *Trends in Gestural Control of Music*, IRCAM – Centre Pompidou, 2000, pp. 41–70.

[23] P. Nikitin, "Leon Theremin (Lev Termen)," *IEEE Antennas and Propagation Magazine*, vol. 54, no. 5, pp. 252–257, Oct. 2012.

[24] G. Torre, K. Andersen, and F. Baldé, "The hands: the making of a digital musical instrument," *Computer Music Journal*, vol. 40, no. 2, pp. 22–34, Jun. 2016.

[25] D. Buchla *et al.*, "Lightning Users' Guide," *Berkeley Buchla Associates*, 1991.

[26] A. L. Brooks, "Enhanced gesture capture in virtual interactive space (VIS)," *Digital Creativity*, vol. 16, no. 1, pp. 43–53, 2005.

[27] S. Jordà, "New musical interfaces and new music-making paradigms," in *Proceedings of the NIME Conference*, Seattle, USA, Apr. 2001.

[28] M. M. Wanderley and P. Depalle, "Gestural control of sound synthesis," *Proceedings IEEE*, vol. 92, no. 4, pp. 632–644, Apr. 2004.

[29] A. G. V. Martin, *Touchless Gestural Control of Concatenative Sound Synthesis*, M.Tech. Thesis, McGill University, Montreal, Canada, 2011.

[30] W. A. S. Buxton, "Input devices: an illustrated tour," in *Haptic Input*, 2016.

[31] C. Goudeseune, "Interpolated mappings for musical instruments," *Organised Sound*, vol. 7, no. 2, pp. 85–96, Aug. 2002.

[32] J. Rovan and V. Hayward, "Typology of tactile sounds and their synthesis in gesture-driven computer music performance," in *Trends in Gestural Control of Music*, IRCAM – Centre Pompidou, 2000, pp. 355–368.

[33] W. A. S. Buxton, "More to interaction than meets the eye," in *User Centered System Design: New Perspectives on Human-Computer Interaction*, Hillsdale, New Jersey: Lawrence Erlbaum Associates, 1986, pp. 319–337.

[34] S. Zhai and P. Milgram, "Human performance evaluation of isometric and elastic rate controllers in a six-degree-of-freedom tracking task,"in *Telemanipulator Technology and Space Telerobotics*, SPIE, 1993, vol. 2057, pp. 130–141.

[35] R. W. Lindeman, "Classifying 3D input devices," WPI Dept. of Computer Science.

[36] Dear Reality GmbH, *Dear VR Spatial Connect Manual*, DearReality, Manual, May 2019. www.dearvr.com/products/dearvr-spatial-connect.

[37] Leap Motion, "North Star Project by Leap Motion," *North Star Project*, Jul. 2020. https://developer.leapmotion.com/northstar.

[38] W. A. S. Buxton, "The haptic channel," in *Readings in Human-Computer Interaction: Toward the Year 2000, 2nd edition*, Morgan Kaufmann Publishers, 1995.

[39] R. J. K. Jacob, L. E. Sibert, D. C. McFarlane, and M. P. Mullen, "Integrality and separability of input devices," *ACM Transactions on Computer-Human Interaction*, vol. 1, no. 1, pp. 3–26, Mar. 1994.

[40] M. T. Marshall, J. Malloch, and M. M. Wanderley, "Gesture control of sound spatialization for live musical performance," in *Gesture-Based Human-Computer Interaction and Simulation*, vol. 5085, M. Sales Dias, S. Gibet, M. M. Wanderley, and R. Bastos, Eds. Berlin, Heidelberg: Springer Berlin Heidelberg, 2009, pp. 227–238.

[41] S. Zhai, *Human Performance in Six Degree of Freedom Input Control*, IE Thesis, University of Toronto, Toronto, Canada, 1994.

[42] D. Quiroz, "A mid-air gestural controller for the Pyramix® 3D panner," in *Proceedings of 2018 AES International Conference on Spatial Reproduction – Aesthetics and Science*, Tokyo, Japan, Jul. 2018, www.aes.org/e-lib/browse.cfm?elib=19643.

[43] J. C. Riedmiller, "Recent advancements in audio – how a paradigm shift in audio spatial representation & delivery will change the future of consumer audio experiences," *Spring Technical Forum Proceedings*, 2015.

[44] V. Pulkki, "Virtual sound source positioning using vector base amplitude panning," *Journal of the AES*, vol. 45, no. 6, pp. 456–466, 1997.

[45] L. Dominjon, A. Lécuyer, J.-M. Burkhardt, and S. Richir, "A comparison of three techniques to interact in large virtual environments using haptic devices with limited workspace," in *Advances in Computer Graphics*, vol. 4035, T. Nishita, Q. Peng, and H.-P. Seidel, Eds. Berlin, Heidelberg: Springer Berlin Heidelberg, 2006, pp. 288–299.

[46] T. Lossius, P. Balthazar, and T. de la Hogue, "DBAP – distance based amplitude panning," in *Proceedings of the International Computer Music Conference* (ICMC), Montreal, 2009.

[47] D. Van Nort, M. M. Wanderley, and P. Depalle, "Mapping control structures for sound synthesis: functional and topological perspectives," *Computer Music Journal*, vol. 38, no. 3, pp. 6–22, Sep. 2014.

[48] A. Hunt and R. Kirk, "Mapping strategies for musical performance," in *Trends in Gestural Control of Music*, IRCAM – Centre Pompidou, 2000, pp. 231–258.

[49] B. Caramiaux, J. Françoise, N. Schnell, and F. Bevilacqua, "Mapping through listening," *Computer Music Journal*, vol. 38, no. 3, pp. 34–48, Sep. 2014.

[50] F. Melchior, C. Pike, M. Brooks, and S. Grace, "On the use of a haptic feedback device for sound source control in spatial audio systems," in *Proceedings of 134th AES Convention*, Rome, Italy, 2013.

[51] W. A. S. Buxton, "Touch, gesture & marking," in *Readings in Human-Computer Interaction: Toward the Year 2000, 2nd ed.*, Morgan Kaufmann Publishers, 1995, Ch. 7.

[52] C. Roland, P. Mobuchon, J.-P. Lambert, F. Gosselin, and X. Rodet, "The phase project: haptic and visual interaction for music exploration," in *Proceedings of Int. Computer Music Conf.*, Barcelona, Spain, Sep. 2005.

[53] F. Melchior, C. Pike, M. Brooks, and S. Grace, "On the use of a haptic feedback device for sound source control in spatial audio systems," presented at the *134th AES Convention*, Rome, Italy, May 2013, www.aes.org/e-lib/browse.cfm?elib=16742.

[54] S. Gelineck, M. Büchert, D. Overholt, and J. Andersen, "Towards an interface for music mixing based on smart tangibles and multitouch," in *Proceedings of NIME Conference*, KAIST, Daejeon, Korea, May 2013.

[55] J. Bennington and D. Ko, "Physical controllers vs. hand-and-gesture tracking: control scheme evaluation for vr audio mixing," in *Proceedings of 147th AES Convention*, New York City, USA, Oct. 2019. www.aes.org/e-lib/browse.cfm?elib=20663.

[56] W. A. S. Buxton, "Gesture based interaction," in *Readings in Human-Computer Interaction: Toward the Year 2000, 2nd ed.*, Morgan Kaufmann Publishers, 1995.

[57] J. Malloch and M. M. Wanderley, "Embodied cognition and digital musical instrument design and performance," in *The Routledge Companion to Embodied Music Interaction*, Routledge, 2017.

[58] M. Boerum, J. Kelly, D. Quiroz, and P. Cheiban, "The effect of virtual environments on localization during a 3d audio production task," in *Proceedings of International Conf. on Spatial Reproduction – Aesthetics and Science*, Tokyo, Japan, Aug. 2018.

[59] P. Kolesnik, *Conducting Gesture Recognition, Analysis and Performance System*, M.Tech. Thesis, McGill University, Montreal, Canada, 2004.

[60] W. Balin and J. Loviscach, "Gestures to operate DAW software," in *Proceedings of 130th AES Convention*, London, UK, May 2011, www.aes.org/e-lib/browse.cfm?elib=15923.

[61] S. Sapir, "Gestural control of digital audio environments," *Journal of New Music Research*, vol. 31, no. 2, pp. 119–129, Jun. 2002.

[62] M. Lech and B. Kostek, "Testing a novel gesture-based mixing interface," *Journal of the AES*, vol. 61, no. 5, 2013.

[63] J. Ratcliffe, "Hand motion-controlled audio mixing interface," in *Proceedings of NIME Conference, London, UK, June–July*, 2014.

[64] R. Selfridge and J. D. Reiss, "Interactive mixing using wii controller," presented at the *130th AES Convention*, London, UK, May 2011. www.aes.org/e-lib/browse.cfm?elib=15863.

[65] J. Kelly and D. Quiroz, "The mixing glove and leap motion controller: exploratory research and development of gesture controllers for audio mixing," in *Proceedings of 142nd AES Convention*, Berlin, Germany, May 2017. www.aes.org/e-lib/browse.cfm?elib=18690.

[66] J. Mycroft, T. Stockman, and J. D. Reiss, "Audio mixing displays: the influence of overviews on information search and critical listening," in *Proceedings of the 11th International Symposium on CMMR*, Plymouth, UK, Jun. 2015, pp. 682–688, http://cmr.soc.plymouth.ac.uk/cmmr2015/downloads.html (accessed Nov. 26, 2015).

[67] J. Wakefield, C. Dewey, and W. Gale, "LAMI: a gesturally controlled three-dimensional stage leap (motion-based) audio mixing interface," in *Proceedings of 142nd AES Convention*, Berlin, Germany, May 2017. www.aes.org/e-lib/browse.cfm?elib=18661.

[68] Auro 3D, "Morten Lindberg On Auro-3D®: 'Recorded Music Is No Longer A Flat Canvas, But A Sculpture You Can Literally Move Around,'" Oct. 2014. www.auro-3d.com/blog/interview-morten-lindberg-on-auro-3d-recorded-music-is-no-longer-a-flat-canvas-but-a-sculpture-you-can-literally-move-around/.

[69] Dolby Laboratories, "Authoring for Dolby Atmos cinema sound manual," 2013. www.dolby.com/uploadedFiles/Assets/US/Doc/Professional/Authoring_for_Dolby_Atmos_Cinema_Sound_Manual(1).pdf.

[70] M. Thornton, "Avid release Pro Tools 12.8HD with complete Dolby Atmos integration | Pro Tools," *Pro Tools Expert*.

[71] Merging Technologies and R. Ryan, "Space Mouse configuration for Pyramix panner window." https://confluence.merging.com/display/PUBLICDOC/3DConnexion+Mouse+-+Configuration+Guide.

[72] D. I. Quiroz and D. Martin, "Exploratory research into the suitability of various 3d input devices for an immersive mixing task," presented at the *147th AES Convention*, New York City, USA, Oct. 2019. www.aes.org/e-lib/browse.cfm?elib=20667.

5

Psychoacoustics of height perception in 3D audio

Hyunkook Lee

5.1 INTRODUCTION

3D multichannel audio formats such as Dolby Atmos®, Auro-3D®, and NHK 22.2 commonly utilize loudspeakers elevated above the listener to add the height dimension to a reproduced sound field. This overcomes the limitations of conventional (horizontal) surround formats such as 5.1 and 7.1, which are limited to the width and depth dimensions. Compared with two-channel stereo, although such formats can enhance perceived spatial impression, they are still unable to accurately represent auditory events happening above us in real life (e.g. birds, helicopters, aircraft, or ceiling reflections in a room). Consider a musician playing an instrument in a concert hall. The direct sound from the musician arrives at one's ears from only a single direction, whereas the early reflections and reverberation arrive randomly from all directions. Humans do not necessarily recognize them as separate auditory events due to the precedence effect [1]. However, they play an important role in the auditory perception of the width, depth, and height of the space, which ultimately influences the quality of the auditory experience of that musical performance.

The so-called "height" channels used in 3D audio formats can help sound engineers provide the listener with a more realistic and immersive experience of sound reproduction. For film, pop, and electroacoustic music content, the height channels could be used to pan individual sound objects to different locations in a 3D space for creative purposes. In the context of classical music, the height channels are typically fed with ambient sounds (early reflections and reverberation) in order to enhance the reproduced sense of both spatial impression and realism, and accordingly, such features often increase the sense of auditory immersion.

At present, technologies for recording, mixing, coding, transmitting, and reproducing surround sound with height channels are rapidly advancing. They increasingly allow consumers to easily access 3D audio content in their home or with mobile devices. However, to utilize the height channels most effectively, sound engineers and researchers should carefully consider the psychoacoustic principles of auditory height perception since they are fundamentally different from those of conventional stereo and surround sound perception.

From this background, this chapter discusses some of the key psychoacoustic principles of vertical stereophony that are relevant to practical applications, such as 3D panning, microphone technique, mixing, and upmixing. In the remainder of this chapter, Section 5.2 first explains the basic mechanism of vertical auditory localization, and then Section 5.3 describes how a vertically orientated stereophonic phantom image is localized. Finally, in Section 5.4, the perception and rendering of the vertical spatial impression are discussed. In each of these sections, key differences between the horizontal and vertical auditory mechanisms are summarized, and practical implications are provided as to how the principles of vertical stereophony influence 3D audio recording, mixing, and upmixing.

5.2 VERTICAL AUDITORY LOCALIZATION

The "duplex theory" suggests that human auditory localization in the horizontal plane mainly relies on the interaural time difference (ITD) at frequencies below around 1 kHz and interaural level difference (ILD) at higher frequencies due to the head-shadow effect [2]. An ITD produces interaural phase differences (IPDs) at different frequencies, but at frequencies above around 1 kHz, the IPD becomes too ambiguous for the brain to resolve due to the short wavelength of the sound. It is important to note that the ITD model of the duplex theory is only valid for pure tones. Research found that for a complex sound like music, the brain tends to use the ITD of the "envelopes" of the ear signals at high frequencies rather than the IPDs of individual frequencies [3].

In the median plane (which is vertical and extends in front of and behind the listener) on the other hand, a sound arrives at both ears at the same time regardless of the elevation angle, and therefore the ITD does not exist. Some ILD would still exist at frequencies above around 8 kHz because the pinnae shapes of the left and right ears tend to differ from each other [4]. However, numerous studies have confirmed that auditory localization in the median plane is mainly governed by spectral notches and peaks present between 3 kHz and 12 kHz in head-related transfer functions (HRTF) [5]–[8]. The HRTF, which can be described as the frequency spectrum of head-related impulse response (HRIR), is unique for every different direction of sound due to the complex shape of the pinnae causing reflections and diffractions at high frequencies. For complex sounds, the brain uses this aural fingerprint as the main cue for resolving any potential front-back confusion, as well as localizing elevated sound sources. Below 3 kHz, a spectral notch is produced in the HRTF due to shoulder/torso reflections, with its frequency varying with the elevation angle of the sound source [9].

It is important to note that the perceived vertical position of a pure tone or a narrow frequency band is inherently determined by the frequency itself rather than the actual vertical position of the sound source. In general, the higher the frequency of a pure tone, the higher its perceived position tends to be. This phenomenon was first reported by Pratt [10] and in the literature is often referred to as the "Pratt" effect or the "pitch-height" effect [11] (in this chapter, the latter term will be used). Roffler and Butler [12] confirmed the existence of this effect using multiple pure tones played from loudspeakers elevated to different heights. They found that a higher tone tended to be localized higher than a lower tone, regardless of the physical height of the loudspeaker that presents the tone. A similar result was reported also for filtered noise signals; a noise that was high-pass filtered at 2 kHz was always localized at a similar position as the presenting loudspeaker, whereas a noise low-passed at 2 kHz was perceived to be below the ear level, regardless of the physical loudspeaker height.

Cabrera and Tilley [11] observed the pitch-height effect for octave-band noise signals. As can be seen in Figure 5.1, bands with a higher centre frequency were generally localized at a higher position than those with a lower one, regardless of the physical height of the loudspeaker. Furthermore, it is noticeable that the perceived positions of negatively elevated sources were higher than the physical heights of the sources, and this difference became larger with higher frequency bands. On the other hand, the full-band noise was accurately localized at every target position, which shows the importance of a broadband spectrum that includes high frequencies for vertical localization.

Blauert [13] found that certain frequency bands are strongly associated with certain perceived regions in the median plane. For example, the 1/3-octave-band centred at 4 kHz would be localized in front, the 8 kHz band overhead, and the 1 kHz band at the back, regardless of the loudspeaker position in the median plane. These particular bands are referred to as "directional bands." This might be explained physiologically by observing relative spectral differences between the HRTFs of the three source positions. For example, the HRTF for 90° elevation has more energy around 8 kHz and less energy around 4 kHz than the HRTF of the front position in the median plane. Similarly, the HRTF of the back position has slightly more energy at around 1 kHz, and that of the front position has more at 4 kHz.

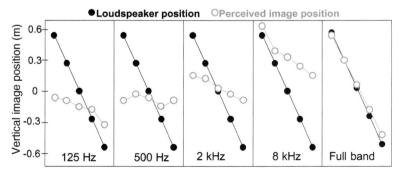

Figure 5.1 The "pitch-height" effect in the localization of octave-band and full-band pink noise in the median plane: after Cabrera and Tilley [11].

While past studies on vertical localization mainly focused on the median plane, Mironovs and Lee [14] investigated sound localization at various angles of elevation – and azimuth – with 30° intervals (Figure 5.2) in an acoustically treated listening room. For a broadband noise source, it was found that the accuracy of elevation localization was generally lower for the sources at the back of the listener than those in front. Furthermore, the accuracy of azimuth localization and front-back confusion became worse at a higher elevation. The fact that the localization of real broadband sources was not accurate for most of the tested positions implies that the accuracy of 3D amplitude panning would also be likely to be poor or even worse, especially both at a higher target-elevation angle and for sound sources with a limited frequency range. In fact, the existing 3D panning method "vector base amplitude panning (VBAP)" [15] is not based on psycho-acoustics, but designed as a computationally convenient method. Further research into a more perceptually optimized method for 3D panning seems to be necessary.

However, it should be acknowledged that our ability to localize highly elevated sounds seems to be inherently limited. In real life, most of the sound sources we encounter are at ear height or below (e.g. footsteps and keyboard typing). Positively elevated sources tend to have a narrow frequency range focused around high-mid frequencies (e.g. bird) or high-frequency roll-off characteristics (e.g. aeroplane), which considering the pitch-height effect, are not ideal for localizing them at their actual position. Reflections and reverberation arriving from above also tend to have reduced high-frequency energy due to absorption. We are not usually familiar with hearing artificial or musical sounds from positively elevated positions. This might explain why elevated real sources are more difficult to localize than horizontally located sources. The vertical-localization ambiguity could be easily resolved by moving the head. However, in multichannel listening, the listener usually faces forward

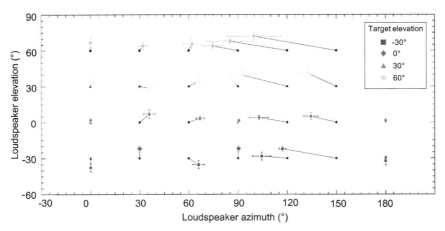

Figure 5.2 Localization of broadband pink noise at different azimuth and elevation angles [14].

constantly while listening to music or watching a film. It may be possible to improve the elevation-localization ability through continuous training and exposure to 3D sound content; head movement gives interaural cues that the brain can use to resolve not just front-back confusion, but also up-down confusion. Furthermore, the brain might be able to learn spectral patterns associated with various types of sound image, when elevated to "typical" positions.

Overall, the literature reviewed above generally suggests that in terms of localization, a 3D reproduction system would not significantly benefit from a height layer that presents a sound that predominantly contains low frequencies. Therefore, in film or popular-music production where the height layer would be used for sound source imaging, it is recommended that in order to maximize the sense of height in reproduction, sources containing sufficient high-frequency content are allocated to the height layer. However, for classical music recorded in an acoustic space, the height channels are mainly used for reproducing diffuse sound such as reverberation, which typically has a high-frequency roll-off characteristic. In this case, the height layer tends to produce a greater depth and openness to the reproduced sound field rather than providing a strong sense of vertical localization of the reverberation.

5.3 VERTICAL STEREOPHONIC IMAGING

In horizontal stereophonic reproduction, phantom-image localization essentially relies on the translation of interchannel level difference (ICLD) or/and an interchannel time difference (ICTD) between loudspeaker signals into the ITD and ILD. In two-channel loudspeaker reproduction, for example, the ipsilateral loudspeaker signal is summed with so-called "acoustic crosstalk" (i.e. the contralateral loudspeaker signal, which is both attenuated at high frequencies due to head shadowing, and delayed relative to the former due to the path difference between the loudspeakers and each ear). This causes both the ITD and ILD to be changed by variations of the ICTD or/and ICLD, leading to an angular shift of the image between the loudspeaker positions. This mechanism is called "summing localization." The minimum ICLD and ICTD required for a full image shift (e.g. ±30° for the standard 60° loudspeaker setup) are around 17 dB and 1 ms, respectively [17].

However, a trade-off between the ICLD and ICTD is possible for certain image positions. A different ratio between the two cues can produce different perceived spatial characteristics while maintaining the image position; for example, the more ICTD was used, the more image spread would be perceived and the less easy it would be to localize the image. If the ICTD is greater than around 1 ms and below the echo threshold (e.g. 30–50 ms for speech), the precedence effect [1] operates, where the phantom image is localized entirely at the position of the earlier loudspeaker.

On the other hand, in vertical stereophony – when two or more loudspeakers are arranged vertically, the ICTD and ICLD cues do not operate in the same manner as in horizontal stereophony. As discussed above, the auditory localization of a single sound source in the vertical plane is largely governed by spectral cues, and this also holds true for the localization of a vertical phantom image. Wallis and Lee [18] examined the influence of the ICTD on vertical stereophonic localization (via vertical time-delay-based panning). The stimuli were octave-band and broadband noise signals, and they were presented from loudspeakers placed at 0° and 30° elevation angles in front of the listener, with the upper loudspeaker delayed with respect to the lower by 0, 0.5, 1, 5, and 10 ms. If the summing localization and precedence effect still operated as in horizontal stereophony, the resulting phantom image would be localized at the position of the lower loudspeaker with the ICTD of 1 ms or greater.

However, as shown in Figure 5.3, this was not found to be the case. The perceived image position of the broadband noise was random between the lower and upper loudspeakers. Furthermore, the localization of octave-band stimuli was mainly governed by the pitch-height effect regardless of the ICTD. The broadband result seemed to be associated with comb filtering between the lower and upper loudspeakers arriving at the ear with an ICTD. The perceived image height might depend upon the positions and magnitudes of frequency notches and the resulting relative spectral energy weighting as per the directional bands theory by Blauert [13].

A vertically introduced ICLD has been found to be a more useful cue for vertical localization than an ICTD, although its perceptual resolution is still limited. Barbour [19] found that, in an experiment using an amplitude panning applied between loudspeakers arranged at 0° and 45° or 60° elevations in the median plane, the perceived image elevation angle of a noise source consistently increased as the ICLD increased. Mironovs and Lee [20] observed a similar phenomenon for various musical sources in an experiment with a pair of loudspeakers at 0° and 30° elevations, placed at either 0° or 30° azimuth. However, in both studies, the error bars (that represent inter-subject variability) of nearby ICLD conditions largely overlapped, suggesting that there were only three groups of localized points: lower speaker, upper speaker, and somewhere in between.

The above findings have an important implication for acoustic recording using microphones. In typical classical-music recording scenarios, the main microphone layer is responsible for source imaging, while the height layer is likely to be used for additional ambience. Since the vertical ICTD alone fails to trigger the precedence effect as discussed above, using vertically spaced omnidirectional microphones might suffer from vertical imaging instability and potential tonal colouration due to a comb-filtering effect at the ear. Therefore, the level of direct sound captured in the height-channel microphone(s) – known as vertical interchannel crosstalk (VIC) – would need to be reduced in order to have the source image aligned with the position of the main layer. Further, Wallis and Lee [21], [22] found that for the phantom image to be thus localized at the position of the main layer, the minimum amount of VIC level-reduction in the height layer should be about 7.5 dB for a vertical ICTD of 1–10 ms, and 9.5 dB for 0 ms ICTD for musical sources. This can be achieved effectively by employing cardioid or supercardioid microphones angled upwards for the height channels.

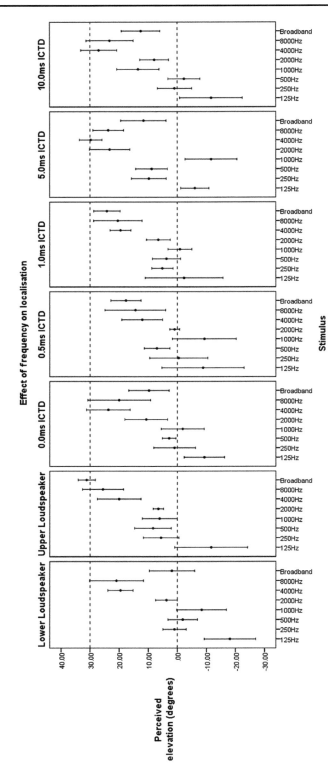

Figure 5.3 Experimental results showing the effect of ICTD applied to a vertically orientated loudspeaker pair [18].

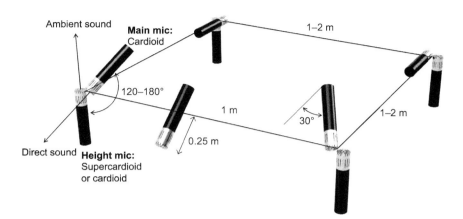

Figure 5.4 An example 3D microphone array configuration based on vertical stereo psychoacoustics: PCMA-3D [29].

An example 3D microphone configuration is shown in Figure 5.4; one cardioid microphone points towards the sound source and the other, either cardioid or supercardioid microphone, points directly upwards or faces away from the source for a maximal suppression of VIC. If the angle subtended between the microphones was just large enough to produce the 9.5 dB ICLD, the upper microphone would not disturb the localization of the phantom image. However, potential perceptual effects of VIC might still exist, for instance, vertical-image spread or tonal colouration.

For a complete separation between the main and height-channel signals, a larger ICLD is required. Using supercardioid or cardioid microphones for the height channels, this can be easily achieved by pointing the height microphone directly towards the ceiling or away from the source (i.e. a 180° "back-to-back" cardioid pair). The channel separation between the lower and upper microphones allows for greater headroom to change the level of ambience in the height layer. With omnidirectional microphones used for the height layer, raising the signal level of the height layer also increases the loudness of the source, as well as shifting the image upwards. On the other hand, with unidirectional height microphones and sufficient VIC suppression, the height ambience can be boosted without affecting the source image – which could be useful for controlling the vertically orientated spatial impression. The vertical spacing between the main and height microphones is discussed in Section 5.4.

Furthermore, it is worth mentioning that a phantom-centre image created between two loudspeakers tends to be inherently elevated upwards, with its angle being dependent upon the base angle of the loudspeaker pair. In 1947, de Boer [23] found a striking phenomenon; when two symmetrically arranged loudspeakers present identical signals, as the loudspeaker base angle increased, the perceived phantom-centre image also tended to be elevated to a higher position, and this effect was maximal when the base angle reached 180° (i.e. the loudspeakers at the listener's sides). This "phantom image elevation effect" depends upon the type of sound sources; sources with a flatter frequency response and a more transient nature tend to be elevated higher and with a greater certainty [24], as can be seen in Figure 5.5.

As mentioned above, HRTFs above around 6 kHz play a particularly important role in the auditory localization of a vertically positioned sound source. However, the perceptual mechanism of the elevation of a "phantom" source resulting from two loudspeakers at ear level relies on absolute cues at frequencies below 1 kHz, as well as relative cues at higher frequencies. The conventional explanation of this effect by Blauert is based on his directional boost theory [13], which suggests that the spectral energy weighting of the ear-input signal dictates the perceived elevation of a sound image. This was demonstrated in [21]; as the loudspeaker base angle increased from 0° to 180°, the resulting ear-input signal had more energy at frequencies around 8 kHz and less around 4 kHz (the directional bands theory proposes that 8 kHz and 4 kHz are mapped to above and front), thus leading to a more elevated image position.

However, the present author proposed a new theory – that the elevation effect is also associated with a spectral notch (at a lower frequency) caused by acoustic crosstalk [21]. Consider a sound radiating from a real

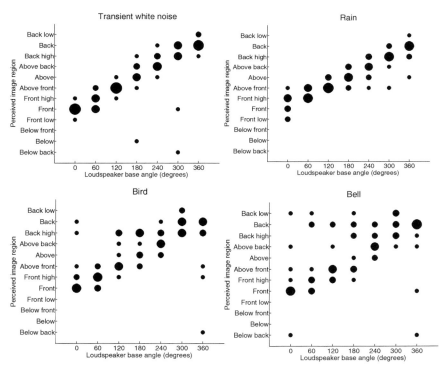

Figure 5.5 Experimental results showing sound source and loudspeaker base angle dependency of the phantom image elevation effect [24].

loudspeaker placed directly above the head. The sound will produce a shoulder reflection, which would arrive at the ear about 0.7 ms later than the direct signal. Now, when two laterally placed loudspeakers radiate the same signals, the time delay between the direct sound and the acoustic crosstalk would be almost the same as that of the shoulder reflection of the signal from the overhead loudspeaker. Both of these real and phantom-image conditions produce the first spectral notch at around 650 Hz. Therefore, the brain might use the notch as an elevation-localization cue for the phantom-centre image. This psychoacoustic effect has been exploited for a virtual elevation-panning method called "virtual hemispherical amplitude panning (VHAP)" [25], [26]. As illustrated in Figure 5.6, by using only four loudspeakers placed at ±90°, 0°, and 180° at ear height, VHAP is able to render an elevated phantom image in the upper hemisphere. This method has been found to produce perceptually convincing virtual height loudspeakers for practical Dolby Atmos® and Auro-3D® 9.1 content [27].

5.4 VERTICALLY ORIENTATED SPATIAL IMPRESSION

In the context of concert hall acoustics, the term "spatial impression" is often described to have two main dimensions: apparent source width (ASW) and listener envelopment (LEV). ASW is defined as the perceived spread of the sound source that is caused by reflections arriving within about 80 ms after the direct sound, whereas LEV is a sensation of being surrounded by a reverberant sound field, associated with reflections arriving more than 80 ms after the direct sound [28]. Although this might be considered to be a somewhat oversimplified paradigm, these two sub-attributes of spatial impression have been used as successful measures for evaluating the perceived qualities of concert halls.

In the context of sound recording and reproduction, the concept of ASW is described as phantom-image broadening or spread. The horizontal image spread (HIS) can be rendered to different degrees by applying ICTD-based panning [17] or interchannel decorrelation of stereophonic signals [29], or by changing the spacing between microphones at the recording stage [30]. In general, a greater amount of decorrelation (i.e. normalized cross-correlation coefficient being close to zero) or wider microphone spacing leads to the perception of a greater HIS. In surround recording and reproduction of acoustic music, the rear channels are typically used for reproducing diffuse ambient sounds for LEV, while the frontal channels provide directional imaging and HIS. The perception of LEV is also dependent on the magnitude of decorrelation, especially at low frequencies [31]; the more decorrelated, the more enveloping. In acoustic recording, the wider the spacing between two microphones, the lower the frequency at which a full decorrelation can be achieved. For example, based on the diffuse field coherence theory proposed in [32], a full decorrelation between two omnidirectional microphone signals is achieved at frequencies down to about 100 Hz for a 2 m spacing, but to about 300 Hz for a 50 cm spacing.

However, in 3D sound recording and reproduction, the effect of vertically applied decorrelation on vertical image spread (VIS) and LEV tends to be

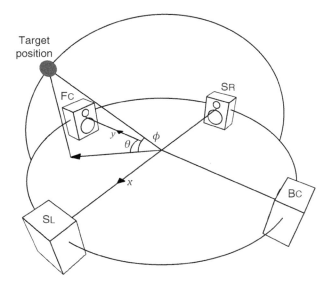

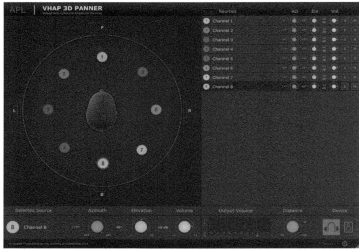

Figure 5.6 (top) VHAP for 3D panning without height loudspeakers: a concept diagram and (bottom) a bespoke VST plugin to control such panning [25, 26].

minimal. Gribben and Lee have conducted extensive investigations into the effect of vertical decorrelation on VIS perception. In [33], they showed that for the same degree of decorrelation applied to identical signals, the magnitude of VIS was significantly smaller than that of HIS. Further, they found in [34] that the effect of vertical decorrelation on VIS was largely negligible, although it did cause significant amounts of tone colouration below about 500 Hz. It was also reported that vertical decorrelation only applied above about 500 Hz between loudspeakers at 0° and 30° elevation angles did not have a significant difference from a full-band decorrelation in perceived VIS, regardless of the tested azimuth angles of the loudspeaker pair (0°, 30°, and 110°) [35].

The present author also proposed an alternative way for controlling VIS, named "perceptual band allocation (PBA)" [36]. Exploiting the pitch-height principle, PBA flexibly maps frequency bands decomposed from an original signal to either the main or height loudspeaker layer depending on the inherent perceived vertical positions of the bands. By applying different band allocation schemes, it is possible to control the perceived magnitude of VIS. With this method, the signals of each band are summed together at the ear without any of the phase cancellation that would likely occur with phase-based decorrelation methods. The simplest scenario of PBA is a split between low and high frequencies around 1 kHz, where the low-passed signals are routed to the main loudspeakers, while the high-passed ones are presented from the height loudspeakers. This has been applied to a 2D-to-3D upmixing of four-channel surround reverberation signals captured in a concert hall [37]. The result showed that, in some conditions, PBA-upmixed eight-channel reverberation produced higher listener ratings for spatial impression and subjective preference than the original reverberation captured using an eight-channel 3D microphone array.

PBA was also used in a low-complexity stereo-to-3D upmixing system proposed by Kraft and Zölzer [38]. In that system, direct and ambience signals are first separated from the original two-channel stereo content, and then the ambient signals upmixed to create surround channel signals using a decorrelation filter. Then, the resulting four-channel ambient signals are upmixed to feed four height channels using the two-band PBA method with a crossover frequency of 1 kHz.

As with the effect of vertical decorrelation, the effect of vertical microphone spacing on the overall spatial impression is minimal. Lee and Gribben [39] compared four different spacings between the main and height microphone layers (0, 0.5, 1, and 1.5 m) and found that there was no significant difference among 0.5, 1, and 1.5 m, while 0 m produced a slightly greater magnitude of spatial impression, as well as a higher preference rating. This finding led to a new design concept of the "horizontally spaced and vertically coincident (HSVC)" 3D microphone array, which has been adopted for PCMA-3D [29] (Figure 5.4), ORTF-3D [40], and ESMA-3D [41]. Different types of 3D microphone techniques are discussed in detail by Geluso in Chapter 12. In addition, an extensive open-access database of recordings and multichannel room impulse responses captured using a number of 3D microphone arrays, named "3D-MARCo," can be found in [42].

A recent study [43] objectively compared the 3D microphone arrays included in the 3D-MARCo database in order to consider the interaural cross-correlation coefficient (IACC). When measured for main layer reproduction with a five-channel microphone array, the IACC hardly changed after a four-channel height-microphone layer was added to provide extra ambient sounds, regardless of the types of microphone arrays. This suggests that the height channels might not increase the perceived

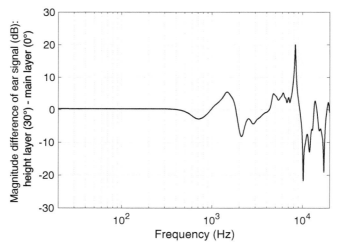

Figure 5.7 Delta-magnitude spectrum of the ear-input signal of loudspeakers at ±30° azimuth and 30° elevation to loudspeakers at ±30° azimuth and 0° elevation. Positive plots mean that the elevated loudspeakers produced greater energy at the corresponding frequencies in the ear signal.

magnitude of horizontal LEV, although they may still contribute to the perception of vertical LEV as Furuya *et al.* [44] suggested. However, the perceptual mechanism of vertical LEV and its role in the perception of both the overall spatial impression and the sense of immersion, have not yet been confirmed in the context of practical sound recording and reproduction. Considering that typical reverberation has a roll-off at high frequencies and a lack of high frequencies would cause the image to be perceived at a lower elevation than the presenting loudspeaker (due to the pitch-height effect discussed earlier), it can be considered that reverberation presented from a height loudspeaker might not produce a great sense of auditory height. As also mentioned earlier, in a room or a concert hall, we do not necessarily localize individual reflections or reverberation arriving from above, apart from distinct echoes, but they still contribute to the perceived size of a space, as well as adding to the realism and naturalness of a sound. This also reflects the present author's personal experience with height channels in 3D sound recording and reproduction.

Another benefit of the use of height channels for reproduction of reflections and reverberation might be that of perceived tonal clarity and balance. As shown in Figure 5.7, different loudspeaker positions produce different spectral balances of the ear-input signals due to HRTF. At typical height-loudspeaker position (e.g. 30° azimuth, 30°-45° elevation), the ear-input signal would have more energy between 4 kHz and 8 kHz and less energy between 2 kHz and 4 kHz compared with the ear signal at the corresponding main loudspeaker position (30° azimuth, 0° elevation). This means that a signal routed to the height loudspeaker would sound brighter than the same signal presented from the main loudspeaker, whereas the main loudspeaker would make it sound harder and more intelligible. When capturing reverberant sounds with main and height microphones, the spectral difference between the two loudspeaker layers could make the overall tonality of the reverberant sounds more balanced and "open" compared with when both signals are only routed to the main loudspeaker layer.

5.5 CONCLUSION

This chapter discussed psychoacoustic principles related to vertical auditory localization, vertical stereophonic imaging, and vertically orientated spatial impression – in the context of multichannel audio with height channels. From the research reviewed, it can be generally suggested that the addition of height channels in 3D audio systems would not automatically enhance the perceived spatial impression. The way we perceive a vertically orientated phantom image is fundamentally different from how we perceive a horizontal image in the width and depth dimensions. The former is mainly governed by spectral cues, whereas the latter is largely about interaural cues. For that reason, conventional recording and mixing techniques for stereo and surround sound would not work as effectively when they are applied to the height channels. Therefore, in order to be able to create perceptually effective 3D audio content, sound designers and engineers must carefully consider the relevant psychoacoustic principles, and adopt the most appropriate recording and mixing techniques for the height channels. Furthermore, researchers and developers working in the field of 3D audio are encouraged to consider psychoacoustics in the development of new techniques and tools for 3D audio content production.

REFERENCES

[1]   R. Y. Litovsky, H. S. Colburn, W. A. Yost, and S. J. Guzman, "The precedence effect," *The Journal of the Acoustical Society of America*, vol. 106, pp. 1633–1654, Oct. 1999.

[2]   J. Blauert and R. A. Butler, "Spatial hearing: the psychophysics of human sound localization by Jens Blauert," *The Journal of the Acoustical Society of America*, vol. 77, no. 1, pp. 334–335, 1985.

[3]   E. R. Hafter and R. H. Dye, "Detection of interaural differences of time in trains of high-frequency clicks as a function of interclick interval and number," *The Journal of the Acoustical Society of America*, vol. 73, no. 2, pp. 644–651, 1983.

[4]   C. L. Searle, L. D. Braida, D. R. Cuddy, and M. F. Davis, "Binaural pinna disparity: another auditory localization cue," *The Journal of the Acoustical Society of America*, vol. 57, no. 2, pp. 448–455, 1975.

[5]   R. A. Butler and K. Belendiuk, "Spectral cues utilized in the localization of sound in the median sagittal plane," *The Journal of the Acoustical Society of America*, vol. 61, no. 5, pp. 1264–1269, May 1977.

[6]   F. Asano, Y. Suzuki, and T. Sone, "Role of spectral cues in median plane localization," *The Journal of the Acoustical Society of America*, vol. 88, no. 1, pp. 159–168, 1990.

[7]   V. C. Raykar, R. Duraiswami, and B. Yegnanarayana, "Extracting the frequencies of the pinna spectral notches from measured head-related impulse responses," *The Journal of the Acoustical Society of America*, vol. 116, no. 4, pp. 2625–2625, Oct. 2004.

[8]   D. J. Kistler and F. L. Wightman, "A model of head-related transfer functions based on principal components analysis and minimum-phase reconstruction," *The Journal of the Acoustical Society of America*, vol. 91, no. 3, pp. 1637–1647, 1992.

[9]   V. R. Algazi, C. Avendano, and R. O. Duda, "Elevation localization and head-related transfer function analysis at low frequencies," *The Journal of the Acoustical Society of America*, vol. 109, no. 3, pp. 1110–1122, 2001.

[10] C. C. Pratt, "The spatial character of high and low tones," *The Journal of Experimental Psychology*, vol. 13, no. 3, pp. 278–285, Jun. 1930.

[11] D. Cabrera and S. Tilley, "Vertical localization and image size effects in loudspeaker reproduction," *24th AES International Conference on Multichannel Audio, The New Reality*, pp. 1–8, 2003.

[12] S. K. Roffler and R. A. Butler, "Localization of tonal stimuli in the vertical plane," *The Journal of the Acoustical Society of America*, vol. 43, no. 6, pp. 1260–1266, Jun. 1968.

[13] J. Blauert, "Sound localization in the median plane," *Acta Acustica united with Acustica*, vol. 22, no. 4, pp. 205–213, 1969. www.ingentaconnect.com/content/dav/aaua/1969/00000022/00000004/art00004.

[14] M. Mironovs and H. Lee, "On the accuracy and consistency of sound localization at various azimuth and elevation angles," *144th Audio Engineering Society Convention*, May 14, 2018.

[15] V. Pulkki and T. Lokki, "Creating auditory displays to multiple loudspeakers using VBAP: a case study with DIVA project," *International Conference on Auditory Display*, pp. 1–5, 1998. www.icad.org/Proceedings/1998/PulkkiLokki1998.pdf.

[16] C. J. Chun, H. K. Kim, S. H. Choi, S. J. Jang, and S. P. Lee, "Sound source elevation using spectral notch filtering and directional band boosting in stereo loudspeaker reproduction," *IEEE Transactions on Consumer Electronic*, vol. 57, no. 4, pp. 1915–1920, Nov. 2011.

[17] H. Lee and F. Rumsey, "Level and time panning of phantom images for musical sources," *Journal of the Audio Engineering Society*, vol. 61, no. 12, pp. 978–988, 2013.

[18] R. Wallis and H. Lee, "The effect of interchannel time difference on localization in vertical stereophony," *Journal of the Audio Engineering Society*, vol. 63, no. 10, 2015.

[19] J. Barbour, "Elevation perception: phantom images in the vertical hemisphere," *AES 24th International Conference Multichannel Audio*, pp. 1–8, 2003. www.aes.org/e-lib/browse.cfm?elib=12301.

[20] M. Mironov and H. Lee, "The influence of source spectrum and loudspeaker azimuth on vertical amplitude panning," *142nd Audio Engineering Society Convention*, 2017.

[21] R. Wallis and H. Lee, "Vertical stereophonic localization in the presence of interchannel crosstalk: the analysis of frequency-dependent localization thresholds," *Journal of the Audio Engineering Society*, vol. 64, no. 10, 2016.

[22] R. Wallis and H. Lee, "The reduction of vertical interchannel crosstalk: the analysis of localisation thresholds for natural sound sources," *Applied Sciences*, vol. 7, no. 3, 2017.

[23] K. de Boer, "A remarkable phenomenon with stereophonic sound reproduction," *Philips Technical Review*, vol. 9, pp. 8–13, 1947.

[24] H. Lee, "Sound source and loudspeaker base angle dependency of phantom image elevation effect," *Journal of the Audio Engineering Society*, vol. 65, no. 9, 2017.

[25] H. Lee, D. Johnson, and M. Mironovs, "Virtual hemispherical amplitude panning (VHAP): a method for 3D panning without elevated loudspeakers," *144th Audio Engineering Society Convention*, pp. 1–13, 2018.

[26] H. Lee, M. Mironovs, and D. Johnson, "Binaural rendering of phantom image elevation using VHAP," *AES 146th International Convention*, 2019.

[27] H. Lee and K. Borzym, "Creating virtual height loudspeakers for Dolby Atmos and Auro-3D using VHAP." *Audio Engineering Society*, May 28, 2020.

[28] T. Hidaka, L. L. Beranek, and T. Okano, "Interaural cross-correlation, lateral fraction, and low- and high-frequency sound levels as measures of acoustical quality in concert halls," *The Journal of the Acoustical Society of America*, vol. 98, no. 2, pp. 988–1007, Aug. 1995.

[29] F. Zotter and M. Frank, "Efficient phantom source widening," *Archives of Acoustics*, vol. 38, no. 1, pp. 27–37, 2013.

[30] C. Millns and H. Lee, "An investigation into spatial attributes of 360° microphone techniques for virtual reality," *144th Audio Engineering Society Convention*, pp. 1–9, 2018. https://github.com/APL-Huddersfield/MAIR-.

[31] D. Griesinger, "Objective measures of spaciousness and envelopment," *Audio Engineering Society 16th International Conference: Spatial Sound Reproduction*, pp. 27–41, 1999.

[32] R. K. Cook, R. V. Waterhouse, R. D. Berendt, S. Edelman, and M. C. Thompson, "Measurement of correlation coefficients in reverberant sound fields," *The Journal of the Acoustical Society of America*, vol. 27, no. 6, pp. 1072–1077, Nov. 1955.

[33] C. Gribben and H. Lee, "A comparison between horizontal and vertical interchannel decorrelation," *Applied Sciences*, vol. 7, no. 11, 2017.

[34] C. Gribben and H. Lee, "Vertical interchannel decorrelation on the vertical spread of an auditory image," *Journal of the Audio Engineering Society*, vol. 66, no. 8, pp. 537–555, 2018.

[35] C. Gribben and H. Lee, "The perception of band-limited decorrelation between vertically oriented loudspeakers," *IEEE/ACM Transactions on Audio, Speech and Language Processing*, vol. 28, pp. 876–888, 2020.

[36] H. Lee, "Perceptual band allocation (PBA) for the rendering of vertical image spread with a vertical 2D loudspeaker array," *Journal of the Audio Engineering Society*, vol. 64, no. 12, pp. 1003–1013, 2016.

[37] H. Lee, "2D-to-3D ambience upmixing based on perceptual band allocation," *Journal of the Audio Engineering Society*, vol. 63, no. 10, 2015.

[38] S. Kraft and U. Zölzer, "Low-complexity stereo signal decomposition and source separation for application in stereo to 3D upmixing." *140th Audio Engineering Society Convention*, May 26, 2016.

[39] H. Lee and C. Gribben, "Effect of vertical microphone layer spacing for a 3D microphone array," *Journal of the Audio Engineering Society*, vol. 62, no. 12, 2014.

[40] H. Wittek and G. Theile, "Development and application of a stereophonic multichannel recording technique for 3D audio and VR." *143rd Audio Engineering Society Convention*, Oct. 08, 2017.

[41] H. Lee, "Capturing 360° audio using an equal segment microphone array (ESMA)," *Journal of the Audio Engineering Society*, vol. 67, no. 1, pp. 13–26, 2019.

[42] H. Lee and D. Johnson, "An open-access database of 3D microphone array recordings," *147th Audio Engineering Society Convention*, 2019.

[43] H. Lee and D. Johnson, "3D microphone array comparison: objective measurements," *Journal of the Audio Engineering Society*, accepted for publication, 2021.

[44] A. Wakuda, H. Furuya, K. Fujimoto, K. Isogai, and K. Anai, "Effects of arrival direction of late sound on listener envelopment," *Acoustical Science and Technology*, vol. 24, no. 4, pp. 179–185, 2003.

6

Ambisonics understood

Cal Armstrong and Gavin Kearney

6.1 BACKGROUND

Ambisonics is an expandable, mathematics-based approach to spatial audio reproduction. It encompasses the encoding, storage, and rendering of directional auditory data by formulating the spatial sampling of an infinitesimal sound field such that it may be resynthesized by a finite number of point sources.

Developed throughout the 1970s, Ambisonics was first conceptualized by Michael Gerzon [1] and then formalized [2], [3], with further work that discussed microphone capture techniques [4] and practical reproduction methods [5]. Unlike traditional surround sound techniques (such as Dolby's 5.1 [6]), Ambisonics does not require any knowledge of the playback system (e.g. loudspeaker layout) during the encoding process [7]. Ambisonics provides a generic representation of a sound field such that it may be reproduced, with limitations, over almost any loudspeaker configuration.

Ambisonic format data is made up of a series of time-domain mono audio streams known as channels. Ambisonic channels represent specific orthogonal scaled portions of a sound field as defined by multidimensional trigonometric-based harmonic functions. The number of Ambisonic channels depends upon the *order* of Ambisonics being used. Higher orders require a greater number of channels and generally provide an increased spatial resolution at the cost of data storage, computation, and increased complexity [8].

Using almost identical principles to the mid-side microphone technique [9] Ambisonic channels may be weighted and summed together to spatially sample the sound field with a known directivity and directionality. The resulting polar pick-up patterns are often referred to as "virtual microphones". The algorithms used to calculate these channel weightings are known as Ambisonic decoders and are used to determine the loudspeaker signals of an arbitrary playback array.

6.2 SPHERICAL HARMONIC FUNCTIONS

6.2.1 Overview

Spherical harmonic functions are 3D functions derived from the solution of Laplace's equation [10]. They are used as the basis for sampling the sound field in a particular direction. Their 2D alternatives are the cylindrical harmonic functions; however, as it is most relevant to modern Ambisonic applications, the 3D case will be prioritized in this chapter.

Spherical harmonic functions are defined over the sphere by the multiplication of sinusoidal functions (defined by an azimuthal component) with associated Legendre polynomial functions (defined by an elevatory component) [11]. They are depicted in Figure 6.1.

The harmonics are grouped by spatial sampling complexity and defined by parameters passed to the associated Legendre polynomials and sinusoidal functions during their derivation. The parameters are defined here as degree, $m \geq 0$, and index, $i = 0,1,\ldots m$. A spin is also defined:

$$\sigma = \begin{cases} +1 & \text{if} \quad i = 0 \\ \pm 1 & \text{if} \quad i > 0 \end{cases} \qquad (6.1)$$

that is related to the orientation of each function. Elsewhere in the literature, it is common to omit σ and define $-m \leq i \leq m$. However, the current notation is preferable since it highlights the symmetrical properties of the spherical harmonics. In Figure 6.1, an increasing degree is represented by row. Within each degree there are $2m+1$ harmonic functions. Functions

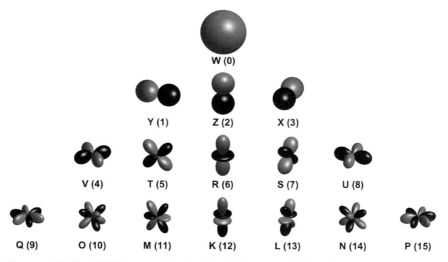

Figure 6.1 The SN3D normalized spherical harmonic functions as used in up to third-order Ambisonics labelled with alphabetical (FuMa) and ACN notation. The absolute value of the function is depicted as the radii of each surface. The sign is indicated by shade: grey positive, black negative.

on the left observe a 90° rotation about the z-axis (vertical) compared to those on the right (e.g. ACN 1, and 3, 10, and 14). These harmonics are defined by the same index but opposite spin. Ambisonic representations are defined by order, M, and include $(M+1)^2$ spherical harmonic components of degree $m \leq M$.

6.2.2 Notation

The notation of these functions is far from standardized. Mathematical texts tend to favour the terms degree, l, then order, m, referring to the rows and columns of Figure 6.1 respectively [10], [12]. However, this became confusing when Ambisonic literature began to use the term "order" to mean the order of Ambisonics (which in fact relates more directly to the mathematical "degrees" or rows of spherical harmonics). To avoid confusion, the following nomenclature is proposed:

- degree (m): refers to an individual row of Figure 6.1
- order (M): refers to an order of Ambisonics that includes a collection of rows from Figure 6.1
- index: (i): refers to the columns of Figure 6.1

6.2.3 Derivations

The infinitesimal collection of spherical harmonics, $Y_{mi}^{\sigma \text{O3D}}$, of degree $m = 0,1, \ldots \infty$ form a complete orthogonal basis set such that

$$\int Y_{mi}^{\sigma}(\varphi, \vartheta) \cdot Y_{m'i'}^{\sigma'}(\varphi, \vartheta) \, d\Omega = \delta_{mn} \cdot \delta_{ii'} \delta_{\sigma \sigma'} \qquad (6.2)$$

where δ_{ab} denotes the Kronecker delta function, which is one for $a = b$. Simply put, this means that the result of any function times another will integrate to zero over the sphere.

The spherical harmonics are defined as follows:

$$Y_{mi}^{\sigma} = N_{mi}^{\{\ldots 3D\}} P_{mi}' \left(\sin(\vartheta)\right) \times \begin{cases} \cos(i\varphi) & \text{if } \sigma = 1 \\ \sin(i\varphi) & \text{if } \sigma = -1 \end{cases}$$

$$= P_{mi}^{\{\ldots 3D\}} \left(\sin(\vartheta)\right) \times \ldots \qquad (6.3)$$

where a normalized associated Legendre function is written as

$$P_{mi}^{\{\ldots 3D\}} = N_{mi}^{\{\ldots 3D\}} P_{mi}' \qquad (6.4)$$

and $N_{mi}^{\{\ldots\}}$ denotes the normalization factor and P_{mi}' denotes the unnormalized Legendre polynomial. Common normalization factors include SN3D (Schmidt semi-normalized), N3D (an orthogonal normalization), and what will be referred to as O3D (an orthonormal normalization):

$$N_{mi}^{\text{SN3D}} = \sqrt{\varepsilon_m \frac{(m-i)!}{(m+i)!}} \qquad (6.5)$$

$$N_{mi}^{\text{N3D}} = \sqrt{\varepsilon_m \cdot (2m+1) \cdot \frac{(m-i)!}{(m+i)!}} \qquad (6.6)$$

$$N_{mi}^{\text{O3D}} = \sqrt{\varepsilon_m \cdot \frac{(2M+1)}{4\pi} \cdot \frac{(m-i)!}{(m+i)!}} \qquad (6.7)$$

where

$$\varepsilon_m = (2 - \delta_{i0}) \qquad (6.8)$$

Note that the Schmidt semi-normalization omits the Condon–Shortley phase factor of $(-1)^m$ which is commonly implemented in quantum mechanics. Generally speaking, SN3D normalization reduces the weighting of higher-order spherical harmonics. In practice, this may be preferred as it results in a similar maximum value across all functions.

6.3 AMBISONIC ENCODING

6.3.1 Overview

Ambisonic encoding covers the conversion of real-world source and direction information to the intermediate storage format of Ambisonics (Ambisonic format). In this form, sound fields are decomposed into their spherical harmonic components and saved as Ambisonic channels.

6.3.2 Manual encoding

To manually encode data into Ambisonic format, sources are independently weighted onto each Ambisonic Channel by a spherical harmonic coefficient. The value of each weight is determined by the value of the spherical harmonic function represented by that channel, at the angle one wishes to encode the sound source at. This is done for all sources and all channels with multiple sources simply summed together on each channel:

$$\beta_{mi}^{\sigma} = \sum_{\text{for all } s} s \cdot Y_{mi}^{\sigma}(\vec{v}_s), \qquad (6.9)$$

where

- β_{mi}^{σ} is the Ambisonic Channel representing the spherical harmonic Y_{mi}^{σ}
- s is a source signal
- $\bar{v}_s$ is the angle at which that source is being encoded

6.3.3 Spatial recording

To record directly into Ambisonic format, one must aim to separately capture different regions of the sound field with weightings that match each of the relevant spherical harmonic functions.

For first-order Ambisonics (FOA), the relevant functions are omnidirectional and figure-of-eight – as shown in Figure 6.1 – and so can be captured directly using omnidirectional and figure-of-eight microphones. Alternatively, a tetrahedral array of microphones may be used to record the sound field [4]. This is known as an A-format recording, and is preferred as the configuration allows the microphones to be placed in a more coincident fashion. An example layout is presented in Figure 6.2.

A simple set of equations are then used to convert the A-format recordings into a more standard Ambisonic format:

$$\beta_{00}^{1} = FLU + FRD - BLD + BRU \quad (W)$$

$$\beta_{00}^{1} = FLU + FRD - BLD - BRU \quad (X)$$

$$\beta_{00}^{-1} = FLU - FRD + BLD - BRU \quad (Y)$$

$$\beta_{00}^{1} = FLU - FRD - BLD + BRU \quad (Z) \qquad (6.10)$$

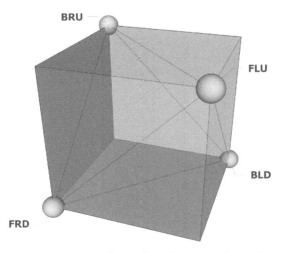

Figure 6.2 Example of a tetrahedral microphone configuration with positions (F)ront, (B)ack, (L)eft, (R)ight, (U)p, and (D)own.

where β_{mi}^{σ} is the Ambisonic Channel representing the spherical harmonic. Y_{mi}^{σ}. The Ambisonic signals should then be refined by means of frequency/phase correction filters to account for the spatial separation of the microphones [13], [14].

Unfortunately, for higher-order Ambisonics (HOA), it is not possible to directly record the sound field with simple microphones since pick-up patterns of sufficient complexity do not exist. Instead, arrays of microphones must be used with beamforming techniques to approximate the correct pick-up patterns. One such example is MH Acoustics' Eigenmike® (mhacoustics.com/home), which holds 32 microphone capsules and can approximate up to fourth-order Ambisonic recordings.

6.3.4 Competing standards

There are a number of competing standards when it comes to Ambisonic format data that relate to the sequential arrangement in which harmonic components are stored and the interchannel normalization. Originally, Gerzon proposed normalizing the first-order harmonics such that X, Y, and Z each had a gain of $\sqrt{2}$ in their directions of peak sensitivity (W is omnidirectional and so has a gain of one in all directions). This was in order that all four channels [W, X, Y, Z] each carried approximately equal average energy for any given sound field recording [5]. In the days of magnetic tape recording, this was an important consideration to maximize the signal-to-noise ratios. This standard was defined as B-format.

A MaxN normalization scheme was similarly but more generally defined such that each channel is scaled to have a maximum gain of ± 1 [15]. Furse and Malham later proposed the generalized FuMa normalization scheme, which follows on from the MaxN normalization scheme with an additional scaling of $\dfrac{1}{\sqrt{2}}$ on the W channel to provide backward compatibility with B-format.

As HOA became more common, a scalable method for channel ordering was required. The Ambisonic channel number (ACN) is now fairly standardized. A simple formula calculates the correct ACN of a harmonic based on its degree, m, and index, i:

$$\text{ACN} = m \cdot (m+1) + \sigma i \qquad (6.11)$$

6.4 AMBISONIC DECODING

6.4.1 Overview

Ambisonic decoding covers the conversion of data from Ambisonic format to a set of loudspeaker signals. The key principle is in generating a decoding matrix, $\boldsymbol{D}$, to weight and sum each Ambisonic Channel independently for each loudspeaker feed.

The "standard" decoder is often assumed to be a *mode-matching pseudo-inverse* decoder, as in [16], which aims to exactly restore the original sound field. There are also alternative types of decoder that each aim to preserve particular auditory qualities when presented with non-ideal reproduction arrays. It therefore follows that by enforcing particular normalization schemes and regular loudspeaker layouts, each method presented here will simplify to an identical decoding matrix. Figure 6.3 is given at this time for reference and its content should become clearer throughout this chapter.

6.4.2 Definitions

When discussing the decoding of Ambisonics it is common to consider the following:

$$\boldsymbol{\beta} = \begin{bmatrix} \beta_1 \\ \beta_2 \\ \vdots \\ \beta_K \end{bmatrix} \quad \boldsymbol{\rho} = \begin{bmatrix} \rho_1 \\ \rho_2 \\ \vdots \\ \rho_L \end{bmatrix}$$

$$\boldsymbol{C} = \begin{bmatrix} Y_{00}^1(\varphi_1, \vartheta_1) & Y_{00}^1(\varphi_l, \vartheta_l) & \cdots & Y_{00}^1(\varphi_L, \vartheta_L) \\ Y_{mi}^\sigma(\varphi_1, \vartheta_1) & Y_{mi}^\sigma(\varphi_l, \vartheta_l) & \cdots & Y_{mi}^\sigma(\varphi_L, \vartheta_L) \\ \vdots & \vdots & \ddots & \vdots \\ Y_{Mi}^\sigma(\varphi_1, \vartheta_1) & Y_{Mi}^\sigma(\varphi_l, \vartheta_l) & \cdots & Y_{Mi}^\sigma(\varphi_L, \vartheta_L) \end{bmatrix} \quad (6.12)$$

where

- $\boldsymbol{\beta}$ is a column vector of the individual Ambisonic channels (W, Y, Z, X, etc.) of length

$$K = \begin{cases} 2M+1 & (2D) \\ (M+1)^2 & (3D) \end{cases}, \quad (6.13)$$

where

  - M is the order of Ambisonics
  - K is the number of Ambisonic channels
- $\boldsymbol{\rho}$ is a column vector, length L, of the decoded loudspeaker signals, where L is the number of loudspeakers
- $\boldsymbol{C}$ is a $K \times L$ matrix of the spherical-harmonic value coefficients for each Ambisonic Channel in the direction of each loudspeaker. It is known as the *re-encoding* matrix

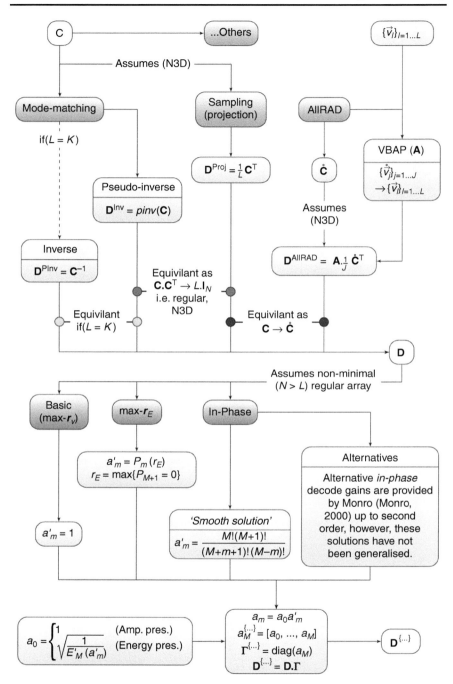

Figure 6.3 Summary and workflow of common Ambisonic decoding techniques and matrix weighting solutions.

An $L \times K$ *decoding* matrix, $\boldsymbol{D}$, is also defined that comprises the Ambisonic Channel weightings for each loudspeaker signal for any given decoding strategy such that

$$\rho = \boldsymbol{D} \cdot \beta \qquad (6.14)$$

6.4.3 Loudspeaker regularity

The regularity of a loudspeaker array will, in general, affect a decoders' performance. A regular array will have loudspeakers that are evenly spaced. However, for 3D reproduction, finding appropriate configurations is a significant and ongoing challenge. Only five distributions (the platonic solids) are known to satisfy this criterion.

Further, geometric regularity does not necessarily guarantee regularity in an Ambisonic sense. Daniel defines the requirement of regularity more precisely in [15, p. 175]:

$$\frac{1}{L} \boldsymbol{C}^{\text{N3D}} \cdot \left(\boldsymbol{C}^{\text{N3D}} \right)^{\text{T}} = I_K , \qquad (6.15)$$

where I_K is the $K \times K$ identity matrix. As a result, there is a dependency upon the spherical-harmonic coefficients [17]; that is, just because a distribution is regular for first order does not make it regular for second or third order. This is summarized for the platonic solids in Table 6.1.

Alternatively, t-designs have been shown to be Ambsonically regular up to order M given the condition $t \geq 2M + 1$ [18].

6.4.4 Velocity and energy vectors

Gerzon introduced two objective measures of sound field restoration: the Makita (or velocity) vector, r_V, and the energy vector r_E. The location of the Makita vector is "the direction in which the head has to face in order that the interaural phase difference is zero." Similarly, the direction of the energy vector is "the direction the head has to face in order that there be no interaural amplitude difference at high frequencies" [5]. These two measures relate directly to the human auditory system's localization

Table 6.1 A summary of the Ambisonic orders (M) to which the vertices of the platonic solids may be considered to be of regular distribution.

Name	Vertices	Regularity ($M \leq \dots$)
Tetrahedron	4	1
Octahedron	6	1
Cube	8	1
Icosahedron	12	2
Dodecahedron	20	2

cues: Interaural Time Difference (ITD) and Interaural Level Difference (ILD), respectively.

The vectors may be calculated by summing together independent vectors pointing in the directions of each loudspeaker within the playback array. The length of each of these independent vectors should be proportional to the amplitude gain of that specific loudspeaker in the case of the Makita vector, or the energy gain in the case of the energy vector :

$$r_v = \sum_{l=1}^{l=L} g_l^{\text{Amplitude}} \cdot \hat{v}_l \qquad (6.16)$$

$$r_E = \sum_{l=1}^{l=L} g_l^{\text{Energy}} \cdot \hat{v}_l, \qquad (6.17)$$

where

- L is the number of loudspeakers
- g is the loudspeaker gain
- $\hat{v}$ is the unity vector in the direction of a loudspeaker

Gerzon also defined a total playback amplitude/energy gain as the sum of the magnitude of each of the independent vectors. The length of the Makita and energy vectors should ideally equal the total playback gain, as would be the case for reproduction with a single loudspeaker. In the case where the length of the vectors does not equal the total playback gain, this can lead to instability in the source imaging.

6.4.5 Mode matching

The mode-matching methodology aims to restore the original sound field within the reproduction array. The loudspeaker signals, ρ, are calculated by equating the summed spherical-harmonic mode excitations of the L loudspeakers to the spherical-harmonic mode excitations of the original sound field (Ambisonic channels), β [19], [16]. The error is then minimized in a least-squared sense by a matrix inversion/pseudo-inversion.

The ideal reconstruction (known as the *re-encoding equation*) may be written as

$$\beta = C \cdot \rho \qquad (6.18)$$

It represents the summed *re-encoding* of each loudspeaker signal as a source in the direction $\vec{v}_l$ into Ambisonic format via the re-encoding matrix, C, equated to the original Ambisonic format signal. By simple rearrangement, it follows that the loudspeaker signals may be derived by

$$\rho = C^{-1} \cdot \beta \qquad (6.19)$$

such that

$$\mathbf{D}^{\text{Inv}} = \mathbf{C}^{-1} \tag{6.20}$$

However, this is only possible if $\mathbf{C}$ is square, that is, if $L = K$. Alternatively, one should consider the Moore–Penrose pseudo-inverse of $\mathbf{C}$ [20]:

$$\mathbf{D}^{\text{PInv}} = \mathbf{C}^{\dagger}, \tag{6.21}$$

where $\mathbf{C}^{\dagger}$, the pseudo-inverse of $\mathbf{C}$, is defined as

$$\mathbf{C}^{\dagger} = \begin{cases} \mathbf{C}^{\text{T}} \cdot \left(\mathbf{C} \cdot \mathbf{C}^{\text{T}} \right)^{-1} & L \geq K \\ \left(\mathbf{C}^{\text{T}} \cdot \mathbf{C} \right)^{-1} \cdot \mathbf{C}^{\text{T}} & L < K \end{cases} \tag{6.22}$$

Conveniently, for square matrices, $\mathbf{D}^{\text{FInv}} \equiv \mathbf{D}^{\text{Inv}}$.

The perceptual quality of mode-matching decoding depends in part on the condition number of $\mathbf{C}$ which can be related to the regularity of the loudspeaker layout. Hollerweger describes how the energy vector of an Ambisonic decode will only align with the source's encoded direction if $[L > K]$ and if the loudspeaker array is either regular, or semi-regular (i.e. the more general case that $\left(\mathbf{C} \cdot \mathbf{C}^{\text{T}} \right)$ is at least diagonal) [15], [21].

6.4.6 Projection

For regular loudspeaker configurations and N3D (or N2D (2D)) normalization, it is shown that where $L \geq K$, the pseudo-inverse of matrix $\mathbf{C}$ is trivialized as the term $\left(\mathbf{C} \cdot \mathbf{C}^{\text{T}} \right)$ becomes the diagonal matrix:

$$\left(\mathbf{C} \cdot \mathbf{C}^{\text{T}} \right) = \begin{cases} L \cdot I_{2m+1} & (2\text{D}) \\ L \cdot I_{(m+1)^2} & (3\text{D}) \end{cases}, \tag{6.23}$$

where I_k is the $(K \times K)$ identity matrix. The result is a simplified derivation of the decoding matrix:

$$\mathbf{D}^{\text{Proj}} = \mathbf{C}^{\text{T}} \cdot \left(\mathbf{C} \cdot \mathbf{C}^{\text{T}} \right)^{-1} = \mathbf{C}^{\text{T}} \cdot \begin{bmatrix} L_{[1,1]} & \cdots & L_{[1,K]} \\ \vdots & \ddots & \vdots \\ L_{[K,1]} & \cdots & L_{[K,K]} \end{bmatrix}^{-1} = \frac{1}{L} \cdot \mathbf{C}^{\text{T}} \tag{6.24}$$

Projection decoding represents the spatial sampling of the Ambisonic channels without any consideration of the placement of any other loudspeaker. In some ways, this makes it a more robust technique in the face of irregular loudspeaker layouts. However, the advantages of matrix stability must be weighed up against the possible directional distortions and energy-balance issues of a rendered source [21], [22].

6.4.7 Alternative techniques

Ambisonic decoding is by no means limited to inversion techniques, and some alternatives are listed for the reader's convenience:

- All-round Ambisonic decoding (AllRAD) [23]
- Energy-preserving decoding [19]
- Constant-angular spread decoding [24]
- Decoding for irregular loudspeaker arrays using interaural cues [25]
- Decoding for irregular arrays of loudspeakers by nonlinear optimization [26], [27]

6.4.8 Near-field compensation

Ambisonics is based on the assumption of plane wave theory. Mathematically encoding a source into spherical harmonics (as described in Section 6.2) in fact assumes that the source has a planar wavefront. Likewise, loudspeaker sources are also often assumed to be planar. For this to be true in the case of a point source, it must be of infinite distance. It was shown in Equation 6.9 that the Ambisonic component, β_{mi}^{σ}, of a *plane* wave signal, s, of incidence (φ, ϑ) may be defined as

$$\beta_{mi}^{\sigma} = s \cdot Y_{mi}^{\sigma}(\varphi, \vartheta). \qquad (6.25)$$

For a radial point source of position $(\varphi, \vartheta, r_s)$, it is necessary to consider the near-field effect filters [28], Γ_m, such that

$$\beta_{mi}^{\sigma} = s \cdot \Gamma_m(r_s) \cdot Y_{mi}^{\sigma}(\varphi, \vartheta)$$

$$\Gamma_m(r_s) = k \cdot d_{\text{ref}} \cdot h_m^-(kr_s) \cdot j^{-(m+1)}, \qquad (6.26)$$

where

- $k = \dfrac{2\pi f}{c} = \dfrac{\omega}{c}$ is the wave number
- d_{ref} is the distance at which the source, s, was measured – it is a compensation factor that derives from the equation (pressure = 1/distance)
- $h_m^-(kr_s)$ are the spherical Hankel functions of the second kind (divergent)
- $j = \sqrt{-1}$
- $\Gamma_m(r_s)$ is the degree-dependent filter that simulates the effect of a non-planar source

The filter is simplified by considering s to be the pressure field measured at the origin, that is, $d_{ref} = r_s$. By doing this, the need to compensate for the attenuation, $\dfrac{1}{r_s}$, and delay, $\tau = \dfrac{r_s}{c}$, of the signal may be ignored. These features may be defined by the zeroth-order (pressure only) near-field effect filter Γ_0. Hence, this factor may be removed from the general filter:

$$\beta_{mi}^{\sigma} = s \cdot F_m \cdot Y_{mi}^{\sigma}(\varphi, \vartheta)$$

$$F_m = \frac{\Gamma_m}{\Gamma_0} = \sum_{i=0}^{m} \frac{(m+i)!}{(m-i)!\,i!}\left(\frac{-jc}{\omega r_s}\right)^i, \tag{6.27}$$

where F_m are the filters that model the effects of wavefront curvature in the spherical-harmonic domain for a source $(\varphi, \vartheta, r_S)$ having been measured from the origin. Unfortunately, for high-order representations and sources close to the origin, these filters become unstable, as shown in Figure 6.4.

Fortunately, in Ambisonics, one must consider the near-field properties of both the original source and the reproduction loudspeakers. Near-field *compensation* filters must also be applied to the loudspeakers (simulating an infinite radius). Near-field compensation filters may be derived as simply the inverse of the near-field effect filters. By combining the filters,

$$H_m = \frac{F_m^{\text{source}}}{F_m^{\text{loudspeakers}}}, \tag{6.28}$$

the unstable nature of the low-frequency boost/cut cancel to produce a stable bass boost/cut shelf filter, as shown in Figure 6.5.

In the case that the source is being rendered at the same radius as the loudspeakers, the two filters will cancel entirely. This is why an Ambisonic system that does not consider near-field compensation will render a source as if "on the loudspeaker array" at the same radius as the loudspeakers.

Note that in the case where a sound field has been recorded directly, near-field effect filters need not be applied. However, near-field compensation filters may still be required if the loudspeaker signals cannot be assumed to be planar.

6.5 SPATIAL ALIASING AND THE SWEET SPOT

6.5.1 Overview

Ambisonic rendering generally defines a process of reproducing a sound field from a finite number of fixed points about a sphere with a particular angular resolution. The angular resolution is dependent on the Ambisonic order and narrows as the number of spherical harmonics that have been utilized increases. This is shown in Figure 6.6.

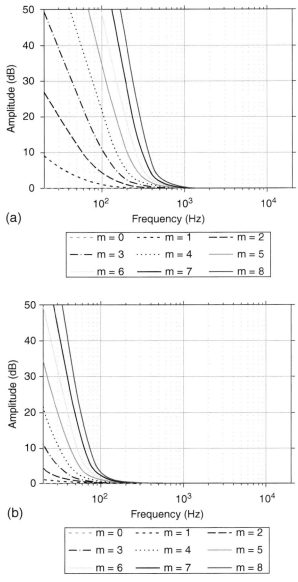

Figure 6.4 Bass-boosting amplitude response of the near-field effect filter for degree, $m = 0,1,\ldots 8$, for a source simulated at (a) 5 m and (b) 1 m.

Therefore, depending on the resolution, it is common for a single point source to be sampled/output from multiple loudspeakers within a reproduction array. This leads to destructive comb filtering when the path difference between the different speakers and the sampling point varies. The outcome is that accurate reconstruction of a sound field is limited to a small area of space in the centre of the reproduction array – known as the sweet spot.

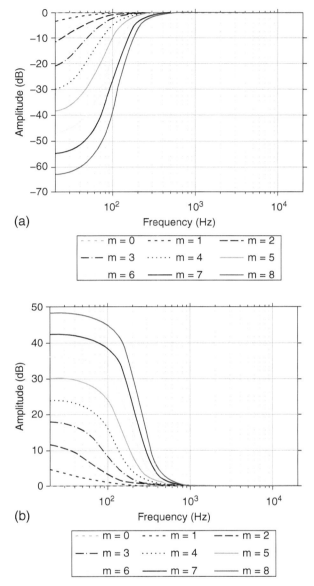

Figure 6.5 Shelf filtering amplitude response of the combined near-field effect and compensation filters for degree, $m = 0,1,\ldots 8$, for a loudspeaker radius of 2 m and a source simulated at (a) 5 m and (b) 1 m.

6.5.2 Effects of discrete sampling

In the ideal case, an infinite number of sample points (loudspeakers) reproduces a sound field stored with infinite angular resolution ($M = \infty$). However, as the number of loudspeakers is reduced, spaces are created in between the sample points. Sparsely sampling a sound field with a high

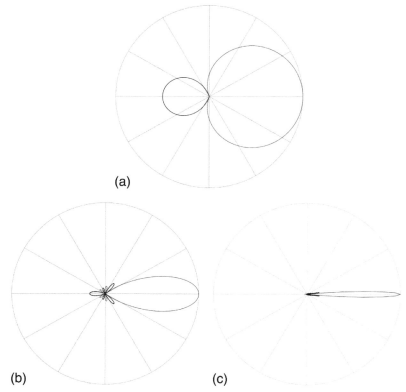

Figure 6.6 Order 1 (a), Order 5 (b), and Order 36 (c). Ambisonic representations of a point source.

angular resolution will lead to gaps where sources are panned in-between loudspeakers. The angular resolution should then be reduced to match to the sampling resolution by ensuring a regular distribution and setting the following criteria:

$$L = K = \begin{cases} 2M+1 \ \ (2D) \\ (M+1)^2 \ \ (3D) \end{cases} \quad (6.29)$$

Figure 6.7 shows a fifth-order Ambisonic source, rendered over three regularly distributed 2D loudspeaker arrays (3, 11, and 100 speakers). Two source positions are shown: exactly in-between two loudspeakers and directly in-line with a loudspeaker.

Figure 6.7a shows that, in the under-sampled case where a source is panned between the loudspeakers, an accurate source signal is not reproduced. The correct wave pattern is not generated and a gap is seen in the reproduced sound field. This is resolved in Figure 6.7c by increasing

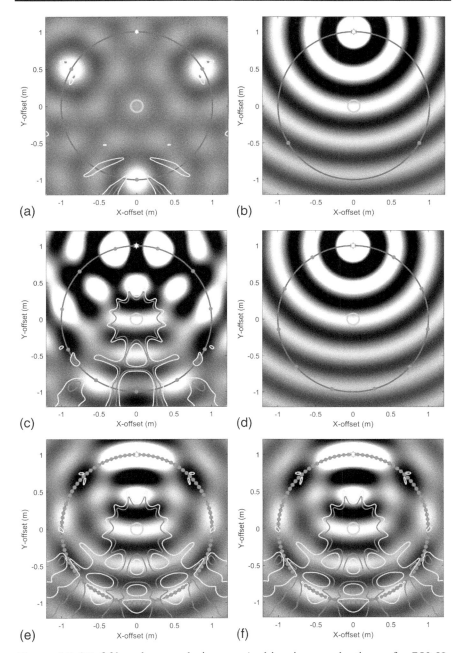

Figure 6.7 2D fifth-order pseudo-inverse Ambisonic reproductions of a 750 Hz sinusoidal point-source represented by a star ($\varphi = 0$, $\vartheta = 0, r = 1$) with a zero-to-peak amplitude of 1 at the centre of the array. Loudspeakers (dots on the circle) have a radius of 1 m. Amplitude is plotted on a capped colour scale: −1-to-1 black-to-white, respectively. The ring at the centre of the circle of radius 8 cm indicates the approximate size of a human head. Areas of accurate reproduction are highlighted by contours: < 20% error (light grey); < 10% error (dark grey). Reproduction is performed on under, ideally, and over sampled arrays.

the number of loudspeakers to close the gap. Here it is seen that the signal is primarily output from the two neighbouring loudspeakers (bright white spots) and an area of accurate reconstruction is shown at the centre of the array. As the number of loudspeakers is increased (Figure 6.7e), more are shown to output the source signal (solid white curve), and significantly higher levels of destructive interference are seen outside the central position.

Figures 6.7b and 6.8d show the reproduction of an under and ideally sampled source panned directly in line with a loudspeaker. In both cases, perfect reconstruction is evident as only the frontal loudspeaker (the one aligned with the source) is outputting any signal (no comb filtering). The other loudspeakers are aligned with the nodes of the virtual microphone pick-up pattern. In the case of the ideally sampled array, this is a result of defining the number of loudspeakers in accordance with Equation 6.29. In the case of the under-sampled array, this is the result of the pseudo-inverse decoder. In the case of the two oversampled arrays, very little source-alignment-dependent difference is apparent since the difference in the number of contributing loudspeakers remains small.

At the precise centre of the array, the path length to every loudspeaker is identical and all signals arrive perfectly in phase. However, from an off-centre position, the path lengths to each loudspeaker vary. This is true for the positions of a person's ears even if their head is centred within the array. As the total solid angle encompassing a collection of loudspeakers increases, so does the variance in their path length. By reducing the differences in path-lengths, the minimum frequency that is subject to interference is increased. Therefore, higher-order Ambisonic reproductions with less spread result in more accurate higher-frequency reproduction.

6.6 DECODER MATRIX WEIGHTINGS

6.6.1 Overview

At high frequencies, which is where spatial aliasing occurs, it is possible to optimize decoders based on alternative psychoacoustic parameters by manipulating the virtual microphone pick-up patterns. To do this, the decoding matrix is modified by a set of weights, $a_m^{\{...\}}$, such that

$$\boldsymbol{D}^{\{...\}} = \boldsymbol{D}' \cdot \Gamma_m^{\{...\}}$$

$$\Gamma_m^{\{...\}} = \mathrm{diag}\left\{\left[a_0^{\{...\}} \quad \cdots \quad \underbrace{a_m^{\{...\}} \quad \cdots \quad a_m^{\{...\}}}_{2m+1} \quad \cdots \quad \underbrace{a_M^{\{...\}} \quad \cdots \quad a_M^{\{...\}}}_{2M+1}\right]\right\}.$$

(6.30)

In this way, the decoders no longer attempt to accurately reconstruct a sound field, but instead are designed to improve the reconstruction of

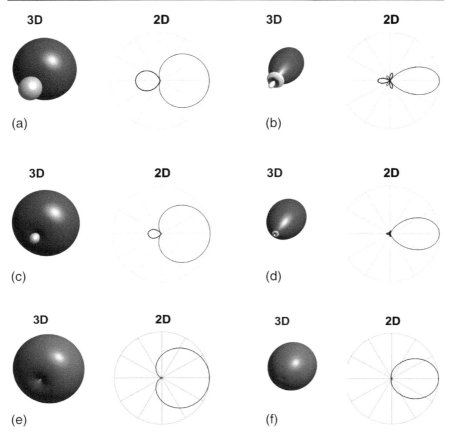

Figure 6.8 Virtual microphone pick-up patterns for first (left) and third (right)-order basic (top), max-r_E (middle) and in-phase (bottom) decoders. Positive values are shown in grey. Negative (out of phase) values are shown in black.

perceptual cues such as ILD [15], [29]. Graphical comparisons of two popular schemes are presented in Figure 6.8 for first and third-order decoders.

6.6.2 Amplitude/energy preservation

Before weighting the Ambisonic channels, consideration must be given as to whether the overall set of weights should preserve either the amplitude or energy levels of the restored sound field. By altering the shape of the pick-up pattern, the spread of a source between neighbouring loudspeakers is altered and therefore the total summation of amplitude/energy levels changes. This may be accounted for with an additional normalization coefficient, a_0.

For amplitude preservation,

$$a_0 = 1. \tag{6.31}$$

For energy preservation, Daniel derives simple linear scaling factors in the case of non-minimal $(L > K)$ regular arrays [30]. Total energy gain as a result of the weights, $\boldsymbol{a}_M$, is defined [15] [Eqn. A.62]:

$$E_M\left(\boldsymbol{a}_M\right) = \frac{1}{L} \sum_{m=0}^{M} (2m+1)\left(a_m\right)^2 \tag{6.32}$$

For playback energy to be preserved, unity gain is required:

$$E_M\left(\boldsymbol{a}_M\right) = 1 \tag{6.33}$$

The normalization factor, a_0, may be considered such that

$$a_m = a_0 \cdot a'_m \tag{6.34}$$

By factoring the term $\left(a_m\right)^2$ in Equation 6.32, it may be written as

$$E_M\left(\boldsymbol{a}_M\right) = \left(a_0\right)^2 \cdot \frac{1}{L} \sum_{m=0}^{M} (2M+1)\left(a'_m\right)^2 . = \left(a_0\right)^2 \cdot E'_M \tag{6.35}$$

Hence, by defining the term,

$$a_0 = \sqrt{\frac{1}{E'_M}} = \sqrt{L \frac{1}{\sum_{m=0}^{M}(2m+1)\left(a_m\right)^2}} \tag{6.36}$$

Equation 6.33 is satisfied by a cancellation of terms in Equations 6.35 and 6.36.

In practice, the absolute normalization of decoding matrices is often redundant and instead, playback levels are managed by the end user; however, normalization must be considered if combining the output from multiple decoding matrices, for example, within a dual-band decoder. In this case it is common to simply normalize the level of the second matrix to match the first:

$$a_0^{\{A\}} = 1$$

$$a_0^{\{B\}} = a_0^{\left\{\frac{B}{A}\right\}} = \frac{a_0^{\{B\}}}{a_0^{\{A\}}} \tag{6.37}$$

where A and B refer to the gains applied to individual decoding matrices. This may be expanded by

$$a_0^{\left\{\begin{smallmatrix} B \\ A \end{smallmatrix}\right\}} = \frac{\sqrt{\dfrac{1}{E'_M\left(a'^{\{B\}}_M\right)}}}{\sqrt{\dfrac{1}{E'_M\left(a'^{\{A\}}_M\right)}}} = \sqrt{\frac{E'_M\left(a'^{\{A\}}_M\right)}{E'_M\left(a'^{\{B\}}_M\right)}}$$

$$= \sqrt{\frac{\sum_{m=0}^{M}(2m+1)\left(a'^{\{A\}}_m\right)^2}{\sum_{m=0}^{M}(2m+1)\left(a'^{\{B\}}_m\right)^2}} \qquad (6.38)$$

In the case that

$$a'^{\{A\}}_m = 1 \qquad m = 0,1,\ldots M \qquad (6.39)$$

(as is true for a basic decoding matrix), this may be simplified further:

$$a_0^{\left\{\begin{smallmatrix} B \\ A \end{smallmatrix}\right\}} = \sqrt{\frac{\sum_{m=0}^{M}(2m+1)(1)^2}{\sum_{m=0}^{M}(2m+1)\left(a'^{\{B\}}_m\right)^2}} \qquad (6.40)$$

and rewritten as

$$a_0^{\left\{\begin{smallmatrix} B \\ A \end{smallmatrix}\right\}} =$$

$$\frac{1}{\sqrt{\text{mean}\left[\left[\left(a'^{\{B\}}_0\right)^2 \quad \cdots \quad \underbrace{\left(a'^{\{B\}}_m\right)^2 \quad \cdots \quad \left(a'^{\{B\}}_m\right)^2}_{2m+1} \quad \cdots \quad \underbrace{\left(a'^{\{B\}}_m\right)^2 \quad \cdots \quad \left(a'^{\{B\}}_m\right)^2}_{2M+1}\right]\right]}}$$

$$(6.41)$$

where normalization is performed with respect to the mean value of the weights applied to each spherical-harmonic coefficient.

6.6.3 Basic (Max-r_V)

Basic weighting refers to either unity or to no additional weighting across all channels. It is suited for holophonic/mode-matching decoders where one wishes to abide by the mathematically founded techniques. By nature, basic weighting maximizes the velocity vector and, as such, is occasionally referred to as max-r_V decoding. This technique is optimally utilized at low frequencies where sound-field reconstruction is at its most accurate and ITD cues may be preserved.

6.6.4 Max-r_E

The high-frequency directional quality of a panned source may be somewhat measured by the energy vector, r_E. Gerzon makes particular reference to how it "is indicative of stability of localisation of images" in the region 700–4000 Hz [31]. It is beneficial to maximize this measure for high-frequency sources, where energy cues dominate timing cues with regard to perceptual localization. Daniel gives a general derivation for the weights, a_m, which maximize r_E over regular loudspeaker layouts [15]. It is common, however, to also implement these weights for irregular arrays – for which they remain approximately correct.

The energy vector, r_E, is first derived, and then differentiated, with respect to a_m. The result is set to zero to find the maximum of the function (i.e. when r_E is maximized). This may be shown to give us the following equation (3D) [15]:

$$(2m+1) \cdot r_E \cdot a_m = (m+1) \cdot a_{m+1} + m \cdot a_{m-1} \tag{6.42}$$

The recurrence relationship bares significant resemblance to one of the Legendre functions – Bonnet's recursion formula [32]:

$$(2m+1) \cdot \eta \cdot P_m(\eta) = (m+1) \cdot P_{m+1}(\eta) + m \cdot P_{m-1}(\eta) \tag{6.43}$$

if $a_m = P_m(\eta)$ and $\eta = r_E$ is defined. The vector r_E is therefore maximized by the relationship

$$a_m = P_m(r_E) \quad m = 0, 1, \ldots M \tag{6.44}$$

Equation 6.44 defines $a_0 = 1$ and $a_1 = r_E$. As it is required that $a_{-1} = 0$ and $a_{M+1} = 0$, r_E may be defined as the largest root of P_{M+1} such that

$$a_{M+1} = P_{M+1}(r_E) = 0 \tag{6.45}$$

is satisfied.

A max-r_E weighting scheme will reduce the presence of side lobes in the pick-up pattern at the expense of a slight widening of the frontal lobe, as shown in Figure 6.8. In practice, this increases the concentration of energy in the direction of the source.

6.6.5 In-phase

In-phase decoding weights the Ambisonic channels such that the outputs from all loudspeakers remain in phase. Unlike the max-r_E weighting scheme there are multiple solutions.

A number of examples are presented up to second order by Monro in [33]. A general Ambisonic panning function, $g(\vec{v}_s)$, is derived and a set of weights, a_m, $m = 0, 1, \ldots M$ are subsequently introduced to the function.

Table 6.2 Example crossover frequencies for basic - max-r_E dual-band decoding [30].

Order	1	2	3	4	5
Freq. (Hz)	700	1250	1850	2500	3000

A multidimensional region may then be defined in terms of a_m such that the output to $g(\vec{v}_s)$ is positive for all $\vec{v}$.

Although any solutions within this region may be considered as in-phase, the solutions that are of most interest lie on the boundary. One such solution is the "smooth solution" that presents a single forward-facing lobe and is defined such that $g(\vec{v}_s)$ must decrease over the range $\vec{v} = \begin{bmatrix} 0 & \pi \end{bmatrix}$ and is zero when $\vec{v} = \pi$ [15]. The weights may be calculated:

$$a_m = \frac{M!(M+!)!}{(M+m-1)!(M-m)!} \tag{6.46}$$

Smooth in-phase decoding has shown particular promise in rendering Ambisonics for large audiences [34]–[36] by removing the output of potentially overpowering side lobes that may become more prominent due to an individual listener's proximity to a particular loudspeaker.

6.6.6 Dual-band decoding

Dual-band decoding is a technique commonly implemented to alter the decoder weighting-scheme over frequency [26], [37], [38]. Typically, a basic weighting would be used at low frequencies with a max-r_E weighting at high frequencies. A crossover frequency is usually selected at the approximate frequency at which holophonic reproduction becomes invalid for a human listener as a result of a shrinking sweet spot [39]. After this point, energy localization is prioritized by the max-r_E weighting scheme to best preserve ILD localization cues. Some example frequencies are shown in Table 6.2 but are by no means definitive [30].

The technique may be applied either by computing two separate decode matrices and combining the individual outputs using a crossover filter, or by applying a series of shelf filters to the Ambisonic channels (in essence, creating a single frequency-dependent weighting) [5], [31], [40], [41]. The methods are entirely equivalent.

6.7 Binaural rendering of Ambisonics

6.7.1 Overview

Spherical loudspeaker rigs suitable for Ambisonic reproduction, for example, Figure 6.9, are expensive, dominating, and hard to come by. With the recent uptake of Ambisonics in the field of virtual reality, portable

Figure 6.9 The 50-point spherical loudspeaker array at the University of York.

head-mounted devices are instead becoming a popular source of content delivery. It is therefore beneficial to consider the binauralization of Ambisonics.

There are two leading methods when it comes to rendering an Ambisonic signal binaurally; via a set of virtual loudspeakers or within the spherical-harmonic domain.

6.7.2 Virtual loudspeakers

The virtual-loudspeaker approach [23], [42]–[44] (Figure 6.10a) most directly replicates the real-world method for rendering Ambisonics over a loudspeaker array and is the simplest to understand. The initial rendering process is identical to real-world reproduction. A loudspeaker configuration is chosen and loudspeaker signals are generated using a decoding matrix. A set of head-related transfer functions (HRTFs) are measured that correspond to the positions of the loudspeakers. The loudspeaker signals are then convolved with the corresponding HRTFs and the results are summed together for each ear into a single stereo file.

6.7.3 Spherical-harmonic domain

The spherical-harmonic domain approach [45] (see Figure 6.10b) differs from the virtual-loudspeaker approach in that the convolutions are now undertaken in the spherical-harmonic domain. The HRTFs are first encoded into spherical harmonics using a transposed decoding matrix. Given the same loudspeaker configuration, the two approaches may be shown to be numerically identical. However, by first encoding the HRTFs into spherical harmonics, the number of convolutions required now depends on the number of spherical harmonics, and not on the number of loudspeakers. As

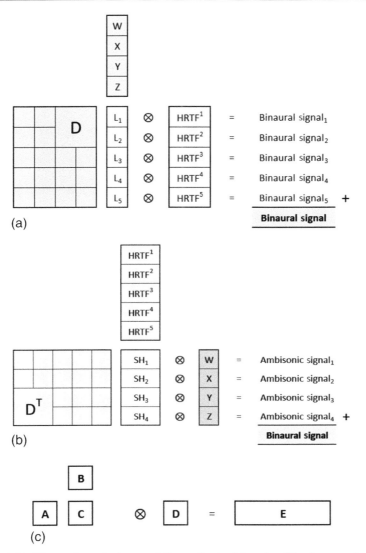

Figure 6.10 Binaural rendering workflows for Ambisonics. The figures should be read left to right and generally depict a matrix multiplication, a series of stereo convolutions, and a final summation. a) Virtual-loudspeaker approach. D is the decoding matrix; [W, X, Y, Z] is the Ambisonic input file. L are the virtual loud-speaker signals. b) Spherical-harmonic domain approach. D^T is the transposed decoding matrix; [W, X, Y, Z] is the Ambisonic input file. SH are the spherical harmonic format HRTFs. c) A is multiplied by B to give C which is convolved with D to give E.

Ambisonic decoding generally requires $L > K$, this reduces the number of convolutions required and therefore the complexity of the renderer. One of the major advantages of this technique is that a dual-band decoder may be implemented by pre-processing the encoded HRTFs, rather than the input signal.

(a)

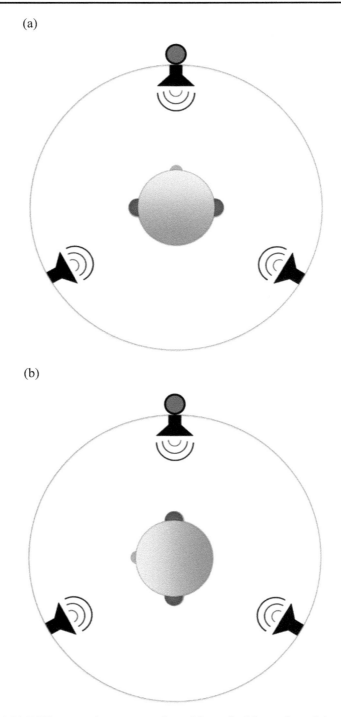

(b)

Figure 6.11 Differences between real-world, static binaural, and head-tracked binaural reproductions of Ambisonic sound fields: (a) (real and virtual) forward facing; (b) (real) left facing; (c) (static binaural) left facing; (d) (head-tracked binaural) left facing. Note the rotation of loudspeakers in (c) and (d), and the counter-rotation of the source in (d).

(c)

(d)

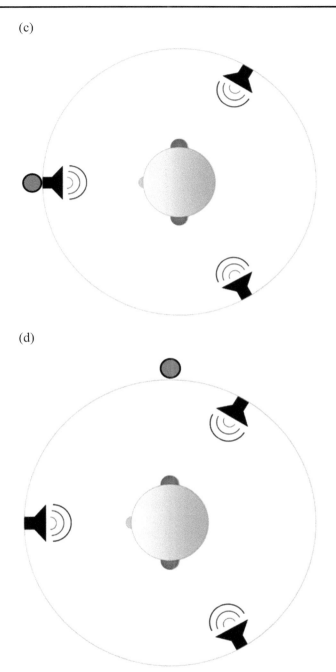

Figure 6.11 Continued

6.7.4 Head-tracked binaural audio

A significant consideration of binaural Ambisonics relates to allowing listeners to "move" or "rotate" within the virtual array (however it is rendered). In real life, this action is trivial. As listeners turn their heads, the surrounding loudspeakers remain stationary. Sources that have been rendered to the loudspeakers therefore also remain stationary and stable. However, for a loudspeaker array that has been rendered virtually over a pair of headphones, this is not the case. As listeners move, so do their headphones and therefore also the virtual array that is carrying the rendered sources.

The solution employs head tracking to monitor subjects' head movements. Simple linear transformation matrices are applied in the spherical-harmonic domain to counter-rotate the Ambisonic sound field [46]. For example, as users rotate their heads to the left, the sound field is counter-rotated to the right. Sound-field rotations are lossless and exhibit a relatively inexpensive computational load.

One consideration is that this technique does not exactly mimic a real-world situation. A comparison of the methods is given in Figure 6.11. In Ambisonics, the accuracy with which a source is reproduced may depend on its alignment with a loudspeaker (or sample point). For a source that is aligned with a loudspeaker in real life, its position never changes. However, for a source that is initially aligned with a loudspeaker and rendered binaurally, its position may shift as users turn their heads and the sound field is rotated. The effects of this discrepancy are more often than not imperceptible, but can be mitigated by opting for optimally regular loudspeaker configurations or dense arrays.

6.8 SUMMARY

A review of Ambisonic principles, mathematical foundations, and encoding and decoding strategies has been presented. Psychoacoustics-based solutions to the spatial aliasing problem outside of the sweet spot (e.g. max-r_E, in-phase) are shown to promote perceptually relevant spatial cues such as interaural time and level differences. It is shown that such manipulations must be carefully considered and that normalization between decoding matrices based upon amplitude or energy preservation must be properly implemented. Methods for presenting Ambisonics binaurally over a virtual loudspeaker array have also been explained. Further reading may be found in the following [15], [21], [35], [47]–[50].

REFERENCES

[1]   M. A. Gerzon, "Periphony: with-height sound reproduction," *Journal of the Audio Engineering Society,* vol. 21, no. 1, pp. 2–10, 1973.

[2]   P. Fellgett, "Ambisonics. Part one: general system description," *Studio Sound,* vol. 17, no. 8, pp. 20–22, 1975.

[3] M. A. Gerzon, "Ambisonics. Part two: studio techniques," *Studio Sound*, vol. 17, no. 8, pp. 24–26, 1975.

[4] M. A. Gerzon, "The design of precisely coincident microphone arrays for stereo and surround sound," in *AES 50th Convention*, 1975.

[5] M. A. Gerzon, "Practical periphony: the reproduction of full-sphere sound," in *AES 65th Convention*, London, 1980.

[6] Dolby, "Dolby Atmos Home Theater Installation Guidelines," 2016. www.dolby.com/us/en/technologies/dolby-atmos/dolby-atmos-home-theater-installation-guidelines.pdf.

[7] J. Herre, J. Hilpert, A. Kuntz, and J. Plogsties, "MPEG-H audio – the new standard for universal spatial/3d audio coding," *Journal of the Audio Engineering Society*, vol. 62, pp. 821–830, 2014.

[8] S. Bertet, J. Daniel, E. Parizet, and O. Warusfel, "Investigation on localisation accuracy for first and higher order Ambisonics reproduced sound sources," *Acta Acustica united with Acustica*, vol. 99, pp. 642–657, 2013.

[9] W. L. Dooley and R. D. Streicher, "M-S stereo: a powerful technique for working in stereo," *Journal of the Audio Engineering Society*, vol. 30, pp. 707–718, 1982.

[10] H. Haber, "The Spherical Harmonics," 2012. http://scipp.ucsc.edu/~haber/ph116C/SphericalHarmonics_12.pdf.

[11] P. Ceperley, "Resonances, waves and fields: spherical harmonics," 2016. http://resonanceswavesandfields.blogspot.co.uk/2016/05/spherical-harmonics.html.

[12] J. N. Goldberg, A. J. Macfarlane, E. T. Newman, F. Rohrlich, and E. C. G. Sudarshan, "Spin-s spherical harmonics and ð," *Journal of Mathematical Physics*, vol. 8, p. 2155, 1967.

[13] A. Farina, "A-format to B-format conversion," 2006. http://pcfarina.eng.unipr.it/Public/B-format/A2B-conversion/A2B.htm.

[14] P. G. Graham and M. A. Gerzon, "Coincident microphone simulation covering three dimensional space and yielding various directional outputs," 4042779, Aug. 16, 1977.

[15] J. Daniel, *Représentation de champs acoustiques, application à la transmission et à la reproduction de scènes sonores complexes dans un contexte multimédia*, These de doctorat, Paris 6, 2001. www.theses.fr/2000PA066581.

[16] M. Poletti, "A unified theory of horizontal holographic sound systems," *Journal of the Audio Engineering Society*, vol. 48, pp. 1155–1182, 2000.

[17] S. Moreau, J. Daniel, and S. Bertet, "3D sound field recording with higher order Ambisonics – objective measurements and validation of a 4th order spherical microphone," in *AES 120th Convention*, 2006.

[18] F. Zotter, M. Frank, and A. Sontacchi, "The virtual t-design Ambisonics-rig using VBAP," in *Proceedings of the 1st EAA-EuroRegio*, Ljubljana, 2010.

[19] F. Zotter, H. Pomberger, and M. Noisternig, "Energy-preserving ambisonic decoding," *Acta Acustica united with Acustica*, vol. 98, pp. 37–47, 2012.

[20] R. Penrose and J. A. Todd, "A generalized inverse for matrices," *Mathematical Proceedings of the Cambridge Philosophical Society*, vol. 51, p. 406, 1955.

[21] F. Hollerweger, *Periphonic Sound Spatialization in Multi-User Virtual Environments*, MSc Thesis, Institute of Electronic Music and Acoustics

(IEM), Graz, Austria, 2006. https://iem.kug.ac.at/en/projects/workspace/projekte-bis-2008/acoustics/awt/periphonic.html.

[22] J. M. Zmölnig, *Entwurf und Implementierung einer Mehrkanal-Beschallungsanlage*, Master's Thesis, Institute of Electronic Music and Acoustics, Graz, Austria, 2002.

[23] F. Zotter and M. Frank, "All-round ambisonic panning and decoding," *Journal of the Audio Engineering Society,* vol. 60, pp. 807–820, 2012.

[24] N. Epain, C. T. Jin, and F. Zotter, "Ambisonic decoding with constant angular spread," *Acta Acustica united with Acustica,* vol. 100, pp. 928–936, 2014.

[25] J. Treviño, T. Okamoto, Y. Iwaya, and Y. Suzuki, "Evaluation of a new ambisonic decoder for irregular loudspeaker arrays using interaural cues," in *Ambisonics Symposium*, Lexington, 2011.

[26] A. J. Heller, E. M. Benjamin, and R. Lee, "A toolkit for the design of ambisonic decoders," in *Linux Audio Conference*, Stanford, 2012.

[27] B. Wiggins, *An Investigation into the Real-Time Manipulation and Control of Three-Dimensional Sound Fields*, PhD Thesis, University of Derby, Derby, UK, 2004. https://derby.openrepository.com/bitstream/handle/10545/217795/BW_PhDThesis.pdf?sequence=1.

[28] J. Daniel, "Spatial sound encoding including near field effect: introducing distance coding filters and a viable, new ambisonic format," in *AES 23rd International Conference*, Copenhagen, 2003.

[29] M. Gorzel, G. Kearney, and F. Boland, "Investigation of ambisonic rendering of elevated sound sources," in *AES 55th International Conference*, 2014.

[30] J. Daniel, J.-B. Rault, and J.-D. Polack, "Ambisonics encoding of other audio formats for multiple listening conditions," in *AES 105th Convention*, California, 1998.

[31] M. A. Gerzon and G. J. Barton, "Ambisonic decoders for HDTV," in *AES 92nd Convention*, Vienna, 1992.

[32] P. M. Morse and U. Ingard, *Theoretical Acoustics*, Princeton University Press, 1968, p. 949.

[33] G. Monro, "In-phase corrections for Ambisonics," in *International Computer Music Conference*, Berlin, 2000.

[34] P. Stitt, *Ambisonics and Higher-Order Ambisonics for Off-Centre Listeners: Evaluation of Perceived and Predicted Image Direction*, PhD Thesis, Queen's University Belfast, 2015. https://ethos.bl.uk/OrderDetails.do?uin=uk.bl.ethos.677279.

[35] G. Kearney, *Auditory Scene Synthesis using Virtual Acoustic Recording and Reproduction*, PhD Thesis, Trinity College, Dublin, Ireland, 2010.

[36] D. G. Malham, "Experience with large area 3-D ambisonic sound systems," in *Institute of Acoustics Autumn Conference Reproduced Sound 8*, 1992.

[37] G. Kearney and T. Doyle, "Height perception in ambisonic based binaural decoding," in *AES 139th Convention*, New York, 2015.

[38] A. J. Heller, R. Lee, and E. M. Benjamin, "Is my decoder Ambisonic?," in *AES 125th Convention*, San Francisco, 2008.

[39] S.-N. Yao, T. Collins, and P. Jancovic, "Timbral and spatial fidelity improvement in Ambisonics," *Applied Acoustics,* vol. 93, pp. 1–8, 2015.

[40] G. Kearney, M. Gorzel, H. Rice, and F. Boland, "Distance perception in interactive virtual acoustic environments using first and higher order ambisonic sound fields," *Acta Acustica united with Acustica,* vol. 98, pp. 61–71, 2012.

[41] M. A. Gerzon, "Psychoacoustic decoders for multispeaker stereo and surround sound," in *AES 93rd Convention*, San Francisco, 1992.

[42] G. Kearney and T. Doyle, "A HRTF database for virtual loudspeaker rendering," in *AES 139th Convention*, New York, 2015.

[43] M. Smyth and S. Smyth, "Headphone virtualisation for immersive audio monitoring," in *AES 140th Convention*, Paris, 2016.

[44] D. Menzies and M. Al-Akaidi, "Nearfield binaural synthesis and Ambisonics," *The Journal of the Acoustical Society of America,* vol. 121, p. 1559, 2007.

[45] A. Avni, J. Ahrens, M. Geier, S. Spors, H. Wierstorf, and B. Rafaely, "Spatial perception of sound fields recorded by spherical microphone arrays with varying spatial resolution," *The Journal of the Acoustical Society of America,* vol. 133, 2013.

[46] M. Kronlachner and F. Zotter, "Spatial transformations for the enhancement of Ambisonic recordings," Graz, 2014.

[47] C. Armstrong, "Improvements in the measurement and optimisation of head related transfer functions for binaural Ambisonics," University of York, 2019.

[48] D. G. Malham, *Space in Music – Music in Space*, MPhil Thesis, University of York, UK, 2003. www.yumpu.com/en/document/view/10492846/higher-order-ambisonic-systems-university-of-york.

[49] F. Hollerweger, "An introduction to higher order ambisonic," University of Graz, Graz, Austria, 2005. http://decoy.iki.fi/dsound/ambisonic/motherlode/source/HOA_intro.pdf.

[50] F. Ortolani, "Introduction to ambisonics (Rev. 2015)," Ironbridge Electronics, 2015.

7

Binaural audio engineering

Kaushik Sunder

7.1 BINAURAL AUDIO TECHNOLOGIES – AN INTRODUCTION

With the advent of virtual reality (VR), augmented reality (AR), and mixed reality (MR) devices, spatial audio – especially presented over headphones – has become indispensable in providing a truly immersive sound experience. We have also seen a surge of "hearables" in the audio market with a variety of smart features. Spatial audio enables such devices to improve the listeners' situational awareness by constantly monitoring real-world events and giving additional directional cues [1]. Ambisonic and binaural integrations in both audio middleware engines such as Wwise, FMOD, Google Resonance, and also plugins such as Facebook 360 and Immerse VST Ambidecoder (Steinberg) have also increased the popularity of spatial music.

Humans have the ability to localize sound objects with great ease. The human brain decodes the directional information from the sound pressure level received at the eardrum. A realistic virtual auditory space can thus be created by capturing the signals at a given recording point (a "subject's" eardrum or blocked ear canal entrance) and reproducing this "binaural signal" at the same or some other point within the ear canal of another listener's ear. This forms the basis of "binaural technology."

Other techniques like Ambisonics and wave field synthesis (WFS) exploit the inherent property of sound wave propagation in space [2]. Ambisonics – originally invented by Gerzon in the mid 1970s [3] – is a recording and playback technique using multichannel mixing technology in live and studio applications that recreates a true realistic sound field, as compared to traditional surround systems. WFS was invented by Berkhout in 1988 [4]. A typical WFS system aims to synthesize the true sound field within the entire listening space such that the sweet spot area covers the entire listening space.

Head-related transfer functions (HRTFs) are a critical component of binaural audio, and they determine much of the quality of any binaural 3D sound that we experience. Human pinnae are as unique as fingerprints and therefore, HRTFs are highly idiosyncratic as well. The need for personalized

130

HRTFs is very well documented in the spatial audio literature and this topic will also be discussed further below.

Most state-of-the-art spatial audio rendering systems use generic HRTFs that do not completely give the best spatial sound experience. The use of generic HRTFs often results in both front-back and up-down confusion, and in-head localization that destroys the veracity of the spatial image. The use of personalized HRTFs helps to eliminate some of these challenges. One of the key issues with obtaining personalized HRTFs is that it is an extremely tedious process. In order to measure a person's HRTFs, the subject has to often sit in an anechoic chamber for several hours with a probe microphone placed at the entrance of the blocked ear canal without any head movement. This process may take several hours depending on the spatial grid that is to be measured. More recent techniques that have shown a lot of promise use 2D images/3D scans of the head to synthesize personalized HRTFs using techniques like the boundary element method (BEM) or finite element method (FEM). These image-based techniques, along with machine learning solutions, allow us to utilize and experience personalized HRTFs for VR, AR, and MR gaming, as well as spatial music applications. We are also seeing an increasing number of hardware devices like Apple Airpods Pro, Audeze Mobius, Oculus, and HTC VIVE that allow head-tracked spatial audio. Especially when head tracking is combined with visual cues – for instance, in AR/VR/MR applications – it significantly affects spatial perception.

Olson *et al.* [5] summarized the four necessary conditions to achieve a realistic perception of spatial audio:

1.   The frequency range of the device should be wide enough to include all the audible components
2.   The dynamic range should be large to permit a distortionless reproduction of spatial audio
3.   The spectral features that correspond to the propagation of the acoustic waves from the source position to the human ear need to be accurately reproduced
4.   The true acoustic space of the environment should be accurately preserved in the reproduced sound

Most state-of-the-art systems in today's world satisfy the first two conditions; however, the main challenges related to spatial audio reproduction originate from the third and the fourth conditions – which form the core of some spatial audio technologies.

7.2 BINAURAL TECHNOLOGY

In a free field, the sound radiated from a point source reaches the ears after undergoing complex interactions, like diffraction and reflection with the anatomical structures (head, torso, and pinnae) of the listener. The HRTF can be described as a linear time invariant process, and is defined as the transfer function that describes the acoustic propagation between the sound source and the listener's ears. An HRTF includes both the magnitude and phase information of the resultant sound wave after its complex interaction with the head, torso, shoulder, and pinnae.

Several definitions of HRTFs have been reported in the literature, as highlighted in [6]:

1. The ratio of the output of a probe microphone located at 1–2 mm from the eardrum of a human subject to the input of the loudspeaker [7]
2. With regard to a mannequin (with an artificial head that mimics the acoustical properties of a human head; see Section 7.2.5) – the ratio of the output pressure of a probe microphone near the eardrum to the free-field pressure at the location of the probe microphone with the head removed [7]–[9]
3. With regard to a mannequin – the ratio of the sound pressure at the opening of a blocked ear canal to the free-field sound pressure as captured by a microphone placed at the position of the "centre of the head" with the head removed [10]
4. With regard to a mannequin – the ratio of the sound pressure at the eardrum to the free-field sound pressure at the centre of the head, with the head removed [11]
5. The ratio of the sound pressure in the ear canal to the sound pressure in the canal with the source directly in front of the listener [12]
6. The ratio of the sound pressure in the left ear canal to the sound pressure in the right ear canal (interaural HRTF) [9]
7. The ratio of the sound pressure in the ear canal to the maximum sound pressure over all locations at that frequency [13]

Among the above list of HRTF definitions, the second, third, and fourth definitions are most widely used. For any arbitrary source position, $HRTF_L$ and $HRTF_R$ for the left and right ears can be defined as:

$$HRTF_L = HRTF_L\left(r,\theta,\phi,f,a\right)=\frac{P_L\left(r,\theta,\phi,f,a\right)}{P_0\left(r,f\right)} \text{ and} \quad (7.1)$$

$$HRTF_R = HRTF_R(r,\theta,\phi,f,a)=\frac{P_R(r,\theta,\phi,f,a)}{P_0(r,f)},$$

where P_L and P_R represent the complex-valued pressures in the frequency domain at the left and right ears, respectively. The symbols r,θ,ϕ,f,a represent the distance from the centre of the head, azimuth, elevation, frequency, and radius of the head, respectively. P_0 represents the complex-valued free-field sound pressure in the frequency domain at the centre of the head with the head absent, which is defined as follows [14]:

$$P_0(r,f)= j\frac{k\rho_0 cQ_0}{4\pi r}e^{-jkr}, \quad (7.2)$$

where ρ_0 is the density of the medium, c is the speed of sound, Q_0 denotes the intensity of the point sound source, and $k = 2\pi f / c$ represents the wave number. The transmission of outward waves with distance is denoted by a factor of e.

The HRTF signal obtained at the eardrum contains several important cues, such as the interaural time differences (ITDs), interaural level differences (ILDs), and the spectral cues (SC) that the auditory system uses to locate a sound source. These cues are explained in detail in the following sections.

7.2.1 Interaural cues

"Duplex theory" [14], [15] states that low-frequency sounds are localized by temporal cues and high-frequency sounds are localized by intensity cues. ILD cues are more prominent at higher frequencies since the head-shadow effect becomes significant for low frequencies unless the source is close to the head (< 50 cm). This is because longer wavelengths can bend around the head, thereby minimizing the intensity differences. On the other hand, the ITD is effective in localizing low frequencies typically below 1.5 kHz. As the ILD and ITD values are increased, the sound image begins to shift towards the ear leading in time or greater in amplitude. The virtual image, however, stops moving along the interaural axis and remains at the leading ear once a critical value of ILD or ITD is reached. Researchers found that the effective ranges of ITD and ILD are approximately 0.005 to 1.5 ms and 1 to 10 dB, respectively [16], [17].

Wightman *et al.* [18] studied the relative salience between ILD and ITD cues in experiments using synthesized stimuli with conflicting ILD and ITD cues. These interaural cues also play an important role in the externalization of the sound image – especially when played back through headphones. Weinrich [19] suggested that ITD is the dominant factor in creating a sensation of spaciousness, and the sound image is localized within the head when there is a discrepancy between the ITD and ILD cues. Thus, only the correct ratio between the ITD and ILD can achieve optimal externalization. Both ILD and ITD are frequency dependent. The interaural cues can be computed using the "spherical-head model" developed by Algazi *et al.* [20].

7.2.2 Spectral cues for localization role of pinnae

The pinna cues are the most widely studied of all localization cues. The pinna acts as a complex resonance cavity, which introduces unique spectral features that are characteristic of a particular source location. The morphology of the pinna has high inter-individual differences, which greatly affect the frequency spectrum at the eardrum. Due to this high variability of pinna morphology, the spectral characteristics become highly idiosyncratic. Thus, if a deployed HRTF is not the listener's own, the spatial image of the perceived sound is distorted [22]–[25]. The extent of the similarity between the features of the individual and the HRTF required to avoid

Table 7.1 Directional cues reported in [21].

Frontal directions spectral cues	Overhead spectral cues	Rear spectral cues
1-octave BW notch, with a lower cut-off between 4 kHz and 8 kHz	¼-octave BW peak between 7 kHz and 9 kHz	Small peak between 10 kHz and 12 kHz
Increased energy above 13 kHz		Decrease of energy above and below this peak

degradation of the spatial image is still unknown. Some researchers [21], [26]–[28] proposed that the similarity covers single features, while others [24], [29], [30] proposed that broad ranges of frequencies are necessary for accurate localization. See Table 7.1.

Some researchers [27], [31]–[33] have observed unique spectral notches generated by the interference of direct sound entering the external auditory canal and the time-delayed reflections off the posterior concha wall. Several other researchers have suggested spectral peaks as important cues [9], [26], [34], [35]. Research has also suggested that there are other spectral features covering a larger frequency range that are relevant for sound localization [24], [28], [36], [37]. Some evidence suggests that the perception of direction can be influenced by spectral cues independently of the location of the sound source. Blauert [26] carried out seminal work through detailed studies of the localization of 1/3-octave-band noise played back through loudspeakers. He first divided the hemisphere around the listener into different regions in the horizontal plane as "front, above, and behind." It was observed that the responses of the subjects were clustered in certain regions depending on the centre frequency of the noise stimuli, which led to the concept of "directional bands." Blauert proposed that each direction in space is associated with a particular frequency band. A later study reported that the directional bands showed individual differences [38].

Butler and Belendiuk [22] conducted experiments, where sound source localization performance in the median sagittal plane was tested under different conditions for real and virtual sound scenes (generic binaural audio playback over headphones) using broadband stimuli as test signals. They conducted 1/3-octave-band measurements centred at 4 kHz, 5 kHz, 6.3 kHz, 8 kHz, and 10 kHz. It was found that as the source moved up in elevation, a notch that coded the frontal region in the median plane moved up in frequency and became narrower in bandwidth. The researchers also observed that subjects did not always perform better using their own personalized recordings. These results were similar to those of Hebrank *et al.* [21], who conducted a number of median-plane-localization experiments.

Bloom [27] studied the hearing's monaural sensitivity curves (by restricting head movement in a near-anechoic chamber to minimize dynamic localization cues), and found that the external ear produces specific distinguishable features in the sound spectra that were dependent on elevation. These spectral features were appropriately modified, and electronically implemented to create an illusion of continuously variable source elevation. The notches in the simplified versions of HRTFs were studied and emulated using a 1-octave-band filter centred at 8 kHz with a varying f_c. Roffler and Butler [39] strongly suggested that in order to localize correctly in the vertical plane, the stimulus must be complex, it must contain frequencies above 7 kHz, and the listener must have the pinnae unoccluded.

Some researchers [40], [41] studied the detection thresholds of peaks and notches. It was seen that the spectral notches were perceptually less salient than corresponding peaks. Gardner et al. [42] performed localization experiments with a series of loudspeakers with and without the pinnae occluded. The employed stimuli were filtered broadband noise of 1/2-octave narrow-band and full-octave bandwidth centred at 2, 3, 4, 6, 8, and 10 kHz. They found that each of the three main parts of the pinna – scapha, fossa, and concha – were important for accurate localization. The occlusion of concha degraded the localization performance the most. They found that localization ability improved if the stimuli had high-frequency content. The localization performance was best for broadband stimuli regardless of the degree of pinna occlusion. The performance degraded with increasing occlusion of the pinna cavities. Localization performance with narrow-band noise stimuli (centre frequency 8–10 kHz) was similar to that with broadband noise stimuli without pinna occlusion. Overall trends in the rear direction were similar, except that the localization performance was in general poorer than in the frontal directions.

7.2.3 Binaural reproduction over loudspeakers

Gardner et al. [43] originally constructed a spatial audio system using conventional stereo loudspeakers. This was accomplished by combining a binaural spatializer with crosstalk cancellation (XTC) technology. A critical condition in binaural rendering is that the left and right recorded/synthesized signals have to be routed exactly to the left and right ears, respectively.

In an XTC-based spatial audio system (Figure 7.1), acoustic crosstalk occurs. This means that the left signal that is targeted at the left ear is perceived not only by the left ear but also by the right ear, and vice versa. The presence of crosstalk inherently distorts the spatial image as the virtual image collapses to a narrow and frontal image with poor elevation perception. Thus, pre-processing filters (H in Figure 7.1) have to be implemented to cancel the unwanted acoustic crosstalk between the left and the right channels [43], [44]. The pre-processing filter has the following functions:

1. Cancels the acoustic crosstalk between left and the right ears
2. Corrects for the transfer function between each loudspeaker and its ipsilateral ear (ear nearer to the source) to achieve a very transparent reproduction of the binaural signals
3. Removes the inherent response of the loudspeaker transducer and preserves only the acoustic propagation effects

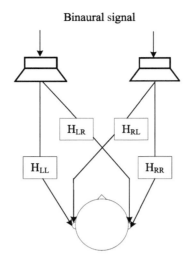

Figure 7.1 Binaural rendering over loudspeakers.

Mathematically, the loudspeaker signals (X) can be derived using the equations:

$$X_L(f) = \frac{B_L(f) \times H_{LL}(f) - B_R(f) \times H_{RL}(f)}{H_{LL}(f) \times H_{RR}(f) - H_{LR}(f) \times H_{RL}(f)} \quad \text{and} \quad (7.3)$$

$$X_R(f) = \frac{B_R(f) \times H_{RR}(f) - B_L(f) \times H_{LR}(f)}{H_{LL}(f) \times H_{RR}(f) - H_{LR}(f) \times H_{RL}(f)},$$

where B_L and B_R are the binaural signals; and H_{LL} and H_{RR} refer to the direct path between each loudspeaker and its corresponding ipsilateral ear; and H_{LR} and H_{RL} correspond to the crosstalk path between the left and right loudspeakers and their contralateral (further from the source) ears. One of the serious limitations in binaural playback over loudspeakers is that crosstalk cancellation works well only in a narrow region in between the speakers, resulting in a limited sweet-spot area. The system is also highly sensitive to listener movement away from the sweet-spot region.

In spite of these challenges, there are some impressive commercial technologies, like BACCH™ filters [45], that deliver convincing binaural audio over loudspeakers. Another example of crosstalk cancellation technology is TRANSAURAL™ technology developed by the Cooper Bauck Corporation.

7.2.4 Binaural audio over headphones

Binaural playback over headphones (Figure 7.2) is the most convenient form of playback mechanism because of its excellent channel separation. Thus, an additional crosstalk cancellation filter is not required. With the

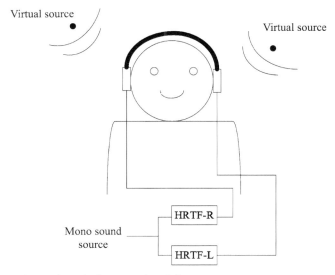

Figure 7.2 Binaural rendering over headphones.

headphones market growing enormously and with the increase in prefer-ence for personal audio, headphone playback method remains the most preferred 3D audio reproduction system. The presence of head trackers in some modern earphones/headphones has also enabled the possibility of head-tracked binaural audio. Binaural playback over headphones, although extremely convenient, is far from ideal, and has several challenges and limitations.

Binaural signals can be typically obtained either by recording the sound scene at the eardrum or the ear canal entrance of a listener/dummy head or by synthesizing the virtual audio using HRTF filters. In the next few sections, the binaural recording and synthesis procedures are elaborated in great detail.

7.2.5 Binaural recording of the sound scene

Binaural signals can be acquired by capturing the signals using a probe microphone placed either at the blocked entrance of the ear canal or very close to the eardrum of the listener. As the signal follows its path to the eardrum, it is modulated by the various diffractions and reflections that are induced by the morphology of the individual, and thus the recording captures this "watermark." Since the anthropometric properties of the ear are unique, the binaural recordings captured at any point within the ear canal are highly idiosyncratic. It is therefore currently impossible to obtain a convincing spatial auditory image that "works" for all listeners. The most effective solution for any given individual is to carry out personalized bin-aural recordings, which is a highly tedious and time-consuming process. Both the positioning of the microphone within the ear and the stability of the head are very important in order to capture the correct information regarding the location of a source; the slightest listener head movement can degrade the integrity of the recorded binaural signal.

For generic purposes, researchers have developed mannequins and dummy heads. These are made of specific materials, and might be fitted with various anatomical structures as well as the head, for example, torso, nose, or pinnae – acoustically similar to those of real humans [46]. The design of a dummy head is based upon the average dimensions from a certain population of listeners or on the anthropometric features of a "typical" human subject and thus a dummy head can simulate the acoustical contributions of the head, torso, and pinnae. These dummy heads typically contain a pair of microphones, often situated at emulated eardrums (or at some point in the ear canal), and so can easily capture binaural recordings with high repeatability [47], [48].

These artificial heads are extremely useful in carrying out a variety of acoustical measurements, binaural recordings, sound-quality evaluation, measurement of headphone responses, evaluation of headsets, and hearing-aid testing. The most important features of various dummy heads are shown in Table 7.2. However, as explained before, dummy head binaural recordings do come with an inherent disadvantage in that the spatial cues present in the recording never exactly match those of the end-listener, which distorts the spatial image considerably.

Table 7.2 Different type of artificial heads and their features.

Type of artificial head	Manufacturer	Features/comments
KEMAR	Knowles/GRAS Sound & Vibration	Commonly used with a human-like appearance, with pairs of both small and large pinnae available
HATS	Brüel & Kjær	Includes head, torso, and detailed pinnae; used for electroacoustic measurements such as HRTF measurements, telephone, and headphone testing
HMS IV	HEAD acoustics GmbH	Consists of head, shoulders, and simplified pinnae; used for evaluation of sound quality and noise
MK2B	01dB-metravib	Used for sound-quality recording in the automobile industry and psychoacoustics
KU 100	Neumann	Consists of only the head; commonly used for binaural recording; diffuse field equalized
VALDEMAR	Aalborg University	Designed for research in binaural and spatial hearing

7.2.6 Binaural synthesis of a sound scene

Instead of recording, the left and right binaural signals (y_L and y_R) [49] can be obtained by filtering a monaural sound source ($x(n)$) with a pair of measured head-related impulse responses (HRIR):

$$y_L(n) = x(n) * HRIR_L \; and \qquad\qquad (7.4)$$

$$y_R(n) = x(n) * HRIR_R$$

When generating binaural signals using HRIRs, the HRIRs first have to be measured at various locations in the 3D space. The responses of the microphones, the headphones, and the coupling between the ear and the headphone ear cup must all be compensated for before rendering the binaural audio for headphones. This equalization process is critical and without it, severe distortions of both magnitude and phase will be induced in the binaural signal. In the following subsection, the process of measuring HRTFs is explained.

7.3 MEASUREMENT OF HRTFS

HRTFs are typically measured by playing an excitation signal (e.g. a "maximum-length sequence," exponential sweep, or white noise) through a loudspeaker and recording the impulse response at the eardrum of the listener using a probe microphone. This suffered from low accuracy and inconvenient data storage in the early years due to the absence of accurate digital measurement techniques, but the advent of digital measurement systems brought quicker and more accurate HRTF measurements with higher spatial resolutions.

Empirical measurements are an important approach in the measurement of HRTFs, and accurately doing so is necessary to create a realistic perception of virtual audio. Generally, the far-field HRTFs at various spatial directions are obtained from the signals captured by the two microphones at both ears by changing the relative direction between the sound source and subject. Such a measuring method is often very lengthy and can suffer from low efficiency. Consequently, a number of methods have been developed to measure the HRTF data at one or a few spatially discrete locations in a single run. One such novel technique of HRTF measurement has been developed using the principle of microphone-loudspeaker reciprocity [50]. The impulse responses are measured simultaneously in a single trial by placing a tiny loudspeaker at the entrance of the ear canal and recording the responses at every microphone placed spatially on a sphere (radius = 0.7 m) around the listener. In this way, the HRTFs of all the microphone positions can be derived at a single time. However, one of the main disadvantages of this method is both the poor low-frequency response of the micro-speaker with a high-frequency range going up to only 16 kHz, and a further disadvantage is the low signal-to-noise ratio of the loudspeaker.

However, HRTFs can also be considered as a continuous function of spatial directions. Fukudome *et al.* [51] proposed a continuous measurement method for obtaining HRTFs continuously in the horizontal plane in a single run using a servo-swivelled chair. HRTFs for any other arbitrary position can then be extracted from the measured signal by applying an appropriate signal processing method. Other researchers used white Gaussian noise and a multichannel adaptive filtering algorithm to measure the continuous HRTF data across all azimuths at different elevations [51]. Ajdler *et al.* [52] used a constantly moving microphone to measure the HRTFs in the horizontal plane, and this method even accounts for the Doppler effect in the recording due to the movement of the microphone relative to a fixed sound source, and cancels it in the reconstructed room impulse response (RIR). An important advantage of this technique is that the HRTFs at all angular positions along the horizontal plane can be measured in less than 1 s. As a result of decades of research and measurements, several HRTF databases (as shown in Table 7.3) are available that help the advancement of spatial audio. Examples of the CIPIC HRTFs can be seen in Figure 7.3.

Table 7.3 List of measured HRTF databases. Examples of the CIPIC HRTFs can be seen in Figure 7.4.

Owner	(Subjects, directions)
AUDIS database [60]	(20; 122)
Austrian Academy of Sciences [56]	(77;1,550)
Bovbjerg [61]	(VALDEMAR; 11,975)
CIPIC, UC Davis [53]	(45;1,250)
Genuit and Xiang [57], [58]	(HMS I and HMS II)
Grassi [63]	(7; 1132)
IRCAM [52]	(51; 187)
Moller *et al.* [65]	(40; 97)
Nagoya University [55]	(96; 72)
Oldenburg University (0.8m,3m) [57]	(HATS; 365)
Reiderer [59]	(51; 7)
SADIE II, University of York [120]	(20; 9,201)
Takane [62]	(3; 454)
TH Köln [121]	(KU100; 2702)
Tohoku University [54]	(3; 454)
TU Berlin (3 m, 2 m, 1 m, 0.5 m) [122]	(KEMAR;360)
University of Maryland [53]	(7; 2,093)
Wightman and Kistler [7]	(10; 144)
Xie *et al.* [64]	(52; 493)
Yu *et al.* [64]	(KEMAR; 3,889)

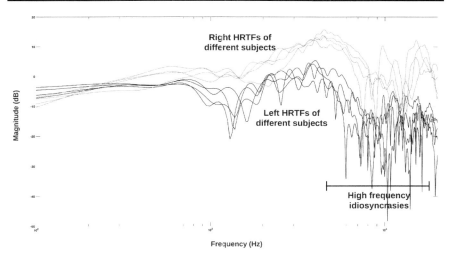

Figure 7.3 Examples of HRTFs (azimuth 45°, elevation 0°) from the CIPIC database for different subjects. Note the unique high-frequency cues in the HRTFs of different subjects.

7.4 PERSONALIZATION OF HRTFS

As mentioned previously, binaural audio is idiosyncratic and highly varies among human subjects. The price to be paid for the use of the HRTFs in Table 7.3 is that the perceived spatial image may be highly distorted. Some of the main challenges that result from the use of generic HRTFs are:

- **Elevation-localization errors:** The use of generic binaural audio has the greatest effect on elevation perception. Confusion in the form of up-down reversals is common with generic HRTFs. This is primarily because of the differences in pinna spectral features between an individual's HRTF and a generic HRTF.
- **Front-back confusion:** Often with generic HRTFs, the frontal directional sources are perceived to be coming from the rear. This is referred to as front-back confusion, and is commonly found in the headphone playback of binaural audio.
- **In-head localization:** The conflict between interaural time and level differences leads to in-head localization of the sound.
- **Timbral colouration:** This is highly noticeable when generic HRTFs are used since using generic HRTFs leads to a spectrum at the eardrum that is vastly different from the listener's true personalized HRTF.

Wenzel *et al.* [23] conducted a detailed NASA study (on the potential efficacy of virtual auditory displays) on perceived localization when using non-individual HRTFs compared to "real" stimuli presented in the free field. In the former, the ITD-based azimuth judgment was very accurate, but elevation and frontal localization were the most affected due to the distortion of the spectral cues on which this is dependent. Further, the rates of front-back and up-down confusion in headphone reproduction using gen-eric HRTFs increased from 19% to 31%, and 6% to 18%, respectively.

The externalization of frontal sound sources was poor and most likely perceived inside the head. In addition, the sound sources appeared to be blurred when non-individual HRTFs were used. The use of dynamic HRTFs – with the help of head tracking – improved the localization per-formance, but confusion was still not completely resolved. An important caveat to remember is that even with personalized HRTFs, the localiza-tion accuracy with virtual sources is poorer than that with real sources [66]. In this study, the researchers observed that the front-back confusion rates almost doubled (11%) for virtual sources compared to those of real sources (6%).

7.4.1 HRTF personalization techniques

The modelling of personal HRTF cues mainly involves the identification of the appropriate parameters that completely define the idiosyncratic features in the spectral cues. Table 7.4 lists and compares all the HRTF personaliza-tion techniques in the literature, and these will now be discussed.

Acoustical measurements: The most straightforward individualiza-tion technique is to measure the personalized HRTFs for every listener at different sound positions [66]. This is the ideal solution, but it is extremely tedious and involves highly precise measurements. These measurements also require subjects to remain seated with a fixed orientation and remain relatively motionless for long periods, which may cause fatigue in subjects. Section 7.3 discussed some fast HRTF measurement systems. An even faster and more accurate method was designed at the Air Force Research Laboratory [67], which allows the entire collection of HRTFs in just 6–10 minutes.

Anthropometric data: Personalized HRTFs can also be modelled as weighted sums of basis functions, which can be performed either in the frequency, time, or spatial domains. The basis functions are usu-ally common to all individuals, and the individualization information is often conveyed by weights. The HRTFs are essentially expressed as a weighted sum of a set of eigenvectors, which can be derived from principal component analysis (PCA) or independent component analysis (ICA) [68]. The individual weights are derived from the anthropometric parameters that are captured by optical descriptors either through direct measurement, derivation from pictures, or by obtaining a 3D mesh of the morphology [69].

The solution to the problem of diffraction of an acoustic wave around the listener's body results in individual HRTFs. This solution may be obtained by analytical or numerical methods, such as the BEM or FEM [69]. The inputs in these methods can be a simple geometrical primitives [70] (a sphere, cylinder, or ellipsoid), a 3D mesh obtained from an mag-netic resonance imaging (MRI) or laser scanner, or a set of 2D images [71]. An important advantage of these techniques is that the relative effects of a particular morphological element (pinna, torso, and head)

Table 7.4 List of personalization techniques.

How to obtain individual features	Techniques	Pros	Cons	Performance and remarks
Acoustical measurements	Individual measurement [65], IRCAM, France [74], CIPIC [53], Uni of Maryland, Tohoku Uni [54], Nagoya Uni [54], Austrian Academy of Sciences [56] etc.	Ideal, accurate	Requires high precision and time-consuming, impractical for every listener	Usually a reference to compare for individualization techniques
Listening/ training	Selection from non-individualized HRTF [75], frequency scaling [73], tune magnitude spectrum [76], active sensory tuning [77], PCA weight tuning [78], critical band tuning [79]	Easy to implement, directly relates to perception	Takes time like a training, might require regular training sessions, can cause fatigue	Making use of the individual listening to obtain the best HRTF – perceptually
Anthropometric data	Select cepstrum parameters [80] Optical descriptors: 3D mesh [71] Analytical or numerical solutions: PCA + multiple linear regression [68], FEM, BEM [69], multiple linear regression [81], multiway array analysis [82], support vector machine[83], artificial neural network [83] Structural model of HRTFs [69] and HRTF database-matching [72]	Based on acoustic principles, can study effects due to independent elements of the morphology	Needs a large database, tedious, requirement for: high resolution of imaging/ expensive equipment/ qualified users	Making use of the correlation between individual HRTFs and anthropometric data
Playback mode	Frontal-projection headphones [84]	No additional measurement, listening training	A new structure – not applicable to normal headphones, special equalization required	Automatic customization, more than 50% reduction on front-back confusion

and their variation in size, location, and shape can be independently investigated [69].

Perceptual feedback: Subjects go through listening tests, where they choose their preferred HRTFs based upon the correct perception of frontal sources and reduced front-back reversals [72]. Listeners can also adapt to non-individual HRTFs by modifying the HRTFs in order to suit their personal perception. Middlebrooks observed both that the peaks and notches of HRTFs are frequency shifted for different individuals, and that the extent of the shift is related to the size of pinna [73]. These perceptually based methods are much simpler in terms of both required resources and effort than personalization methods either obtained by acoustical measurements or derived from the morphology. However, these training sessions can be quite long and can also result in listener fatigue.

Playback mode: Sunder *et al.* [84] developed frontal-projection headphones that resulted in a reduction of more than 50% of front-back confusion. This method does not require any additional measurements, customization, or training.

7.5 AUDITORY DISTANCE PERCEPTION

Auditory distance perception has been an area that has eluded scientists for several years. The ability of human beings to estimate distance is much more complex and also displays less accuracy than directional perception [49]. It was found that listeners often overestimated the distance in the near field and underestimated the distance beyond 1 m in free-field anechoic conditions [85]. The perceived distance increases with the actual distance and then, saturates beyond a point in the far field [86]. This phenomenon was named the "acoustic horizon" by Von Bekesy [87]. Intensity, interaural level differences, spectral cues, and the direct-to-reverberant energy ratio are found to be the most important cues for distance perception [49].

Intensity cues: Intensity as a potential distance cue has long been investigated in various studies [85]. The perceived distance has been found to increase less rapidly than the physical distance, for distances greater than 1 m – when intensity is the only prominent cue; the 6 dB loss in sound pressure for every doubling of distance does not hold in the near field. It was also shown that a difference greater than 6 dB is required for a perceived doubling of distance in the proximal region ("close" to the head, often quoted as a distance less than 1 m). Other studies have investigated loudness as a possible distance cue [88]. Under conditions where intensity is the only prominent cue, loudness varies inversely with perceived distance, and a 10 dB decrement is required for a sensation of "half loudness" or doubling distance [89]. In fact, when there are multiple cues (in the diffuse sound field) at the disposal of the listener, the loudness of the sound source has been found to remain constant under distance-varying conditions. The existence of loudness constancy is attributed to the reverberant sound energy being independent of the distance.

Binaural cues: Binaural cues play a critical role in human distance perception in the near field, in anechoic conditions. At low frequencies, the

ILD and ITD values can be well approximated by the spherical head model [90]. It is noted that the ILD values rapidly increase with a decrease in distance, while the ITD values increase only modestly. This is typically because the ITD value depends upon the difference in distance, while the ILD value depends on the ratio of the distance between the source and the two ears [91]. It is also speculated that listeners may first make use of the distant-invariant ITDs to first determine the azimuth location, and then use the ILD cues to estimate the distance [92]. Furthermore, the ILD of a lateral source increases by 15 just-noticeable difference (JND) or more, when the source distance decreases from 1 m to 12 cm, while the ITD increases only by 2–3 JND. Brungart [93] studied the localization response for different stimulus conditions and found that the distance accuracy in the low-pass condition with a cut-off frequency of 3 kHz was similar to that in broadband conditions. This result – along with poor distance perception in monaural conditions – indicates a strong distance perception dependence on the low-frequency ILD cues in anechoic listening conditions.

7.5.1 Models of auditory distance perception

Based on the perceptual cues identified for distance perception, several distance perception models have been developed, and they can be grouped as follows:

Using binaural cues: Stewart [94] first derived mathematical equations for the ILD and ITD that originate in the near field with regard to a spherical head. Using Stewart's derivations, Hartley *et al.* [95] also manually tabulated ILDs and ITDs at different distances and found that a combination of distance-invariant ITDs and distance-varying ILDs could account for the binaural cues in the near field. Hirsch [96] presented a model of binaural distance perception that allowed a listener to determine the distance of a sound source via its response to the ILD and ITD. However, this model ignores the effects of sound diffraction induced by the head; the ears are represented by point receivers in free space. Tahara and Sakurai [97] proposed a model similar to Hirsch's that simulates the changes in distance perception based on the simultaneity of the changes of the ITD, with the ILD as a function of distance as well as azimuth angle.

Using auditory parallax: Kim *et al.* [98] modelled the distance perception up to 2 m using the parallax angle information available in the HRTFs as previously investigated by Brungart and Rabinowitz [90], [98]. Their "auditory parallax model" was simulated using a simple geometrical model that defined the relation between the angle with respect to the centre of the head – a sort of "average" angle compared to that subtended by each of the two ears. It was found that the perceived distance of the virtual sound simulated by the auditory parallax model was similar to the free-field listening condition.

Simulating the spectral cues directly: Duda and Martens [99] developed an algorithm to compute the sound pressure at the surface of the spherical head for a source at various distances. Rabinowitz *et al.* [100] later studied the frequency scalability of HRTFs for an enlarged head, which maintained

a fixed head size and also varied the distance from the source. Ze-Wei
et al. [101] modified the model developed by Rabinowitz *et al.* [100]
and analysed the proximal-region HRTFs by adding the neck and torso.
Interestingly, it was found that the human neck influences the near-field
HRTFs but its perceptual effects are still unclear. Though these models
give an idea about the spectral variation in proximal-region HRTFs, they
cannot be used to directly synthesize binaural audio as the spectral effects
of the external ear are missing.

Simulation from far-field HRTFs: Duraiswami *et al.* [102] modelled
near-field HRTFs using a range extrapolation of measurements from a
single distance. This was carried out by expressing the HRTF in terms of
a series of multi-pole solutions of the Helmholtz equation. Romblom and
Cook [103] used difference filters between the near and far field derived
from the spherical-head model and used an HRTF look-up table to calculate the near-field HRTFs. However, perceptual results were not reported
in this study. Kan *et al.* [104] used a similar method and simulated the
near-field HRTFs by applying the distance variation function (DVF) to
the personalized far-field HRTFs. Kan *et al.* also showed that the DVF
technique improved the distance perception compared to a simple intensity adjustment. The directional perception in this study was reported to
be good – with less front-back confusion – since personalized HRTFs
were used.

Furthermore, it is interesting to understand the effect of idiosyncrasy
on the perception of distance. Zahorik [105] noted that distance localization when using generic HRTFs is similar to that using personalized
HRTFs in the presence of other distance cues. Brungart's [93] observation of good distance perception using low-pass filtered stimuli further
shows the lack of dependence that distance perception has on the idiosyncratic spectral cues. On the contrary, directional perception is degraded
by the use of generic HRTFs. Thus, even though distance perception is
less dependent on the individual features, personalized characteristics
in the HRTF are extremely important for the accurate reproduction of
spatial audio.

7.5.2 Reverberation

Reverberation plays a critical role in accurate distance perception.
Reflections from surroundings, such as walls, ceilings, the floor, and furniture, have a significant effect on spatial hearing. When a source emits
sound, the direct sound is first to arrive at the listener position and is then
followed by early reflections (see Figure 7.4). These early reflections typically arrive with different time delays with the first reflection (with the
shortest time delay) coming from the reflective source that has the least
difference in the acoustical path relative to the direct sound. What follows
the early reflections is a rapidly increasing number of reflections with a high
temporal density. However, the total energy of these reflections reduces
exponentially due to both surface absorption and other high-frequency air
absorptions.

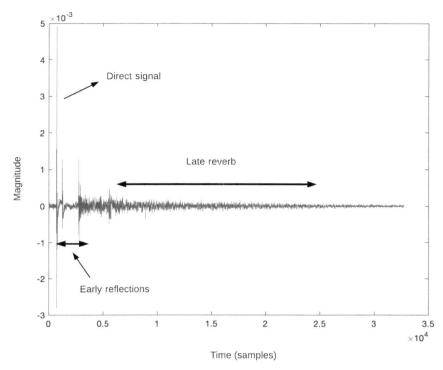

Figure 7.4 A typical breakdown of a BRIR.

The sound transmission between the sound source and a receiver can be modelled as a time-invariant process and can be described as an RIR. An RIR, where the receiver is treated as a point source, is a useful tool for investigating several acoustical parameters, including reverberation time [106]. As mentioned, in the presence of a human listener, the anatomical features of the human ears, head, and torso modify the sound that reaches the two ears through diffraction and reflection. In general, the early reflections are direction dependent and idiosyncratic due to the unique anthropometric features of the human ear. All this information can be encoded in the form of binaural RIRs (BRIRs). BRIRs can either be measured or modelled using anechoic HRIRs and the corresponding RIRs of the acoustic space [107]. Two of the most common classical geometric methods for modelling room reverberation are "ray tracing" and the "image source" method, modifications of which are used extensively today in several physics-based engines for room modelling.

In VR and AR applications, modelling reverberation accurately becomes all the more important. Several middleware audio engines such as Resonance Audio, FMOD, and Visisonics have implemented physics-based room models that simulate real-time acoustics based upon a user's interaction with the virtual world. With AR, such real-time estimation is especially important since the acoustics of the environment are not known a priori. Llewellyn and Paterson discuss this topic further in Chapter 3.

Murgai *et al.* [108] at Magic Leap developed methods for the blind estimation of the reverberation fingerprint using machine learning techniques.

7.6 HEADPHONE EQUALIZATION

In addition to the need for personalized HRTFs, the accurate equalization
of headphone responses is necessary in order to have a true immersive perception of the sound [109]. Headphones are not acoustically transparent as
they modify the sound spectrum of the binaural audio as it is reproduced.
Typically, the headphone response or a headphone transfer function (HPTF)
comprises the headphones transducer response and the acoustic coupling
between the headphones and the listener's ears. This path is indicated in
Figure 7.5.

In order to compensate for the headphone response, the HPTF is first
measured at either the blocked ear entrance or at the eardrum. It is not
a necessary requirement that the "reproduction point" (i.e. the point
where the binaural signals are played back) is the same as the recording
point. However, the location mismatch between the reproduction and the
recording point also has to be compensated for [110]. In general, headphone equalization accounts for the compensation in Figure 7.5.

In addition, equalization may also compensate for the frequency
response of the microphone and the mismatch of the radiation impedance
looking out from the ear entrance. To correctly compensate for the HPTF
response for an individual, again, one needs to measure the HTPF at the
eardrum or the blocked ear entrance of the individual. The headphone
responses are highly individualized since the frequency response involves
the headphone-ear coupling. Moller *et al.* [10] measured the HTPF (at the
blocked ear canal) of 40 individuals for 14 different headphones and found
that the inter-subject variations at high frequencies were much higher than
those in the low-frequency region. Blocked ear-canal measurements tend
to have less inter-individual variations than open ear-canal measurements.

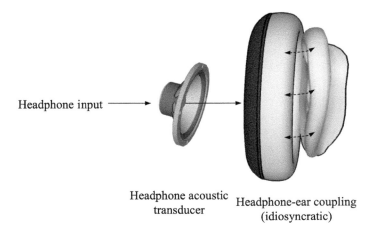

Headphone input

Headphone acoustic
transducer

Headphone-ear coupling
(idiosyncratic)

Figure 7.5 Breakdown of a HPTF.

Pralong and Carlile [111] obtained the HPTFs of the left and right ears of 10 subjects with Sennheiser HD 250 circumaural headphones. It was found that considerable variability in the responses occurred at frequencies above 6 kHz, with an inter-individual standard deviation peaking up to 17 dB at frequencies around 9 kHz for the right ear. Both spectral features and inter-individual differences are also found to be similar to those in the HRTFs. Since high-frequency spectral cues are extremely important for localization perception, personalized equalization is necessary to ensure that headphones do not degrade the spatial perception. HPTFs also vary considerably every time the headphones are removed and repositioned. Kulkarni *et al*. [112] obtained 20 transfer-function measurements on the KEMAR mannequin when removing and repositioning the headphones after each measurement. The variation of the responses at low frequencies is lower than that at higher frequencies since the pressure at low frequencies within the earphone cavity can be approximated to be the same everywhere. Because of this variation in the measurement of headphone responses, effective compensation is not possible, even with personalized equalization, and the spectrum at the eardrum becomes unpredictable.

The mean response (across the different measurements) is still inadequate because the mean inverse filter cannot equalize the filter functions completely. Interestingly, it has been pointed out that the variance produced by the spectral artefacts due to repositioning is lower than the variance of the spectral features in the HRTFs [113].

7.7 HRTF-BASED LOCALIZATION FOR SMART HEARABLES/EARPHONES

In the last few years, we have seen a surge of hearables in the audio market with a variety of smart features. Listeners that wear these devices are increasingly cut off from the outside world and are a potential safety hazard, but augmenting them with a hearing aid equipped with binaural technology can locate the direction of important cues. Sound – unlike vision – can provide us with complete 360° situational awareness, for instance, for a biker to be aware of the possible presence and direction of nearby vehicles, thereby improving safety. The current state of the art in directional sensing uses multiple-microphone techniques, but using only two microphones enormously helps to reduce the cost and computational complexity of this technology. However, a unique approach using just two microphones can still be used to "detect the direction" of an acoustic event. This method makes use of the unique spectral and temporal features of the listeners' HRTF [1].

Such HRTF-based sound-localization methods have traditionally been applied to human-centred robotic systems for telepresence operations [114]. The idea behind most HRTF-based localization algorithms is to identify a pair of HRTFs that corresponds to the emitting position of the source such that the correlation between left and right microphone observations is maximum. There are four different principal techniques in the literature:

- Matched filtering approach

  In the matched filtering approach, the inverse of the HRTFs is first computed and then filtered with the microphone "observations." An index computed to indicate the highest cross-correlation yields the source location of interest. The disadvantage of this approach is that the inversion of the HRTFs can be problematic due to instability. This is primarily because of the linear phase component of the HRTFs that encodes the ITDs. Keyrouz *et al.* [115] proposed a solution that uses outer-inner factorization converting an unstable inverse to an anti-causal and bounded inverse.

- Source cancellation approach

  Keyrouz and Diepold [116] proposed the source cancellation approach, which is an extension of the matched filtering approach. In this method, the cross-correlation between all the pairs of ratios of recorded observations and ratio of HRTFs is taken. The main improvement is that the HRTFs need not be inverted, and can be precomputed and stored in memory.

- Reference-signal approach

  The reference signal [117] method uses four microphones: two for the HRTF-filtered signals and two others outside the ear canal for original sound signals. While this method has several advantages, this method does not fall under the scope of binaural localization as it requires four microphones.

- Cross-convolution approach

  In order to avoid any instability issues, Usman *et al.* [118] exploited the associative property of the convolution operator and proposed the cross-convolution approach. Among all the HRTF-based sound source localization techniques, for example, the three above, the cross-convolutional method is the one that is most robust, delivers the highest accuracy, and is best suited for this application. In this method, the left and right observations or the hearing-aid signals are filtered with a pair of contralateral HRTFs. The filtered signals are then converted to the frequency domain before an index of the maximum cross-correlation is calculated. The direction corresponding to the index is the estimated direction of the source:

$$
\begin{aligned}
\hat{S}_{L,i} &= H_{R,i} \cdot X_L \\
&= H_{R,i} \cdot X_{L,i_0} \cdot S \\
&= H_{L,i} \cdot X_{R,i_0} \cdot S \\
&= H_{L,i} \cdot X_L \\
&= \hat{S}_{R,i} \Leftrightarrow i = i_0 \\
i_0 &= argmax_i \left\{ F\left(\hat{S}_{L,i}\right) \oplus F\left(\hat{S}_{R,i}\right) \right\},
\end{aligned}
\tag{7.5}
$$

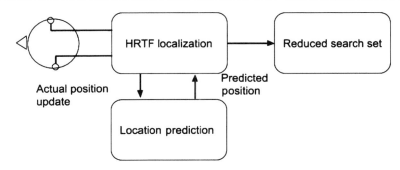

Figure 7.6 Block diagram of a typical HRTF localization system.

where i_0 is the best estimate from this method, X_L and X_R are the left and right observations of the hearing aids, respectively, and $H_{L,i}$ and $H_{R,i}$ are the reference HRTFs obtained after applying the difference filter.

In addition to these methods, it is often advantageous to determine a region of interest – especially for moving sources – to reduce the computational costs of such HRTF-based sound localization. A schematic indicating this can be seen in Figure 7.6. Overall, a multitude of tracking models are used in microphone-based sound localization, and these primarily predict the path of a sound source and offer accurate localization results [119].

7.8 SUMMARY

This chapter is a primer for binaural audio engineering. HRTFs form an integral part of spatial audio that determines the quality of spatial audio rendering. The need for personalized HRTFs and the different techniques of HRTF personalization were explained in detail in Section 7.4. With the advancements in hardware and areas like artificial intelligence and machine learning, we are in a better position to obtain personalized HRTFs and enjoy a realistic perception of spatial audio. The important cues for auditory-distance perception were also discussed in detail. Human ability to estimate distances is extremely poor – even in real life, and thus accurately *simulating* distance perception is even more difficult; its dependence on the direct-to-reverberant energy ratio makes it a tricky cue to reproduce virtually. Headphone equalization is another critical component that is often overlooked in spatial audio. Headphones come in all shapes and sizes, and thus have a huge influence on the perceived quality of sound. When it comes to binaural audio, not equalizing for the headphones results in degradation of the spatial image with poor externalization and tonal quality. There are several studies that discuss user preferences in headphone equalization for stereo content, but this has not been explored in great detail for spatial audio content. Further perspectives on binaural applications can be seen in Chapter 1 by Pike and Chapter 13 by Lord.

In order to create a truthful rendering of binaural audio, it is important to understand and strive to maintain the integrity of an idealized "creator-medium-consumption" ecosystem. Creators must use personalized HRTFs to create and monitor their spatial audio content with maximum fidelity. They must also understand the medium of consumption (e.g. headphones vs. speakers, and headphone equalization) and tailor the content experience accordingly. On the consumption side, listeners should also try to use personalized HRTFs in order to best experience the sound as the creators intended. With the availability of an increasing number of tools for spatial audio, establishing such a creator-medium-consumption ecosystem might yet become a possibility in the near future.

REFERENCES

[1] K. Sunder and Y. Wang, "An HRTF based approach towards binaural sound source localization," *147th Audio Engineering Society Convention*, 2019. www.aes.org/e-lib/online/browse.cfm?elib=20662.

[2] F. Rumsey, *Spatial Audio*. Focal Press, 2001.

[3] M. A. Gerzon, "Ambisonics in multichannel broadcasting and video," *Journal of the Audio Engineering Society*, vol. 33, no. 11, pp. 859–871, 1985.

[4] A. J. Berkhout, "A holographic approach to acoustic control," *Journal of the Audio Engineering Society*, 1988. www.aes.org/e-lib/browse.cfm?elib=5117.

[5] H. F. Olson, "Modern sound reproduction," 1972.

[6] D. S. Brungart and B. D. Simpson, "Auditory localization of nearby sources in a virtual audio display," in *Applications of Signal Processing to Audio and Acoustics, 2001 IEEE Workshop on the*, 2001, pp. 107–110.

[7] F. L. Wightman and D. J. Kistler, "Headphone simulation of free-field listening. I: Stimulus synthesis," *The Journal of the Acoustical Society of America*, vol. 85, no. 2, pp. 858–867, 1989.

[8] S. Mehrgardt and V. Mellert, "Transformation characteristics of the external human ear," *The Journal of the Acoustical Society of America*, vol. 61, no. 6, pp. 1567–1576, Jun. 1977.

[9] S. Carlile and D. Pralong, "The location-dependent nature of perceptually salient features of the human head-related transfer functions," *The Journal of the Acoustical Society of America*, vol. 95, no. 6, pp. 3445–3459, 1994.

[10] H. Møller, D. Hammershøi, C. B. Jensen, and M. F. Sørensen, "Transfer characteristics of headphones measured on human ears," *Journal of the Audio Engineering Society*, vol. 43, no. 4, pp. 203–217, 1995.

[11] W. G. Gardner and K. D. Martin, "HRTF measurements of a KEMAR," *The Journal of the Acoustical Society of America*, vol. 97, no. 6, pp. 3907–3908, 1995.

[12] E. A. G. Shaw and M. M. Vaillancourt, "Transformation of sound-pressure level from the free field to the eardrum presented in numerical form," *The Journal of the Acoustical Society of America*, vol. 78, no. 3, pp. 1120–1123, 1985.

[13] J. C. Middlebrooks, J. C. Makous, and D. M. Green, "Directional sensitivity of sound-pressure levels in the human ear canal," *The Journal of the Acoustical Society of America*, vol. 86, no. 1, pp. 89–108, 1989.

[14] P. M. Morse and K. Ingard, *Uno: Theoretical Acoustics*. McGraw-Hill Book Co., Inc., 1968.

[15] L. Rayleigh and A. Lodge, "On the acoustic shadow of a sphere. with an appendix, giving the values of Legendre's functions from P0 to P20 at intervals of 5 degrees," *Philosophical Transactions of the Royal Society of London. Series A, Containing Papers of a Mathematical or Physical Character*, vol. 203, no. 359–371, pp. 87–110, 1904.

[16] B. M. Sayers, "Acoustic-image lateralization judgments with binaural tones," *The Journal of the Acoustical Society of America*, vol. 36, no. 5, pp. 923–926, 1964.

[17] F. E. Toole and B. M. Sayers, "Lateralization judgments and the nature of binaural acoustic images," *The Journal of the Acoustical Society of America*, vol. 37, no. 2, pp. 319–324, 1965.

[18] F. L. Wightman and D. J. Kistler, "The dominant role of low-frequency interaural time differences in sound localization," *The Journal of the Acoustical Society of America*, vol. 91, no. 3, pp. 1648–1661, Mar. 1992.

[19] S. G. Weinrich, "Improved externalization and frontal perception of head-phone signals," presented at the *92nd Audio Engineering Society Convention*, Vienna, Austria, 1992. www.aes.org/e-lib/browse.cfm?elib=6842.

[20] V. R. Algazi, R. O. Duda, R. Duralswami, N. A. Gumerov, and Z. Tang, "Approximating the head-related transfer function using simple geometric models of the head and torso," *The Journal of the Acoustical Society of America*, vol. 112, no. 5, pt. 1, pp. 2053–2064, Nov. 2002.

[21] J. Hebrank and D. Wright, "Spectral cues used in the localization of sound sources on the median plane," *The Journal of the Acoustical Society of America*, vol. 56, no. 6, pp. 1829–1834, 1974.

[22] R. A. Butler and K. Belendiuk, "Spectral cues utilized in the localization of sound in the median sagittal plane," *The Journal of the Acoustical Society of America*, vol. 61, no. 5, pp. 1264–1269, 1977.

[23] E. M. Wenzel, M. Arruda, D. J. Kistler, and F. L. Wightman, "Localization using nonindividualized head-related transfer functions," *The Journal of the Acoustical Society of America*, vol. 94, p. 111, 1993.

[24] F. Wightman and D. Kistler, "Multidimensional scaling analysis of head-related transfer functions," in *Proc. IEEE Workshop on Applications of Signal Processing to Audio and Acoustics (WASPAA'93)*, pp. 98–101.

[25] D. Begault, E. Wenzel, A. Lee, and M. Anderson, "Direct comparison of the impact of head tracking, reverberation, and individualized head-related transfer functions on the spatial perception of a virtual speech source," *Journal of the Audio Engineering Society*, vol. 49, no. 10, pp. 904–916, 2001.

[26] J. Blauert, "Sound localization in the median plane," *Acta Acustica united with Acustica*, vol. 22, no. 4, pp. 205–213, 1969.

[27] P. J. Bloom, "Creating source elevation illusions by spectral manipulation," *Journal of the Audio Engineering Society*, vol. 25, no. 9, pp. 560–565, 1977.

[28] H. L. Han, "Measuring a dummy head in search of Pinna Cues," *Journal of the Audio Engineering Society*, vol. 42, no. 1–2, pp. 15–37, 1994.

[29] E. A. Macpherson and W. Center, "On the role of head-related transfer function spectral notches in the judgement of sound source elevation," *Auditory Display: Sonification, Audification, and Auditory Interface, SFI Studies in the Sciences of Complexity,* edited by G. Kramer and S. Smith, Addison-Wesley, Reading, MA, pp. 187–194, 1994.

[30] E. H. A. Langendijk and A. W. Bronkhorst, "Contribution of spectral cues to human sound localization," *The Journal of the Acoustical Society of America*, vol. 112, no. 4, pp. 1583–1596, 2002.

[31] E. A. G. Shaw and R. Teranishi, "Sound pressure generated in an external-ear replica and real human ears by a nearby point source," *The Journal of the Acoustical Society of America*, vol. 44, no. 1. pp. 240–249, 1968.

[32] A. J. Watkins, "Psychoacoustical aspects of synthesized vertical locale cues," *The Journal of the Acoustical Society of America*, vol. 63, no. 4, pp. 1152–1165, 1978.

[33] C. A. Rodgers, "Pinna transformations and sound reproduction," *Journal of the Audio Engineering Society*, vol. 29, no. 4, pp. 226–234, 1981.

[34] R. A. Humanski and R. A. Butler, "The contribution of the near and far ear toward localization of sound in the sagittal plane," *The Journal of the Acoustical Society of America*, vol. 83, no. 6, pp. 2300–2310, Jun. 1988.

[35] J. C. Middlebrooks, "Narrow-band sound localization related to external ear acoustics," *The Journal of the Acoustical Society of America*, vol. 92, no. 5, pp. 2607–2624, 1992.

[36] E. H. Langendijk and A. W. Bronkhorst, "The contribution of spectral cues to human sound localization," *The Journal of the Acoustical Society of America*, vol. 105, no. 2. pp. 1036–1036, 1999.

[37] F. Asano, Y. Suzuki, and T. Sone, "Role of spectral cues in median plane localization," *The Journal of the Acoustical Society of America*, vol. 88, no. 1, pp. 159–168, Jul. 1990.

[38] M. Itoh, K. Iida, and M. Morimoto, "Individual differences in directional bands in median plane localization," *Applied Acoustics*, vol. 68, no. 8, pp. 909–915, 2007.

[39] S. K. Roffler and R. A. Butler, "Factors that influence the localization of sound in the vertical plane," *The Journal of the Acoustical Society of America*, vol. 43, no. 6, pp. 1255–1259, Jun. 1968.

[40] R. Bücklein, "The audibility of frequency response irregularities," *Journal of the Audio Engineering Society*, vol. 29, no. 3, pp. 126–131, 1962.

[41] B. C. J. Moore, S. R. Oldfield, and G. J. Dooley, "Detection and discrimination of spectral peaks and notches at 1 and 8 kHz," *The Journal of the Acoustical Society of America*, vol. 85, no. 2, pp. 820–836, 1989.

[42] M. B. Gardner and R. S. Gardner, "Problem of localization in the median plane: effect of pinnae cavity occlusion," *The Journal of the Acoustical Society of America*, vol. 53, no. 2, pp. 400–408, 1973.

[43] W. G. Gardner, *Transaural 3-D Audio*. Citeseer, 1995.

[44] W. G. Gardner, *3-D Audio Using Loudspeakers*, vol. 444. Springer Science & Business Media, 1998.

[45] E. Y. Choueiri, "Optimal crosstalk cancellation for binaural audio with two loudspeakers," *Princeton University*, vol. 28, 2008. www.researchgate.net/profile/Edgar_Choueiri/publication/264886721_Optimal_Crosstalk_Cancellation_for_Binaural_Audio_with_Two_Loudspeakers/links/54bf9bf00cf2f6bf4e05050e/Optimal-Crosstalk-Cancellation-for-Binaural-Audio-with-Two-Loudspeakers.pdf.

[46] M. Vorländer, "Past, present and future of dummy heads," *Proc. Acústica, Guimarães, Portugal*, 2004.

[47] Ku100, "Neumann KU 100 dummy head documentation." https://en-de.neumann.com/ku-100.

[48] F. Christensen, C. B. Jensen, and H. Møller, "The design of VALDEMAR-an artificial head for binaural recording purposes," presented at the *109th Audio Engineering Society Convention*, Los Angeles, USA, 2000. www.aes.org/e-lib/browse.cfm?elib=9085.

[49] D. R. Begault, *3-D Sound for Virtual Reality and Multimedia*. AP Professional, 2000.

[50] D. N. Zotkin, R. Duraiswami, E. Grassi, and N. A. Gumerov, "Fast head-related transfer function measurement via reciprocity," *The Journal of the Acoustical Society of America*, vol. 120, no. 4, pp. 2202–2215, Oct. 2006.

[51] K. Fukudome, T. Suetsugu, T. Ueshin, R. Idegami, and K. Takeya, "The fast measurement of head related impulse responses for all azimuthal directions using the continuous measurement method with a servo-swiveled chair," *Applied Acoustics*, vol. 68, no. 8, pp. 864–884, 2007.

[52] T. Ajdler, L. Sbaiz, and M. Vetterli, "Dynamic measurement of room impulse responses using a moving microphone," *The Journal of the Acoustical Society of America*, vol. 122, no. 3, p. 1636, Sep. 2007.

[53] V. R. Algazi, R. O. Duda, D. M. Thompson, and C. Avendano, "The CIPIC HRTF database," in *Proc. IEEE Workshop on Applications of Signal Processing to Audio and Acoustics (WASPAA'01)*, pp. 99–102.

[54] Tohoku University, "The RIEC HRTF dataset." www.riec.tohoku.ac.jp/pub/hrtf/.

[55] T. Nishino, Y. Nakai, K. Takeda, and F. Itakura, "Estimating head related transfer function using multiple regression analysis," *IEICE Transactions A*, vol. 84, pp. 260–268, 2001.

[56] P. Majdak, M. J. Goupell, and B. Laback, "3-D localization of virtual sound sources: effects of visual environment, pointing method, and training," *Attention Perception & Psychophysics*, vol. 72, no. 2, pp. 454–469, 2010.

[57] H. Kayser, S. D. Ewert, J. Anemüller, T. Rohdenburg, V. Hohmann, and B. Kollmeier, "Database of multichannel in-ear and behind-the-ear head-related and binaural room impulse responses," *EURASIP Journal on Advances in Signal Processing*, vol. 2009, no. 1, p. 298605, Jul. 2009.

[58] K. Genuit and N. Xiang, "Measurements of artificial head transfer functions for auralization and virtual auditory environment," *Proceedings of the 15th ICA, Trondheim*, vol. 2, pp. 469–472, 1995.

[59] K. A. Riederer, *Head-Related Transfer Function Measurements*, Master's Thesis, Aalto University, Aalto, Denmark, 1998.

[60] J. Blauert *et al.*, "The AUDIS catalog of human HRTFs," in *The Journal of the Acoustical Society of America*, vol. 103, pp. 2901–2902, 1998.

[61] B. P. Bovbjerg, F. Christensen, P. Minnaar, and X. Chen, "Measuring the head-related transfer functions of an artificial head with a high-directional resolution," presented at the *109th Audio Engineering Society Convention*, Los Angeles, USA, 2000. www.aes.org/e-lib/browse.cfm?elib=9074.

[62] S. Takane, D. Arai, T. Miyajima, K. Watanabe, Y. Suzuki, and T. Sone, "A database of head-related transfer functions in whole directions on upper hemisphere," *Acoustical Science and Technology*, vol. 23, no. 3, pp. 160–162, 2002.

[63] E. Grassi, J. Tulsi, and S. Shamma, "Measurement of head-related transfer functions based on the empirical transfer function estimate," in *Proceedings 9th International Conference on Auditory Display (ICAD2003)*, pp. 119–121.

[64] B. Xie, X. Zhong, D. Rao, and Z. Liang, "Head-related transfer function database and its analyses," *Science China Series G: Physics, Mechanics & Astronomy*, vol. 50, no. 3, pp. 267–280, 2007.

[65] H. Møller, M. F. Sørensen, D. Hammershøi, and C. B. Jensen, "Head-related transfer functions of human subjects," *Journal of the Audio Engineering Society*, vol. 43, no. 5, pp. 300–321, 1995.

[66] F. L. Wightman and D. J. Kistler, "Headphone simulation of free-field listening. II: psychophysical validation," *The Journal of the Acoustical Society of America*, vol. 85, no. 3, pp. 868–878, 1989.

[67] D. S. Brungart, G. Romigh, and B. D. Simpson, "Rapid collection of head related transfer functions and comparison to free-field listening," *Principles and Applications of Spatial Hearing*, pp. 139–148, 2011.

[68] W. W. Hugeng and D. Gunawan, "Improved method for individualization of head-related transfer functions on horizontal plane using reduced number of anthropometric measurements," *Journal of Telecommunications*, vol. 2, no. 2, pp. 31–41, 2010.

[69] R. Nicol, *Binaural Technology*. AES, 2010.

[70] R. O. Duda, V. R. Algazi, and D. M. Thompson, "The use of head-and-torso models for improved spatial sound synthesis," *113th Audio Engineering Society Convention* 2002. www.aes.org/e-lib/browse.cfm?elib=11294.

[71] M. Dellepiane, N. Pietroni, N. Tsingos, M. Asselot, and R. Scopigno, "Reconstructing head models from photographs for individualized 3D-audio processing," in *Computer Graphics Forum*, vol. 27, pp. 1719–1727.

[72] D. N. Zotkin, J. Hwang, R. Duraiswaini, and L. S. Davis, "HRTF personalization using anthropometric measurements," in *Proc. IEEE Workshop on Applications of Signal Processing to Audio and Acoustics (WASPAA'03)*, pp. 157–160.

[73] J. C. Middlebrooks, "Individual differences in external-ear transfer functions reduced by scaling in frequency," *The Journal of the Acoustical Society of America*, vol. 106, no. 3, pp. 1480–1492, 1999.

[74] IRCAM, "LISTEN HRTF Database." 2003.

[75] B. U. Seeber and H. Fastl, "Subjective selection of non-individual head-related transfer functions," in *Proceedings 9th International Conference on Auditory Display (ICAD 2003)*, vol. 3, pp. 259–262.

[76] C.-J. Tan and W.-S. Gan, "User-defined spectral manipulation of HRTF for improved localisation in 3D sound systems," *Electronics Letters*, vol. 34, no. 25. p. 2387, 1998.

[77] P. Runkle, A. Yendiki, and G. Wakefield, "Active sensory tuning for immersive spatialized audio," presented at the *Conference on Auditory Display*, Atlanta, Georgia, USA, Apr. 2000. https://smartech.gatech.edu/handle/1853/50665 (accessed Mar. 19, 2021).

[78] K. J. Fink and L. Ray, "Tuning principal component weights to individualize HRTFs," in *Proceedings IEEE International Conference Acoustics, Speech and Signal Processing (ICASSP'12)*, pp. 389–392.

[79] Y. Hur, S.-P. Lee, Y.-C. Park, and D.-H. Youn, "Efficient individualization of HRTF using critical-band based spectral cues control," *The Journal of the Acoustical Society of Korea*, vol. 30, no. 4, pp. 167–180, May 2011.

[80] S. Shimada, N. Hayashi, and S. Hayashi, "A clustering method for sound localization transfer functions," *Journal of the Audio Engineering Society*, vol. 42, no. 7/8, pp. 577–584, 1994.

[81] H. Hu, L. Zhou, J. Zhang, H. Ma, and Z. Wu, "Head related transfer function personalization based on multiple regression analysis," in *Proceedings International Conference on Computational Intelligence and Security*, vol. 2, pp. 1829–1832.

[82] M. Rothbucher, M. Durkovic, H. Shen, and K. Diepold, "HRTF customization using multiway array analysis," in *Proceedings 18th European Signal Processing Conference (EUSIPCO'10)*, pp. 229–233.

[83] Q.-H. Huang and Y. Fang, "Modeling personalized head-related impulse response using support vector regression," *Journal of Shanghai University*, vol. 13, no. 6, pp. 428–432, 2009.

[84] K. Sunder, E.-L. Tan, and W.-S. Gan, "Individualization of binaural synthesis using frontal projection headphones," *Journal of the Audio Engineering Society*, vol. 61, no. 12, pp. 989–1000, 2013.

[85] P. Zahorik, "Auditory display of sound source distance," in *Proceedings of 8th International Conference on Auditory Display*, 2002, pp. 1–7.

[86] P. Zahorik, D. S. Brungart, and A. W. Bronkhorst, "Auditory distance perception in humans: a summary of past and present research," *Acta Acustica united with Acustica*, vol. 91, no. 3, pp. 409–420, 2005.

[87] G. von Békésy, "The moon illusion and similar auditory phenomena," *The American Journal of Psychology*, vol. 62, no. 4, pp. 540–552, 1949.

[88] P. Zahorik and F. L. Wightman, "Loudness constancy with varying sound source distance," *Nature Neuroscience*, vol. 4, no. 1, pp. 78–83, 2001.

[89] S. S. Stevens and M. Guirao, "Loudness, reciprocality, and partition scales," *The Journal of the Acoustical Society of America*, vol. 34, no. 9B, pp. 1466–1471, 1962.

[90] D. S. Brungart and W. M. Rabinowitz, "Auditory localization of nearby sources. Head-related transfer functions," *The Journal of the Acoustical Society of America*, vol. 106, no. 3 Pt 1, pp. 1465–1479, Sep. 1999.

[91] J. Blauert, *Spatial Hearing (Revised Edition)*, Massachusetts Institute of Technology, 1997.

[92] B. G. Shinn-Cunningham, S. Santarelli, and N. Kopco, "Tori of confusion: binaural localization cues for sources within reach of a listener," *The Journal of the Acoustical Society of America*, vol. 107, no. 3, pp. 1627–1636, Mar. 2000.

[93] D. S. Brungart, "Auditory localization of nearby sources. III. Stimulus effects," *The Journal of the Acoustical Society of America*, vol. 106, no. 6, pp. 3589–3602, 1999.

[94] G. W. Stewart, "The acoustic shadow of a rigid sphere, with certain applications in architectural acoustics and audition," *Physical Review (Series I)*, vol. 33, no. 6, p. 467, 1911.

[95] R. V. L. Hartley and T. C. Fry, "The binaural location of pure tones," *Physical Review*, vol. 18, no. 6, pp. 431–442, 1921.

[96]   H. R. Hirsch, "Perception of the range of a sound source of unknown strength," *The Journal of the Acoustical Society of America*, vol. 43, no. 2, pp. 373–374, 1968.

[97]   Y. Tahara and H. Sakurai, "A tentative model for the localization of sound based on the simultaneity of time difference and level difference of the sound between ears," in *Proceedings of the Congress of the Acoustical Society of Japan*, pp. 373–374, 1974

[98]   H.-Y. Kim, Y. Suzuki, S. Takane, and T. Sone, "Control of auditory distance perception based on the auditory parallax model," *Applied Acoustics*, vol. 62, no. 3, pp. 245–270, 2001.

[99]   R. O. Duda and W. L. Martens, "Range dependence of the response of a spherical head model," *The Journal of the Acoustical Society of America*, vol. 104, no. 5, pp. 3048–3058, 1998.

[100] W. M. Rabinowitz, J. Maxwell, Y. Shao, and M. Wei, "Sound localization cues for a magnified head – Implications from sound diffraction about a rigid sphere," *Presence: Teleoperators and Virtual Environments*, vol. 2, no. 2, pp. 125–129, 1993.

[101] C. Ze-Wei, Y. Guang-Zheng, X. Bo-Sun, and G. Shan-Qun, "Calculation and analysis of near-field head-related transfer functions from a simplified head-neck-torso model," *Chinese Physics Letters*, vol. 29, no. 3, p. 034302, 2012.

[102] R. Duraiswaini, D. N. Zotkin, and N. A. Gumerov, "Interpolation and range extrapolation of HRTFs [head related transfer functions]," in *2004 IEEE International Conference on Acoustics, Speech, and Signal Processing*, 2004, vol. 4, pp. iv–iv.

[103] D. Romblom and B. Cook, "Near-field compensation for HRTF processing."

[104] A. Kan, C. Jin, and A. van Schaik, "A psychophysical evaluation of near-field head-related transfer functions synthesized using a distance variation function," *The Journal of the Acoustical Society of America*, vol. 125, no. 4, pp. 2233–2242, 2009.

[105] P. Zahorik, "Distance localization using nonindividualized head-related transfer functions," *The Journal of the Acoustical Society of America*, vol. 108, no. 5, pp. 2597–2597, 2000.

[106] M. R. Schroeder, "Natural sounding artificial reverberation," *Journal of the Audio Engineering Society*, vol. 10, no. 3, pp. 219–223, 1962.

[107] G. Kearney, C. Masterson, S. Adams, and F. Boland, "Approximation of Binaural Room Impulse Responses," in *IET Irish Signals and Systems Conference (ISSC 2009)*, 2009, pp. 1–6.

[108] P. Murgai, M. Rau, and J.-M. Jot, "Blind estimation of the reverberation fingerprint of unknown acoustic environments," *143rd Audio Engineering Society Convention*, 2017. www.aes.org/e-lib/online/browse.cfm?elib=19302.

[109] S.-M. Kim and W. Choi, "On the externalization of virtual sound images in headphone reproduction: a Wiener filter approach," *The Journal of the Acoustical Society of America*, vol. 117, no. 6. pp. 3657–3665, 2005.

[110] H. Møller, "Fundamentals of binaural technology," *Applied Acoustics*, vol. 36, no. 3, pp. 171–218, 1992.

[111] D. Pralong and S. Carlile, "The role of individualized headphone calibra-
       tion for the generation of high fidelity virtual auditory space," *The Journal
       of the Acoustical Society of America*, vol. 100, no. 6, pp. 3785–3793, 1996.

[112] A. Kulkarni and H. S. Colburn, "Variability in the characterization of the
       headphone transfer-function," *The Journal of the Acoustical Society of
       America*, vol. 107, no. 2, pp. 1071–1074, Feb. 2000.

[113] K. I. McAnally and R. L. Martin, "Variability in the headphone-to-ear-
       canal transfer function," *Journal of the Audio Engineering Society*, vol. 50,
       no. 4, pp. 263–266, 2002.

[114] M. Rothbucher, D. Kronmüller, M. Durkovic, T. Habigt, and K. Diepold,
       "HRTF sound localization," in *Advances in Sound Localization*, INTECH
       Open Access Publisher, 2011.

[115] F. Keyrouz, K. Diepold, and P. Dewilde, "Robust 3D robotic sound local-
       ization using state-space HRTF inversion," in *2006 IEEE International
       Conference on Robotics and Biomimetics*, Dec. 2006, pp. 245–250.

[116] F. Keyrouz and K. Diepold, "An enhanced binaural 3d sound localization
       algorithm," in *2006 IEEE International Symposium on Signal Processing
       and Information Technology*, Aug. 2006, pp. 662–665.

[117] F. Keyrouz and A. A. Saleh, "Intelligent sound source localization based on
       head-related transfer functions," in *2007 IEEE International Conference
       on Intelligent Computer Communication and Processing*, Sep. 2007, pp.
       97–104.

[118] M. Usman, F. Keyrouz, and K. Diepold, "Real time humanoid sound source
       localization and tracking in a highly reverberant environment," in *2008 9th
       International Conference on Signal Processing*, Oct. 2008, pp. 2661–2664.

[119] D. Bechler, M. Grimm, and K. Kroschel, "Speaker tracking with a micro-
       phone array using Kalman filtering," *Advances in Radio Science*, vol. 1,
       no. B. 3, pp. 113–117, 2003.

[120] University of York, "SADIE II Database," www.york.ac.uk/sadie-project/
       database.html.

[121] TH Koln, http://audiogroup.web.th-koeln.de/.

[122] TU Berlin, "The FABIAN head-related transfer function data base," https://
       depositonce.tu-berlin.de/handle/11303/6153.

8

Contextual factors in judging auditory immersion

Sungyoung Kim

8.1 INTRODUCTION

From the moment I started my PhD studies, I made do with subjective "quantified" data and dealt with the question of how to appropriately interpret reported subjective impressions of auditory experience. Any sensory evaluation takes for granted that assessors can translate their judgment (strength or magnitude) of a salient percept into quantifiable data. A recent pop culture example is the blind taste testing of Coke vs. Pepsi. When people choose Pepsi over Coke, they (or marketing executives) believe the choice is made from a confident judgment, which will be reliably consistent in a different situation. Is this true? French philosopher Jean-Jacques Rousseau seemed to have a different opinion. Rousseau claimed that a person "establishes his own private scale of values" according to his own "*taste*," as quoted by Boulez in his book [1, p. 44]. At least Rousseau took a stance where he did not know the answer (as did Boulez): when a subject reports an amount of affective or hedonic response, how much has one's "taste" influenced this judgment? A more realistic question is, what factors influence "taste"? If the person used "his or her own private scale," would it be possible to compare this person's data with others"?

An assessor's previous experiences, personality, and even mood can modulate perceptual and cognitive judgments; "taste" is not consistent. This topic has caught the attention of the consumer marketing world, and associated investigations aim to explain how humans form new references for judgment when faced with circumstantial changes. Studies have shown that adding a third "asymmetrically dominated" opinion can change people's preferences. In other words, the existence of a simple external reference could influence a subject's internal value or endowment [2], [3]. This is why companies spend billions of dollars on commercials that emphasize not a specific product's merits, but instead appeal to consumers' emotions.

How significant are non-experimental variables in our listening, and therefore to our subjective evaluations of auditory perception? While Boulez's discussion of "taste" in his book refers to listeners' apprehension towards classical music, Rousseau's idea must translate across listening activities: we each use our own scale to evaluate auditory information.

Many researchers (including but not limited to [4]–[6]) have investigated the potential influence of such non-experimental variables, referring to them as psychological error [7] or bias [5]. In particular, Zielinski *et al.* [8] published a comprehensive review of key biases encountered in modern audio-quality listening tests. They defined biases as "systemic errors affecting the results of a listening test" and asserted that those errors "are very difficult to identify as they manifest themselves by a repeatable and consistent shift in the data." By contrast, there are "random errors" that are easy to recognize and easy to minimize. The latter type of error has been treated as negligible – regardless of its presence, which can be controlled by a statistical method, for instance, with the randomized stimuli being presented as a nuisance variable [9]. However, biases or systemic errors might not be small enough to be neglected as being random variables, and can prevent the experimenter from deriving a reliable conclusion.

Zielinski *et al.*'s paper [8] investigated a range of potential biases originating from experimental preparation to a listener's judgment process. Among them, many biases are associated with "quantification" processes (mapping judgment of perceptual magnitudes to a scalable unit, such as an integer number), and influences on the process from internal and external factors. For example, a "contraction bias" refers to an assessor's tendency to be conservative and to avoid the extremes of the scale. As Zielinski *et al.* pointed out, the result is that the "center value becomes an 'attractor' and listeners' responses tend to be biased toward it." Other biases have distinct influences on listeners' responses, which are summarized in Table 1 in [8].

Contextual factors result in a particular form of subjective bias. Zielinski *et al.* mentioned this "situational context bias," and asserted that this kind of bias could happen during affective judgments and might modulate qualitative assessments of a given sound stimulus – depending on the situational context. Begault [10] also asserted that the given context matters for listeners, and that their perceptual judgments for everyday listening versus a psychoacoustic experiment could be different. While this seems reasonable and parallel with other areas of study (which are comprehensively summarized in [11]), Zielinski concluded that it was not possible to find either strong evidence or "direct data" supporting the existence of an influence associated with this kind of contextual effect. Nevertheless, I believe that the contexts investigated in previous studies ([12], [13] referred to in [8]) were simply not strong enough to modulate the assessors' responses. However, our brain instructs the auditory system to give more attention to (subjectively) "significant" or "familiar" acoustic patterns against background and unknown auditory streams – this is known as the "cocktail party effect" [14, p. 107]. Since such auditory attention mediates perception and behaviour, it is a context-dependent phenomenon; listening inherently is context dependent. Therefore, there must be a case in which listeners' high-level perceptual judgments are also modulated by an idiosyncratic and contextual circumstance. One example is when auditory-quality judgment is modulated by a specific musical style.

8.2 A CASE STUDY: INFLUENCE OF MUSICAL CONTENT ON LISTENER PREFERENCE

My colleagues and I have previously investigated the relationship between multichannel recording techniques and musical content [15]. In best practice, when a sound recording engineer captures a surround sound field and reproduces it through a loudspeaker array, consideration is given to all possible variables that might influence the final sonic quality, and fine details are subjectively adjusted before playback. Our study showed that determining the configuration of a specific microphone array is significantly influenced by a producer's or engineer's aesthetic appreciation for a specific style of music. For this study, we selected four solo piano pieces composed in the European concert music tradition and deemed to be approximately representative of various eras: Bach, Schubert, Brahms and a contemporary piece. The musical selections represented a wide variety of acoustical possibilities regularly encountered by audio engineers and producers when recording "classical" piano music. We hypothesized that different musical selections would activate the piano-hall acoustical system differently.

We selected four surround microphone arrays to capture the piano-hall acoustical system: Fukada tree [16], Polyhymnia pentagon [17, p. 197], OCT with Hamasaki square [18], and SoundField MKV with SP451 surround processor. During the recording process, each microphone array was optimally placed in the concert hall according to the expertise of the engineers conducting the study, and each musical piece was simultaneously recorded by all microphones. All musical excerpts were performed in the same concert hall by a single musician and played on a single piano.

After the recording, preference testing was conducted using male and female recording engineers as participants. They were asked to report a subjective preference between two randomly selected stimuli. The data from experiments were subsequently analysed to build a preference model, and the results showed that even though no single microphone technique was chosen as the most preferred throughout all musical selections, listener preferences with regard to microphone technique were modulated by musical selection (this being an aggregate of the composition in question, the performance practice used by the musician, and the performance itself). In other words, musical selection as a context significantly modulated listeners' affective judgment of selecting a "preferred" microphone technique. Context should therefore be understood as an aggregate quantity – or a composite – of various factors that may be perceived as being unified within a given listening scenario.

8.3 ACTIVE CONTEXT IN IMMERSIVE AUDIO?

One subsequent question is whether a contextual factor would be equally meaningful for the new paradigm of immersive audio, for both capture and reproduction technologies. The audio and media industries are witnessing fast growth of this "immersive" audio field, as is indicated by headlines of new and game-changing applications and technologies that will drastically improve the sonic experience. With the advent of virtual reality (VR), augmented reality (AR), mixed reality (MR), and extended reality (XR), companies are aggressively competing to steer the direction of this new market. "Immersive audio" is another context that might foster several unique bases for judgment. How should we cope with this completely new context of immersive and/or 3D audio in a specific circumstance? What could be the hidden underlying factors that influence an understanding of auditory immersion? Kim and Rutkowski [19] found that when listeners are exposed to multichannel sound fields, they understand spatial auditory features using a procedure similar to visualization. When transposed to immersive audio, their findings suggest that in the presence of potentially confusing or unfamiliar spatial cues (a novel listening context), listeners' cognitive processes for an affective judgment would be different from those for a two-channel stereophonic sound field. We need new information regarding perception within the novel spatial experiential scenarios created by VR, AR, and so on, and thus it would be valuable to summarize what has been previously studied around contextual factors that modulate listeners' auditory immersion.

In the following sections, two specific contextual aspects that have influenced perceptual judgment in auditory immersion will be introduced: cultural background and listening environment.

8.4 CULTURAL BACKGROUND

What is culture? While there are many detailed explorations of this idea, a common and dictionary definition would be "the customary beliefs, social forms, and material traits of a racial, religious, or social group," which appears as the first of six definitions in the Merriam-Webster dictionary. Members of a group sharing a specific cultural background would tend to form similar beliefs around common values. It is therefore "context dependent" when a specific belief, value, or practice is shared by the group. For example, watching someone ingesting a living octopus is considered mouth-watering to many Koreans raised on traditional cuisine, while the same culinary practice is viewed as disgusting by many North Americans unfamiliar with this custom. As a cultural psychologist wrote in [20], mind and environment are interdependent.

A number of previous studies have examined how cultural practices shape psychological processes, some showing that North Americans and East Asians differ significantly for certain cognitive processes. A study by Yokosawa *et al.* [21] showed that Japanese and US subjects were different in their colour preferences, influenced by "the affective nature of people's interactions with and beliefs about the objects in the physical and social environments within their culture." Other studies [22]–[24] explained that differences in the cognitive style of thought between North Americans and East Asians could result in differences in perceptual judgments. Nisbett and Miyamoto [24] showed that North Americans are characterized by an analytic cognitive process: "organizing objects by emphasizing rules and categories and to focus on salient objects independently from the context," while East Asians use a holistic cognitive process: attending to the entire "context and to the relationship between the objects and the context."

A recent study by Saulton *et al.* [25] shows that spatial cognitive processing between Germans and Sound Koreans are different; Koreans, with holistic and interdependent tendencies, look around rooms more, attending to the entire context more than Germans, who tend to be focused and independent. Korean subjects showed significantly less bias when judging a room's rectangularity (width-to-depth ratio) than their German counterparts. This prompts a question fundamental to spatial audition: do we differ in our sonic understanding of spatial information? Does cultural background influence listeners' apprehension of an auditory environment or auditory perception of an enclosed space? If the conceived size of an enclosure is differentiated by a specific socio-cultural context, that context would similarly factor into how members of the group understand the auditory environments in which they reside and conduct their daily lives. For example, according to data on the average residential floor space per capita in 2009, a Japanese person occupies 43 m^2 while a Canadian takes up 72 m^2 and an American 77 m^2 [26]. North American listeners pass their home lives in spaces almost twice as large as Japanese listeners. It would be hard to expect identical affective, perceptual, and cognitive understandings of a specific room from these two groups, as might be shown by their evaluations of the question: how big is "big enough" for daily living?

8.5 CULTURAL INFLUENCE ON CONCEPTUALIZATION OF IMMERSIVE AMBIENCES IN MULTICHANNEL MUSIC REPRODUCTION

Previously, researchers from a wide range of studies [27]–[30] reported their findings on cultural influence on auditory perception. For example, a specific "noise" sound that was perceived as annoying for Chinese listeners was not equally annoying for Japanese listeners. Another example was a study that I conducted, which showed that Japanese listeners are different from North American listeners in their preferred spectral balance, favouring reduced gain in the mid-frequency bands [30].

In order to assess whether a similar influence can be observed in the auditory spaciousness and immersion of multichannel-reproduced music, we conducted a controlled listening experiment. Spatial sound theorists Blesser and Salter [31] assert that a listener "selects specific aural attributes of a space based on what is desirable in a particular cultural background to describe" a given auditory environment. Our research question was to identify such "desirable characteristics" – of immersive ambiences in multichannel-reproduced music – for a specific listener group.

I had a personal interest in researching how a listener's immersive auditory experience is modulated by loudspeaker configurations [32]. If we have four loudspeakers for height ambiences, what would be the best configuration for enhanced auditory immersion? Karampourniotis *et al.* piloted an investigation into this question and concluded that the height-loudspeaker *configuration* has a greater influence on sound quality and auditory immersion than types or kinds of height-ambient *signals* [33]. Later I continued the same investigation with three listener groups [34].

The experiment used two horizontal layers of loudspeakers. The lower layer of five was configured as per ITU-R recommendations [35]. This was augmented with a height layer comprising 12 loudspeakers with an elevation of 30°. Since the study sought to explore the influence of various height-channel loudspeaker configurations, the high 12 were positioned with azimuths of ±30°, ±50°, ±70°, ±90°, ±110°, and ±130°, and from this, several subsets of four (each with left-right symmetry) were selected to play alongside the horizontal layer to create a total of 15 possible nine-channel immersive-audio playback configurations. In the end, we only used eight of these configurations due to the practicalities of working in a finite space. The azimuth angles of the four loudspeakers in these eight configurations (C1 through C8) were: (C1) ±30° and ±90°; (C2) ±30° and ±110°; (C3) ±30° and ±130°; (C4) ±50° and ±90°; (C5) ±50° and ±110°; (C6) ±50° and ±130°; (C7) ±70° and ±110°; and (C8) ±70° and ±130°. Listeners were asked to compare these eight configurations (presented randomly) and rank them based upon perceived sound quality. Some listeners found it hard to define the sound quality, and in such cases, I asked them to judge how well one configuration immersed the listener in the music.

The same experiment was conducted in three locations (Canada: Group 1, US: Group 2, and Japan: Group 3). Eleven trained listeners that comprised Group 1 were master's students or faculty members of the Sound Recording Area at McGill University and trained as classical recording engineers/producers with multichannel audio mixing experience. Twelve students from various disciplines formed Group 2 and represented ordinary listeners and consumers. In Japan, 14 students and faculty members of the Tokyo University of the Arts who studied and taught sound recording and psychoacoustics participated as Group 3. Groups 1 and 3 represented experienced listener groups. I expected to observe different results depending on the amount of "training and/or experience" because the degree of training is regarded as an important factor differing listeners' judgment [8], [25]. Thus, responses by members in Groups 1 and 3 would be more similar, and differ from those from Group 2. However, the collected rank data showed different results: the stronger effect correlated with geographical context, as Groups 1 and 2 (North American listeners) were similar and Group 3 differed substantially (Japanese listeners). Figure 8.1 shows the mean rankings associated with a merged North American (solid line with circle symbol) group versus the Japanese group (dotted line with triangle symbol).

According to further analysis of the "descriptions" that each of the two groups used, it appeared that different ranking patterns came from idiosyncrasies in preference towards a given configuration. Across groups, listeners perceived similar descriptive characteristics of a given configuration, but the ranking of preferred immersive conditions was different between

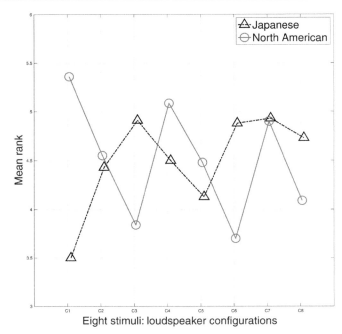

Figure 8.1 Average ranking (*Y*-axis) of perceived sound quality of multichannel-reproduced music with different height-loudspeaker configurations (*X*-axis). Responses from the North American listener groups (Group 1: Canada, and Group 2: US) are merged, and shown in comparison with responses from the Japanese listeners (Group 3).

groups. For example, both groups perceived configuration 1 (C1 with four height loudspeakers at ±30° and ±90°) "frontal" and "narrow." However, the North American group highly favoured this "frontal" and "narrow" configuration, but the Japanese group did not. Whereas although the C6 (±50° and ±130°) configuration conveyed a "wide" and "surrounding" impression to both groups, their hedonic responses were opposite: in ranking the eight configurations, Japanese listeners preferred "enveloping" and "wide" configurations to "frontal" and "narrow" ones (as they were described by all groups).

My study indicates that there are socio-culturally correlated contextual factors in judging auditory immersion. This finding provides support for my premise that when ranking multichannel configurations, members of the North American group, with "analytic" and "independent" cultural values, focus on the integrity of the frontal image. By contrast, members of the Japanese group, with "holistic" and "interdependent" cultural values, focus on the spread of lateral and rear images.

This factor would have little impact if 3D audio was only produced and consumed within the boundaries of a particular socio-cultural situation. However, immersive audio users are not culturally isolated: there is near-ubiquitous access to globally distributed electronic media. Accordingly, both the user experience of culturally diverse content, and also the novel spatial experiences produced by ever-emerging XR technologies propagate across cultural boundaries. Despite such technological innovations, we lack experimental data to support or refute the significance of cultural influences on the shared experience. In particular, auditory information is strongly associated with listeners' emotions and moods [36] and as such will similarly influence listeners' experiences of virtual environments. If an environment is understood differently because of hidden factors, such as the values formed through socio-cultural experience, then a consensus on interactive communication, concepts transmitted via new technologies, and even "satisfaction" with the experience of immersive environments could be limited.

However, there is another important challenge to anticipate and address for user preference in 3D audio design: multichannel audio systems are installed in many different architectural settings. Room dimensions and acoustics greatly influence audio reproduction and its perception [37]. As Howie reports [38]: "well executed listening tests often require […] implementation of the experiment in an environment that is acoustically and technologically appropriate." This statement is appropriate in the context of designing a listening test, and many organizations therefore recommend adherence to particular technical specifications for critical listening (e.g. [39]). Yet, there is a need to assess what is appropriate for everyday listening; particularly in terms of the relevance and impact of the factors that can be designed versus those that will vary. Will end users or consumers experience the same features of 3D audio content regardless of where and how it is delivered? If not, what are the perceptual effects of acoustical variation in different listening environments?

8.6 THE LISTENING ENVIRONMENT: INTERACTIONS BETWEEN ROOMS AND LOUDSPEAKERS

Room acoustics interact with sound sources in many unexpected and uncontrollable ways – they alter both timbral and spatial impressions of produced and reproduced sound field(s), and affect the overall sonic experience. Music in a concert hall is a good example scenario of a "produced field." Historically, European composers recognized the significance of acoustics on music perception both for performers and audiences, and utilized architectural acoustics to shape their music ([40] gives a comprehensive summary of music and room acoustics during the Renaissance period). Also see both Chapter 10 by Rebelo and Chapter 11 by McAlpine *et al*.

As for reproduced sound fields, Toole has devoted his life to researching room interactions with loudspeakers, and published a book [37], a corpus of research findings on the topic, and guidelines adopted by many subsequent researchers. Toole coined the term "circle of confusion" to explain the "never-ending cycle of subjectivity" [37, p.19], yet claimed that one could break this circle through controlled psychoacoustical measurements. He started with a dilemma "that there are no truly remarkable technical standards for control room sound," and asserted that this dilemma (lack of standards and control) brought a penalty that "recordings vary, sometimes quite widely, in their sound quality, spectral balances, and imaging." Nonetheless, through a series of controlled listening experiments and associated psychoacoustical models, he discovered how listeners' affective responses and perceptual judgments were influenced by peripherals in the listening chain. Obviously, there is a strong "interaction of multiple loudspeakers and the listening room when those loudspeakers are reproducing a multichannel recording" [37, p.100]. Toole showed how the combination of loudspeakers' radiation patterns and wall reflections within the listening room affects both timbral and spatial attributes of reproduced music in specific ways.

What happens when we employ multiple loudspeaker layers for 3D immersive audio reproduction? Such a system can virtually render a radiation pattern of 3D reflections in a target space. By manipulating 3D audio content, loudspeaker characteristics and configurations, can we reduce undesired room-loudspeaker interactions to deliver a "transparent" experience, and thus assure that what any listener experiences (coupled with the particular acoustics of their listening room) would be identical to what a content creator previously experienced in his or her environment? To address this question, I conducted a listening experiment that compared four listening rooms [41], [42].

8.7 COMPARISON OF AUDITORY ATTRIBUTES IN FOUR LISTENING ROOMS

Four rooms of varying sonic characteristics, yet having geometric similarity, were chosen for this comparison. The first room was at the Rochester Institute of Technology (Rochester, US), a conference room with no acoustic treatment other than a dropped ceiling as is typical in both classroom and living room settings (volume = 140.4 m^3 and RT$_{60}$ = 640 ms). The room was chosen to represent an everyday listening condition: without acoustical treatment. The second room was the 22.2 Studio (volume = 149.8 m^3 and RT$_{60}$ = 168 ms) at McGill University (Montreal, Canada), and the third was Studio B (volume = 208.1 m^3 and RT$_{60}$ = 340 ms) at Tokyo University of the Arts (Tokyo, Japan). These two rooms were designed to first control acoustical behaviour and also to follow the current standard for multichannel reproduction as specified in [43]. The fourth space was an anechoic chamber located in Japan at RIEC, Tohoku University [44].

As stimuli, two classical music pieces were prepared for this study: the first piece was "Mars" from *The Planets* by Gustav Holst, performed by the National Youth Orchestra of Canada. The second piece was *Kaido-Tose* by Kiyoshi Nobutoki, performed by the Tokyo University of the Arts Orchestra and Choir. These musical pieces were recorded and mixed optimally for the 22-channel reproduction format [45]. We then re-recorded the music as reproduced in each of the four rooms under comparison using a head and torso simulator (HATS, Brüel & Kjær 4100D). The same HATS was located at an equivalent listening position in each of the four listening rooms (Figure 8.2) to binaurally capture 22-channel reproduced sound fields to evaluate in the listening experiment.

Figure 8.2 Binaural capture of 22.2-channel reproduced music using the Brüel & Kjær 4100D HATS (left: STUDIO B at Tokyo University of the Arts; right: RIEC anechoic room at Tohoku University).

A total of 34 listeners were invited to participate in a listening test in which they were asked to quantify both their perceived magnitudes of five attributes, and also one overall preference. The five attributes were: spectral balance, apparent source width (ASW), (spatial) clarity, envelopment, and depth.

The preference ratings were analysed through one-way analysis of variance (ANOVA), showing that the listening environment significantly differentiated the preference of 22-channel reproduced music at the $p <$ .00001 level. This result aligns with previous research results for two-channel and five-channel reproduction: the room-induced differences in reproduced sound fields are indeed a factor in listeners' overall hedonic responses to reproduced music.

In this study, it was assumed that an increased number of reproduction channels could recreate more authentic acoustic information from the target space (where the recording was made) so that listeners could experience similar sonic phenomena regardless of their listening environment. The preference result caused us to reject this hypothesis, yet it enlightened another interesting point. When the 22-channel system was compared with its counterpart two-channel reproduction, listeners responded with less variation across the four listening rooms in terms of preference. This variation was metricized using F-statics values: the 22-channel F-statics value was 7.36 while the two-channel F-statics was 36.5. Listeners noticed a room-induced difference, but the difference reduced as the number of reproduction loudspeakers increased.

What possible perceptual difference could explain such a difference between two and 22-channel reproduced music in various playback rooms? Subsequent analysis shows that the listening environment for two-channel reproduced music did significantly modulate listeners' judgment of all attributes except "depth," the perpendicular extension of a sound image. By contrast, the 22-channel reproduction allowed listeners to consistently judge three spatial attributes in four different playback rooms: ASW, (spatial) clarity, and depth. The result of the "depth" ratings is of interest considering the static nature of binaural capture. A non-dynamic representation of binaural sound (without a head tracker) has been known to fail in generating an "externalized sound image" [46]. Two-channel reproduction and its binaural capture thus failed to deliver a successful perpendicular extension of the sound field for all four rooms. However, the participants perceived an extended depth for the 22-channel reproduced binaural stimuli without head tracking and corresponding dynamic binaural processing. The additional ambiences from height and rear channels possibly helped participants to perceive an externalized image.

Another interesting and remarkable point can be found in the listener envelopment (LEV) ratings. As Figure 8.3 illustrates, the LEV ratings of the 22-channel-reproduced music appear significantly different depending on the listening environment at the $p < .05$ level, unlike the other three spatial attributes previously mentioned. The ordinate indicates reported magnitudes of four attributes, ranging from 0 (min.) to 100 (max.). A custom graphical user interface (GUI) was used to directly convert listeners' evaluations of those spatial characteristics into numbers. For example, a listener could draw the perceived image size in the GUI, in a similar manner to that used by Ford *et al.* [47]. Figure 8.3 indicates that variations in room acoustics may differentiate the sense of being enveloped by the sound field (both horizontally and vertically), even for 22-channel-reproduced music. By contrast, the three other spatial attributes were not influenced by the room.

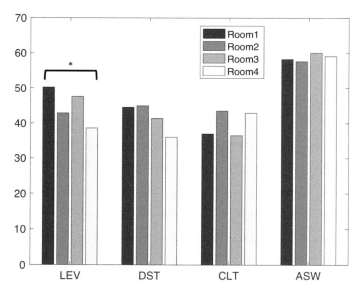

Figure 8.3 Average ratings of four spatial attributes associated with four listening rooms. The LEV ratings are significantly different depending on the room variable at the $p < .05$ level (* symbol in the figure), while the other three ratings remain statistically the same in all four listening rooms. The variable of "where" (the listening environment in which one listens to an immersive sound field) changes the sense of being enveloped.

So, our study indicates that the listening room influences listeners' perceptions of immersive audio reproduction. Different listening environments can differentiate the originally intended sound field and its associated attributes, especially the sense of being enveloped by the sound field (LEV). Importantly, however, this study demonstrates that these influences become smaller when multiple layers of loudspeakers are used to cover extended areas of the listening room, creating greater physical immersion for a listener located within the reproduced sound field. Compared with a two-channel reproduction system, 22 channels delivered a more homogeneous spatial experience to listeners in all four of the listening rooms in the study. Future research might test these findings across a wider range of listening room settings.

8.8 SUMMARY

Auditory immersion enables spatial recognition, induced from a multimodal and intertwined process of visual and auditory sensation within enclosed space. This complex process is also affected by circumstantial contexts. Two experiments have revealed that two contextual factors effectively modulate listeners' judgment of auditory immersion: (1) sociocultural background and (2) listening room acoustics; thus, both the "who" and "where" factors were highly significant. Still, despite the fact that immersive audio technologies are booming, little experimental data exists to support or refute the true significance of these contextual effects on the immersive auditory experience. Interactivity is an important factor in the growth of this market – a market that encourages people to participate, not simply observe. Further investigations into the underlying mechanisms of the contextual factors in auditory perception will support both a new theoretical framework, and also techniques to create "borderless" environments for immersive audio technologies.

REFERENCES

[1]   P. Boulez, *Orientations: Collective Writings*, Harvard University Press, 1985.

[2]   L. Brenner, Y. Rottenstreich, S. Sood, and B. Bilgin, "On the psychology of loss aversion: possession, valence, and reversals of the endowment effect," *Journal of Consumer Research*, vol. 34, 2007, pp. 369–376.

[3]   D. H. Wedell and J. C. Pettibone, "Using judgments to understand decoy effects in choice," *Organizational Behavior and Human Decision Processes*, vol. 67, 1996, pp. 326–344.

[4]   J. P. Guilford, *Psychometric Methods*, 2nd edition. McGraw-Hill, 1954.

[5]   E. C. Poulton, *Bias in Quantifying Judgments*, Lawrence Erlbaum Associates, 1989.

[6]   S. Bech and N. Zacharov, *Perceptual Audio Evaluation – Theory, Method and Application*, Wiley, 2006.

[7]   H. Stone and J. L. Sidel, *Sensory Evaluation Practices*, 3rd edition. Academic Press, 2006, pp. 104–110.

[8]   S. Zielinski, F. Rumsey, and S. Bech, "On some biases encountered in modern listening tests – a review," *Journal of Audio Engineering Society (AES)*, vol. 56, 2008, pp. 427–451.

[9]   D. Basu, "On the elimination of nuisance parameters," *Journal of the American Statistical Association*, vol. 72, 1977, pp. 355–366.

[10] D. Begault, "Preference versus reference: listeners as participants in sound reproduction," in *Proceedings of Spatial Audio & Sensory Evaluation Techniques*, 2006.

[11] H. L. Meiselman (Ed.), *CONTEXT The effect of Environment on Product Design and Evaluation*, Elsevier, 2019.

[12] L. Gros, S. Chateau, and S. Busson, "Effects of context on the subjective assessment of time-varying speech quality: listening/conversation, laboratory/real environment," *Acta Acustica united with Acustica*, vol. 90, 2004, pp. 1037–1051.

[13] K. Beresford, N. Ford, F. Rumsey, and S. Zielinski, "Contextual effects on sound quality judgments: listening room and automotive environments," in *Proceedings of AES 120th International Convention*, Paris, France, 2006.

[14] S. S. Horowitz, *The Universal Sense: How Hearing Shape the Mind*, Bloomsbury, 2012.

[15] S. Kim, M. DeFrancisco, K. Walker, A. Marui, and W.L. Martens, "An examination of the influence of musical selection on listener preferences for multichannel microphone technique," in *Proceedings AES 28th International Conference on The Future of Audio Technology – Surround Sound and Beyond*, Piteå, Sweden, 2006.

[16] A. Fukada, "A challenge in multichannel music recording," in *Proceedings AES 19th International Conference on Surround Sound*, Schloss Elmau, Germany, 2001.

[17] F. Rumsey, *Spatial Audio*, Focal Press, 2001.

[18] K. Hamasaki, "Multichannel recording techniques for reproducing adequate spatial impression," in *Proceedings AES 24th International Conference on Multichannel Audio: The New Reality*, Banff, 2003.

[19] S. Kim and R. M. Rutkowski, "Investigating listener's preference and brain responses of multichannel reproduced piano music," in *Proceedings of SMPC 2011*, Rochester, NY, 2011, p. 95.

[20] R. A. Shweder, *Thinking Through Cultures: Expeditions in Cultural Psychology*, Harvard University Press, 1991.

[21] K. Yokosawa, K. B. Schloss, M. Asano, and S. E. Palmer, "Ecological effects in cross-cultural differences between U.S. and Japanese color preferences," *Cognitive Science*, vol. 40, no. 7, 2016, pp. 1590–1616.

[22] A. Krishna, R. Zhou, and S. Zhang, "The effect of self-construal on spatial judgments," *Journal of Consumer Research*, vol. 35, no. 2, 2008, pp. 337–348.

[23] M. E. W. Varnum, I. Grossmann, S. Kitayama, and R. E. Nisbett, "The origin of cultural differences in cognition: evidence for the social orientation hypothesis," *Current Directions in Psychological Science*, vol. 19, no. 1, 2010, pp. 9–13.

[24] R. E. Nisbett and Y. Miyamoto, "The influence of culture: holistic versus analytic perception," *Trends in Cognitive Sciences*, vol. 9, no. 10, 2005, pp. 467–473.

[25] A. Saulton, H. H. Bülthoff, S. de la Rosa, and T. J. Dodds, "Cultural differences in room size perception, *PLoS One*, vol. 12, no. 4, 2017, p. e0176115.

[26] L. Wilson, How big is a house? average house size by country. http://shrinkthatfootprint.com/how-big-is-a-house/.

[27] S. Iwamiya and M. Zhan, "A comparison between Japanese and Chinese adjectives which express auditory impressions," *The Journal of the Acoustical Society of Japan (E)*, vol. 18, no. 6, 1997, pp. 319–323.

[28] S. E. Olive, T. Welti, and E. McMullin, "The influence of listeners' experience, age, and culture on head- phone sound quality preferences," in Proceedings *AES 137th International Convention*, Los Angeles, CA, 2014.

[29] S. Namba *et al.*, "Verbal expression of emotional impression of sound: a cross-cultural study," *Journal of the Acoustical Society of Japan (E)*, vol. 12, no. 1, 1991, pp. 19–29.

[30] S. Kim, R. Bakker, H. Okumura, and M. Ikeda, "A cross-cultural comparison of preferred spectral balances for headphone-reproduced music," *Acoustical Science and Technology*, vol. 38, no. 5, 2017, pp. 272–273.

[31] B. Blesser and L.-R. Salter, *Spaces Speak, Are You Listening? Experiencing Aural Architecture*, The MIT Press, 2006.

[32] S. Kim, D. Ko, A. Nagendra, and W. Woszczyk, "Subjective evaluation of multichannel sound with surround-height channels," in *Proceedings of AES 135th International Convention*, New York, 2013.

[33] A. Karampourniotis, S. Kim, D. Ko, R. King, and B. Leonard, "Significance of height loudspeaker positioning for perceived immersive sound field reproduction." *Journal of the Acoustical Society of America*, vol. 135, (2014), p. 2282

[34] S. Kim, R. King, and T. Kamekawa, "A cross-cultural comparison of salient perceptual characteristics of height channels for a virtual auditory environment," *Virtual Reality*, vol. 19, (2015), pp. 149–160.

[35] ITU, "Multi-channel stereophonic sound system with or without accompanying picture," Recommendation BS.775-3, Int. Telecommunications Union Radiocommunication Assembly, Geneva, Switzerland, (2012).

[36] L. M. Anderson, B. E. Mulligan, and L. S. Goodman, "Effects of vegetation on human response to sound," *Journal of Arboriculture*, vol. 10, no. 2, 1984, pp. 45–49.

[37] F. E. Toole, *Sound Reproduction: Loudspeakers and Rooms*, 2nd edition. Focal Press, 2010.

[38] W. Howie, D. Martin, S. Kim, T. Kamekawa, and R. King, "Effect of audio production experience, musical training, and age on listener performance in 3D audio evaluation," *Journal of AES*, vol. 67, no. 10, 2019, pp. 782–794.

[39] ITU, "Methods for the subjective assessment of small impairments in audio systems," Recommendation BS.1116-3, Int. Telecommunications Union Radiocommunication Assembly, Geneva, Switzerland, (2015).

[40] D. Howard and L. Moretti, *Sound and Space in Renaissance Venice*, Yale University Press, 2009.

[41] S. Kim, R. King, T. Kamekawa, and S. Sakamoto, "Recognition of an auditory environment: investigating room-induced influences on immersive experience," in *Proceedings AES 28th International Conference on Spatial Reproduction*, Tokyo, Japan, 2018.

[42] S. Kim and S. Sakamoto, "Investigating room-induced influences on immersive experience, Part II: effects associated with listener groups and musical excerpts," in *Proceedings AES 147th International Convention*, New York, 2019.

[43] AES:TD1001, "Multichannel surround systems and operations," Technical Report AESTD1001.1.1-10, AES, (2000).

[44] S. Sakamoto, Y. Suzuki, M. Kakinuma, H. Ohyama, H. Matsuo, and K. Takashima, "Construction of two anechoic rooms with a new experimental floor structure," in *Proceedings Inter-Noise 2016, INTER-NOISE,* Hamburg, Germany, 2016.

[45] K. Hamasaki, K. Hiyama, and R. Okumura, "The 22.2 multichannel sound system and its application," in *Proceedings AES 118th International Conference*, Barcelona, Spain, 2005.

[46] K. Inanaga, Y. Yamada, and H. Koizumi, "Headphone system with out-of-head localization applying dynamic hrtf (head related transfer function)," in *Proceedings AES 98th International Convention*, Paris, France, 1995.

[47] N. Ford, F. Rumsey, and B. de Bruyn, "Graphical elicitation techniques for subjective assessment of the spatial attributes of loudspeaker reproduction – a pilot investigation," in *Proceedings AES 110th International Convention*, Amsterdam, The Netherlands, 2001.

9

Spatial music composition

Natasha Barrett

9.1 INTRODUCTION

Connections between sound, music, and space are prevalent throughout history and society. It is therefore hardly surprising that many modern composers, sound artists, and media artists emphasise the spatial aspects of their work. Recent technological developments allied with academic and artistic research have initiated new spatial musical ideas that expand the traditional boundaries of composition. Spatial experiences, which are primary to our interaction with the environment, are embraced within a musical discourse and serve as an important carrier of structure and meaning.

The topic of spatial music composition is broad. This chapter focuses on 3D sound composition created mainly with the technologies of Ambisonics and wave field synthesis (WFS). The discussion draws on fixed-media electroacoustic music with non-visual elements or, in other words, acousmatic music, and while not intending to exclude other genres, it will allow us to focus on sound. The acousmatic context transmits a powerful sense of spatiality, while in contexts of live electroacoustic music, sound art installations, and 3D visual environments, vision complicates the discussion in areas outside the scope of this chapter.

As composers, we glean knowledge by combining technical instruction, listening, and practice with a reading of history. We can begin by probing this knowledge base in listening to, and engaging with, different auditory environments.

9.2 LISTENING AND SPATIAL COMPOSITION

How do we experience the "real" spatial sound world, and how does this experience differ from that of the composed spatial sound worlds of electroacoustic music?

The way in which we understand the spatial world can be read from the perspectives of science, perception, and anthropology. Science and perception highlight the importance of acoustics, and how we hear and understand the integration of simultaneous components of the auditory scene (e.g. as elaborated by Bregman [1]). Culture, nature, and our environment play equally important roles, which were discussed by Ingold [2]. Of particular interest are Ingold's discussions building on both anthropology and philosophy by embracing perceptual ecology and Gibson's theories that perception is active and linked to movement, and that the information required for building a representation of the world is already contained in the structure of sensation.

For the purpose of the spatial musical discourse, we can first consider sound propagation. Sound waves propagate from a source in a multitude of directions determined by the physical nature of the object. We understand the source's spatial image via a combination of the direct sound and the sound interacting with the environment. Although our perception of the auditory scene is gained from one location, our life experiences shape our expectations as to how it will sound from other locations. Much of the time we are in fact moving, and continuously accumulate information serving to reinforce these expectations. We also gather information about causation, material properties, and behaviour, which are clues informing identity, and of particular interest when ascertaining distance in the absence of impacting environmental cues. We will return to this topic later.

A composed spatial sound world will be reproduced using loudspeakers or headphones. If the sound is played through one loudspeaker, its spatial state is stable, regardless of our listening location, yet is accompanied by restrictions that are insurmountable for many composers: the loudspeaker's spatial-frequency radiation pattern will impose itself on the spatial image, the loudspeaker itself becomes the source, and spatial movement is impossible (unless the loudspeaker is itself in motion). The loudspeaker as a sound object can nevertheless be of potential interest for site-specific sound art installations.

All other speaker-based reproduction methods involve tricking our perception into believing that a source is located at a specific point in a spatial scene, whereby many loudspeakers work together to mimic acoustic cues and project phantom images using methods specified by the chosen spatialization technology. Any information that may imply the source's identity further coaxes our perception into believing the source's spatial location as real.

Spatial images are stable – or representative of the composer's intentions – inside a certain listening area, and as we will see later, the size of this listening area is dependent on both the reproduction method and how the composer chooses to use it. Without exception, there will be a listening position that is too close to any one loudspeaker, where the precedence effect (in which the perceived spatial location is dominated by the direction of the first arriving wavefront) outweighs the phantom image. We can argue that binaural reproduction is more useful, yet this too carries limitations: besides headphones inhibiting the social experience of the concert, we cannot dynamically interact with the sound field unless working from the premise of head tracking and real-time scene rendering. Furthermore, some spatial directions are clearer than others [3], and again, without head tracking and scene rendering, we cannot employ motion-based information gathering to resolve spatial ambiguities.

In summary, regardless of the sound projection method, we cannot expect an "artificial" sound scene to be equivalent to a "real-world" sound scene, nor for the listener to experience consistencies such as parallax between scene elements. WFS partly addresses these challenges while presenting other hurdles, such as the number of loudspeakers required to create the sound field, the frequency ranges that can be reproduced, and that the sweet spot from which it is possible to hear near-field-focused source changes – depending upon both where the source is located and the direction of the moving wavefront. [4]. Furthermore, the visual reference of the loudspeaker may be in direct conflict with the spatial cues the music is aiming to communicate. Nevertheless, tricking our perception appears to have been passable enough for the spatial audio arts to flourish, and the differences between "real" and "constructed" carry an important corollary: they are embedded in artistic investigation.

9.3 HISTORY AND SPATIAL COMPOSITION

Rather than exposing a historical survey recounting names and places, I will delve into two strategies for the performance of fixed-media compositions that have developed over the past decades: electroacoustic music composed in stereo and performed over a loudspeaker orchestra, and multichannel formats played over a fixed loudspeaker configuration. These strategies initially diverged, but have now partially converged.

The development of acousmatic music and the loudspeaker orchestra progressed hand in hand. By placing loudspeakers at different distances and angles from the audience, sound can be spatialized in real time as a live performance. The goal is to project and enhance spatial contrast, movement, scenes, articulations, and other characteristics of the music, and for the chosen interpretation to be for the benefit of the complete audience. In practice, this is achieved by deploying loudspeakers based upon their frequency and radiation characteristics, upon room acoustics, and on the audience area, and in performance by controlling the volume sent to either individual loudspeakers or constellations of them. Loudspeakers of different responses will project spectral variations in the music, and the spatial results are further influenced by room acoustics, and how changes in the precedence effect and directional volume influence our perception of the spatial scene. Performances are coloured by both consistencies and variances: the fixed media sets the main framework for the performance, while the space, the configuration of the loudspeaker system, and the performer's choices, add unique elements to each concert. Loudspeaker orchestras such as the Groupe de Recherches Musicales (GRM) Acousmonium [5] and Birmingham Electroacoustic Sound Theatre (BEAST) [6] have been highly active in Europe since the 1970s.

Practical experience reveals that the loudspeaker orchestra is more than just an instrument, and that the art of sound diffusion is more than just performance practice: it is one stage in the compositional process, where exploring how the music behaves in space enlightens compositional choices as well as serving to express the full potential of the material (a sentiment echoed by Harrison [7], the founder of BEAST). In the spatial sound world of the loudspeaker orchestra, the stereo source can be teased apart to project a living apparition. Sounds can be made to dart around the listener at unnatural speeds, and the variety of immersion has no natural auditory counterpart. From a stereo source, the impossible becomes possible: through performance it becomes 3D. This is no mere chance as the composition has been crafted to afford these possibilities. First-hand experience can be valuable in understanding the spatial expression in general.

Multichannel (channel-based) works are normally composed for a fixed loudspeaker configuration. As hardware became cheaper and more accessible, quadraphonic (Stockahusen's Kontakte from 1960 is often quoted as the first example) gave way to octophonic and larger setups with height. In a perfect world, channel-based works preserve the composer's spatial intentions, yet practical problems prevail. In a concert space, it may not be possible to place loudspeakers in their specified locations, differences in the loudspeakers and room acoustics may upset the spatial picture, and sitting too close to any one speaker can result in a significantly downgraded experience. These problems are often only apparent after the audience is seated, and unlike diffusion performance, little can be adjusted after the concert begins. More recently, loudspeaker orchestras have accommodated channel-based sources. An eight-channel work may be distributed over a number of eight-channel loudspeaker groups and combined with performed diffusion. The benefits of both performance and compositional control may then be combined [8].

9.4 WHAT IS 3D SOUND?

The term "3D sound" has been applied somewhat indiscriminately. Referring to playback systems, 3D has described a variety of loudspeaker arrays, including equally spaced setups arranged horizontally, hemispherical arrays, ad hoc arrays with height, and the "loudspeaker orchestra," where loudspeakers are located at different distances from the listener. All sound is 3D in that it occupies and propagates through space, yet in music and sound art, "3D" has been used to describe sound played back from loudspeakers surrounding the audience. There is however no agreement as to the qualities that merit a definition of 3D. If we focus on composition rather than loudspeakers, we can clarify the sound-based definition by stating that 3D is where musical ideas explore the complete dimensionality of space, and it is from here that the rest of this chapter departs.

Loudspeaker setups are a prominent concern for channel-based and sound diffusion approaches to spatial composition. More recently, we have seen a paradigm shift towards object-based and scene-based approaches. In the object-based method, audio objects or audio stems are coupled to spatial audio description metadata. In the scene-based method, the main features of the auditory scene are preserved and normally encapsulated in one audio file. Both object-based and scene-based approaches allow us to create spatial music without needing to specify, nor in fact consider, the loudspeaker configuration at the outset. As composers, we are liberated from loudspeakers (even if we once again depend on them in performance!).

This next section first presents Ambisonics, and then discusses how compositional choices are influenced by the storage of spatial audio features and how technical approaches might colour our ability to explore an idea in practice.

9.5 AMBISONICS

Ambisonics is a spatialization technology that captures (records or encodes) 360° sound in spherical harmonics [9], [10] and projects the sound field over a loudspeaker array specified in a decoder. Each spherical harmonic is represented in an audio channel, and the number of spherical harmonics determines the maximum precision at which the signal can be recreated. First-order 3D Ambisonics uses the first four spherical harmonics to represent the sound field, and is currently the most commonly recorded format. First-order directionality is blurry and the acceptable audience listening area is small [11]. By contrast, seventh-order 3D Ambisonics, which at the time of writing can only be encoded rather than recorded, consists of 64 spherical harmonics represented in 64 audio channels. The directionality is far more precise than first order, where for each increase in Ambisonic order, directionality increases and blur decreases [12], and the acceptable audience listening area also increases (Hollerweger [13] provides easy-to-read insights, and a technical summary is offered in Chapter 6 by Armstrong and Kearney). Although the area of precise spatial recreation is frequency dependent and mathematically, rather small for higher frequencies, we perceive it as far larger due to how we hear and understand sound. Rather than rely on mathematical models, psychoacoustic experiments using real-world source stimuli are now studying listening experiences in real-world conditions [14].

The projection of higher-order sound fields requires increasingly more loudspeakers than for lower orders [15]. As we shall see later, encoding at a high resolution does not mean we have to decode at that resolution, which is useful if our loudspeaker array is at any time too sparse. There are no reasons to avoid very high orders of encoded Ambisonics other than for practical workflows. Although most commercial digital audio workstations (DAWs) are currently limited to a third-order Ambisonics, alternative software allows unlimited outputs.

9.6 COMPOSITIONAL CONSIDERATIONS OF AMBISONIC FORMATS

The most common way to record a 360° sound field is currently with an A-format microphone. Originally developed in the 1970s, the microphone consists of four cardioid capsules arranged in a tetrahedron from which the first-order Ambisonic components can be derived [16], [17]. Both high-end and cheaper consumer products are now available. Higher-order Ambisonic

(HOA) microphones are less common, and the hardware and technology are reflected in the price [18].

Rather than record a sound field we can also compose (or encode) the sound field at any spatial resolution. With a basic knowledge of encoding and decoding conventions, composers can spatialize their work in HOA using free graphically controlled software. Most of these tools are object based: the composer specifies the positions of mono sounds in space. In a simple configuration, each source is controlled independently, while more advanced approaches may address groups of sources, their spatial behaviour or spatial statistical distribution, as well as sound transformations that emulate characteristics of real-world acoustics. One advantage of the object-based approach is that because spatial features are calculated in real time we can move the virtual listener and change our listening perspective in a way that is not possible for recorded sound fields. We will return to this topic later.

Spatial information is recreated by decoding the Ambisonic signal to loudspeakers or binaural for headphones. Correct decoding involves following simple rules, and as long as the composer follows these rules we should be able to trust the software programmers with the mathematics. A summary of the main rules are as follows: the decoding format should match the encoding format, there is a minimum and maximum number of loudspeakers that should be used for different decoding orders, the speaker locations should be measured and then set in the decoder, and different decoding methods are available for different loudspeaker arrays and for user preferences, such as envelopment and angular accuracy (further explanation can be found in [16]).

The separation of encoding and decoding makes the musical results highly portable as well as being a prudent choice for composers: encoding in a high resolution prepares the work for loudspeaker arrays suitable for very high-order Ambisonic sound fields, which are already becoming more numerous even if our studio systems only offer lower-resolution monitoring. Because the spherical harmonics increase additively where lower-order components are a subset of the complete set, we simply decode the appropriate components to match the loudspeaker configuration. It is also worth remembering that in the brief history of Ambisonics, the encoding principles have hardly changed, whereas the decoding stage has seen the most important developments. By encoding in a high order, we are more likely to benefit from future advances in decoding technology.

Composing in higher order while monitoring a lower-order stream in the studio presents some challenges. In this special situation, we can accurately control our listening location such that the smaller sweet spot of lower-order Ambisonics is of less impact than in a concert situation. Lower orders unfortunately offer less accurate directivity, yet this is less of a problem in the composition studio than it may immediately appear to be: our spatial awareness attends to both relative and absolute spatial information, where the relative separation between two sources or the rate of spatial change provides important information for a critically engaged listener. In lower-order decoding, relative space and motion dynamics will remain somewhat

clear, even though absolute directionality may be blurry. A visual decomposition of the sound field is also useful, such as that shown in the Harpex
display [19]. Composing in 3D while monitoring over a 2D loudspeaker
array is another common compositional conundrum. In these situations,
loudspeaker monitoring can be supplemented by binaural decoding, where
even though our perception of directional elevation is less good than that
in free-field situations, the method serves to convey vertical sound energy
(and significant advances in binaural decoding continue to be made [16]).

Nevertheless, composing "with height" may pose a dilemma: on the one
hand, horizontal Ambisonics significantly reduces the number of channels
in the encoded source but renders the work less interesting for a 3D loudspeaker array, while on the other hand, composing "with height" and
decoding over horizontal arrays may not sound as expected. Horizontal
concert arrays are currently most common, where many music festivals
reach out into eight or 16-channel speaker systems for experimental music.
Hybrid spatial strategies have also been tested, for example, composing a
number of 2D layers to be distributed vertically or overlaid horizontally as
the situation allows [20], or 2D and 3D layers each played through appropriate decoders.

9.7 COMPOSITION INFORMED BY FORMAT, FORMAT INFORMED BY COMPOSITION

The choice of workflow and how the final Ambisonic work is stored may
be determined by the artistic approach, the type of sound materials, or by
certain features of the final composition.

HOA encoding and decoding originally required a composer to implement methods in their own code or work with scripts that controlled code
running in non-real time. The same processes are now steered in real time
using appealing graphical interfaces allowing aural strategies to align
sound transformations with spatialization. Nowadays, both approaches are
accompanied by strengths and weaknesses: in non-real time the number of
individual sources, processes to control those sources and the inclusion of
complex environmental cues are unlimited. In real time, the limits include
both technology (central processing unit (CPU) speed) and human capacity (ability to mentally grasp, explore, and develop a potentially complex
structural vision), yet for many composers, this feels more immediate and
creatively yielding.

Whether a work's final format is object based and played by encoding
and decoding in real time or encoded as a file that is then decoded influences
both the choice of materials and how they are developed. To demonstrate this, we can draw an analogy from the world of stereo composition.
Although many DAWs allow a near-unlimited track count, we often mix
a number of smaller sound fragments and render or export this mix as a
single new audio file. Composition concerns "putting together" sound and
developing ideas through some kind of iterative process. If we decide not to
consolidate many fragments into a new sound, after only a few iterations,
simple tasks such as pitch-shifting explode in complexity – where many

small fragments require individual attention. In Ambisonics, the equivalent to this consolidation process is to render a small object-based mix as an encoded file. Musical experiments in editing, mixing, and other transformations are both easier and more immediate as they operate on consolidated encoded sources rather than on groups of many individual monaural sounds and their spatial data.

Storing the final work in an object-based format defers the choice of spatial rendering, which is indeed appealing. Defined by sources and metadata, a work composed with height can be scaled to horizontal-only, or the spatial dimension of the complete work can be rescaled to emphasize spatial dynamics appropriate for different types of performance space or room acoustics. Minor modifications to the spatial data can offer a suitable input for WFS (an approach I used when composing the work *Urban Melt in Park Palais Meran* [21]), and Dolby Atmos® technology also incorporates the object-based approach, applying amplitude panning rather than Ambisonics. In summary, a work stored in an object-based format can be played through many technologies with only minor modifications rather than a complete remix.

9.8 DISTANCE AND PROXIMITY IN COMPOSITION

The proximity of a sound is conveyed by information consisting of environmental cues, identity, or causality; the curvature of the approaching wavefront (the wavefront of a source located at an infinite distance arrives as a plane wave, while near-field sources produce an increasingly curved wavefront); and changes in sound intensity. Wavefront curvature and intensity are elements of what is called acoustic space. Although mathematically possible, it is currently not practical to control the shape of artificial wavefronts appearing to originate from a virtual source [22] in Ambisonics, and although WFS indeed achieves this result, the effect is only audible from specific listening locations and requires an extremely high density of loudspeakers.

The other proximity cues can be addressed in the compositional process, and are normally implemented prior to entering the Ambisonic domain. When composing distance information, we will be interested in the following features: variations in spectral detail where closer sounds reveal small details, changes in amplitude and high frequencies mimicking the real-world acoustics of distance attenuation and air absorption, changes in the ratio between the direct sound and environmental cues such as reverberation, environmental reflections calculated as images with respect to each source location, changes in the relationship between sources in the scene, and changes in the size of the sound's image, where closer sounds will appear larger. Doppler pitch shifts also imply the speed of a sound approaching or receding. In the Ambisonic domain, reverberation may be applied to the complete sound field using impulse response convolution. The extent (or spread) of the image can also be partially adjusted by scaling the ratio of the spherical harmonics, and most commonly, by adjusting the gain of the omnidirectional component [23]. By controlling these cues to mimic real-world experiences, we can trick our perception into believing the sound is approaching, departing, or moving through the listener's space.

The efficacy of sound identity to our reception of distance is less measurable but compositionally significant, especially when reverberation is subdued: if we believe we recognize a sound's identity, we combine that knowledge with the spectral and spatial information of the signal and deduce the distance of the sound. If the sound lacks a clear identity, we will not know if the sound is close up and dull in the spectral makeup, or located further away and bright and active (and this may be intentionally addressed as part of a compositional method). Recorded Ambisonics captures distance information fairly well, yet with the caveat of a low spatial resolution and limited post-recording modification possibilities, discussed below.

9.9 COMPOSING WITH RECORDED SOUND FIELDS

Editing and collaging Ambisonic recordings is compositionally somewhat limiting. Before exploring how we may *transform* Ambisonic source recordings, one of the simplest ways to add variation is by repositioning the location of the microphone in relation to the source or within the sound field that we are interested in. Recording a source from both near and far will yield contrasting spatial perspectives as the microphone captures correct proximal changes in the signal. Near-field recordings may also appear more enveloping, which is a topic we shall return to later.

Hybrid recording techniques offer further possibilities for variation. We can for example capture the complete sound field in Ambisonics and simultaneously record specific sounds using a number of close-up monaural microphones. The latter can then be spatially aligned with information in the Ambisonic recording, then synthesized in HOA (with or without musical transformations), and the two streams – recorded and synthesized – played simultaneously to create a hyper-realistic landscape. (This is a technique that I have applied in a number of compositions, including *Hidden Values* described in [4] available on *Peat & Polymer* [24]).

Ambisonic transformations occupy four categories: rotational operations, transformations of individual spherical harmonics, the temporal and spectral transformations normally used by composers in the stereo domain, and spatial decomposition. Rotational operations transform the complete sound sphere and preserve the information in the sound field. By contrast, transformations of individual spherical harmonics distort the spatial information. Scaling the W-channel is one distortion already mentioned in Section 9.7. More complex modifications may widen, spread, or emphasize the information in one area of the sound field [25]. Some transformations go as far as to remove the dimensionality of the original and instead materialize as a surround panning function [26]. Many spectral transformations tend to introduce phase differences or other inconsistencies between audio channels, resulting in distortions that tend to degrade rather than transform spatiality. Transformations that maintain channel consistency include resampling or tape transposition, which in addition to altering the pitch or spectral centroid, also serve to increase or decrease the implied size of space due to a change in the time between the spatial reflections.

None of these transformation techniques allow us to focus on one sound in a complex picture, nor move the virtual listener through the scene. Here we need to turn our attention to a method that can isolate sources and their directional information such that the features of the sound field can be repopulated in an object-orientated encoding. One such method for spatial decomposition is beamforming: the signals from a microphone array are combined to focus on information in a specific direction [27]. Similar results are achieved by alternative methods such as Harpex, which applies a parametric B-format model [19] and new methods offering greater accuracy are in development [28]. With these approaches, directional sources can be discovered by a "lighthouse" type scan, by tracking a known sound's movement (especially if the software offers a visualization of the sound field), or by defining multiple beams in all directions.

9.10 COMPOSING IMAGES IN SPACE

Most sound objects create complex directivity patterns where frequencies propagate from surfaces and through cavities, and the sound waves interact with the environment and with the listener in distance and in orientation. A sound object is rarely a point source and instead generates an image over which the spectrotemporal distribution is often non-uniform and may even be in motion. By contrast, Ambisonic synthesis tools encourage a point-in-space approach to spatialization.

How can we compose images in space? The first consideration is a technical trait – a point source encoded in Ambisonic space spreads when decoded. The extent of this spread is determined by the Ambisonic order and is called "angular blur." One advantage of encoding and decoding in HOA is directional precision or low angular blur, yet this very feature emphasizes the mono source as a point in space. Not wanting to forsake either the large audience listening area of HOA or the possibility to achieve clear directionality, a number of technical solutions are available that manipulate the apparent source extent by spreading or blurring in the Ambisonic domain. Some of the techniques include phantom-source widening that disperses the direction of arrival of the sound with frequency [29], Ambisonic spatial blur that takes advantage of lower-order blur [30], manipulations of the interaural cross-correlation coefficient [31], W-panning [23], and other weighting functions to change source extent. Granular techniques have also been applied to source distribution pre-encoding [32]. Although these methods indeed spread the extent of the source, they less-adequately address key features of a spatial image. With these methods, spectrotemporal details are relatively constant as the source changes in width, and there is little consideration of the change in spectral-temporal information with distance.

The key features of sound images can instead be addressed as part of a compositional method [33]. Rather than populate the spatial scene with independent monaural sounds, images can be assembled from many sources, each carrying information that together allude to the features of life-like perceptual objects. These component sources should retain some common features for coherence, as well as reflect temporal and spectral differences representing one aspect or angle of the complete image (real or fictional) interacting correctly with environmental cues. Software such as IRCAM's Spat [34] offers control over the directivity of a source and its interaction inside a simple acoustic model.

Gauging how many component sources to use, how they correlate, and how they are controlled, engages both technical criteria and compositional aims. Figure 9.1 visually represents a sound image assembled from eight components. A simple Cartesian translation illustrated by the image placed at two different distances from the virtual listening location, serves to correctly change the perceived width and depth with distance. Also, when the image is further away, Ambisonic blur naturally reduces the clarity of microdetails as the angular difference between components decreases. When the image is close up, greater spectrotemporal variation and depth of geometry is revealed and the qualities of the image will change in accordance with the relative angle to the virtual listener. (This technique is employed in *Dusk's Gait* [35]).

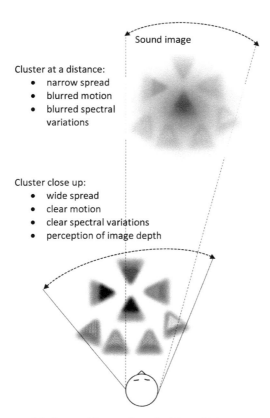

Figure 9.1 An assembled sound image heard at two distances.

Choosing the number of components entails balancing Ambisonic order (or the blur) with the qualities of the sounds used and the proximity of the image to the listener. For very high-order Ambisonics (low blur), the number needs to be greater than for lower-order Ambisonics in order to reduce the likelihood of hearing individual components. Images that are intended to approach the virtual listener likewise require more components than if the image were held further away. Moreover, the number of components depends on the qualities of the sounds and the way in which the image is assembled: sounds that are easier to localize, such as those of high spectrotemporal variation are likely to need to be of a greater number, and a balance between spectrotemporal coherence and possible artefacts such as phasing must be evaluated in each context.

Spectrotemporal variations in the components may be prepared either by sound transformation or in recording. Signals recorded from a number of microphones surrounding a real acoustic object capture natural variations in the sound, and these recordings then serve as components in the synthesized spatial image. Adding local micromovements to either the components or to the global image can further enrich the results and allude to animate qualities. These movements may be controlled by hand or be data driven, where data in turn may be derived from the spectral or amplitude variations in the sources themselves.

9.11 INSIDE A SOUND AND LOOKING ONTO A SCENE

The experience of hearing a stereo composition played over two loudspeakers can be thought of as "looking" onto a scene. From the correct location (in a 60° triangle) we may even experience being inside the scene, where interesting phase and frequency phenomena appear to envelop us in sound. In a diffusion concert we really *enter* the scene, but we may not automatically feel *part of it* and rather still look *out onto it*. The same is true for Ambisonics, where although we may be surrounded by sources from all directions we may still experience a sensation of looking out onto the scene. Placing listeners *inside* the scene to feel as if they are part of the action concerns decoding methods as well as composition. Some decoders tend to highlight sound at the distance of the loudspeaker, for example, the all-round Ambisonic decoding (AllRAD) decoder that is based on amplitude panning, while others sound more spatial, such as the mode-matching decoder with the caveat of strict decoding conventions, or the energy-preserving decoder (refer to [16] for a summary of different decoders). Selecting the decoder and its parameters may entail considerations such as the size of the room, the acoustics, limitations of the loudspeaker array such as symmetry, the size of the audience, and preferences based upon musical content and personal priorities.

Elaborating the decoder discussion is outside the scope of this chapter, so instead I once again focus on composition. Although we cannot touch sound in the same way as physical objects, spatial images may be experienced as occupying our local space, and even if of abstract identity, may still appear tangible. Contrasting the experience of an object as a tangible spatial image is the sensation of being inside the body of sound. Such sensations are less to do with images in space and more to do with what has been called immersion and envelopment – terms that have been applied ambiguously and interchangeably. Listener envelopment has been discussed at length in connection to room acoustics where lateral early reflections affect the apparent source width, while the late reverberant sound influences the feeling of immersion. Berg *et al.* [36] make a distinction between room envelopment or the extent to which we feel surrounded by the reflected sound, and source envelopment or the extent to which we feel surrounded by the sound source(s). For source envelopment, the listener is aware of directional cues and is either looking out "towards" or is "inside" a scene, which may also include a complex array of directional and diffuse information from sounding objects and reflections from the environment.

Considering musical-theoretical opinions, Emmerson supports the idea that, "immersion relates to the scientific insight that the observer is always an implicit participant in the observation being made which translates in compositional terms into a concern with the sensory immersion of the subject (listener) in the work" [37]. He also suggests that immersion is an experience of being inside the sound field without clearly localizing objects [38]. Lynch and Sazdov suggest the opposite: that to be immersed in a space means localizing objects and events from within that space [39].

In light of these differing views, we can consider the following tactics when composing immersive sound fields: an awareness of lateral reflections and late reverberation (from room acoustics), an avoidance of directionally emphasized variations in spectral content (from perception), an understanding of the space as both listener and composer, and if we wish to surround the listener with sources, the number should be significant so that our attention cannot fall on any one direction for any length of time. The second movement of the composition *Hidden Values* called "Optical Tubes" [24] approaches immersion in this way, weaving into the composition a granular cloud consisting of a high density of individually treated events projected in seventh-order 3D Ambisonics.

9.12 COMPOSING SPATIAL MOTION AND THE SPATIAL BEHAVIOUR OF SOUND

I have so far endeavored to unfold the spatial offerings of sound-field recordings, synthesized scenes, sound images, and enveloping sound fields. This section attends to spatial motion and the spatial behaviour of sound.

Channel-based, object-based, and scene-based approaches to composition have developed in parallel to spatialization software. The most common tools concern panning methods either for distributing sound directly over multiple loudspeakers or for Ambisonic encoding. Easy-to-use tools cannot, however, circumvent some central questions: why control spatial motion in the composition? What are the intended results? How to reach those goals? These three questions are interconnected, and the tools we choose should permit a flexible and unfettered workflow.

Although much can be achieved by improvising spatial movement by hand and ear, we are often interested in less ad hoc approaches and instead wish to explore spatial data from other sources as a means to control the behaviour of our sound. Suitable data consists of at least a space-time series and may also include other parameters useful for mapping sound trans-formations. Data may stem from algorithmic processes or be extracted and abstracted from the real world. The latter provides a wealthy supply that is particularly relevant when incorporating real-world sounds in the com-position and where the sounds themselves may drive the spatialization. For example, large and rapid variations in amplitude may be mapped to large and rapid spatial movements in what we can describe as a direct extraction of data from the source. Alternatively, the sound may imply a data source, such as the sound of cascading gravel suggesting the use of an avalanche spatial data model in what we can describe as an indirect extraction from the source.

Many real-world processes and patterns make no sound at all. Using data from these types of source to control space and sound reveals the nature of the source in what is called sonification [40]. For example, phys-ical movements can be recorded in 3D using devices that track multiple points or markers placed over the moving subject. Hardware possibilities range from professional high-speed camera systems (e.g. the Qualisys optical motion-capture system) to consumer devices (e.g. the Leap Motion sensors). A dataset is recorded from each marker, and together they describe correlated points in space and time. Datasets such as these are beneficial when sonifying coherent, 3D coarticulated sound images.

For all such approaches, mapping is a primary concern: a mapping appropriate for the sonification of the micromovements of a shivering body may not serve so well the data from the movement of a flag billowing in a summer breeze.

9.13 DISCUSSION

Ambisonics as a technical method for spatial audio composition features in both contemporary electroacoustic music and in commercial outlets, such as virtual reality and 360° multimedia. Tools such as VST panners for encoding and decoding Ambisonics are easy to use and require minimal knowledge. When approaches to spatial composition become more detailed or hope to align with the information content of the real world, easy-to-use plugins may be less useful. Software should instead offer a means to freely manipulate space, time, and spectrum, and allow a broad approach that may include datasets or custom-made processes. Many composers now build their spatial-ization methods in software such as Max and SuperCollider, using common DAW software simply as a convenient container for the audio.

A thread running through this chapter concerns the balance between recorded and synthesized spatial sound. Although synthesis offers greater control over spatial ideas, acoustic sources offer a wealth of useful infor-mation relevant to all of the topics that have been discussed. Acoustic recordings may motivate an object-orientated synthesis process and vice versa: a synthesized sound field, generated, for example, from sonification, can inform the composition of a real-world acoustic allusion to fold back into the work.

REFERENCES

[1]   A. S. Bregman, *Auditory Scene Analysis: The Perceptual Organization of Sound*, The MIT Press, 1994, pp. 213–393.

[2]   T. Ingold, *The Perception of the Environment: Essays on Livelihood, Dwelling and Skill*, Routledge, 2002, pp. 153–243.

[3]   J. Blauert, *Spatial Hearing, Cambridge, Massachusetts*, The MIT Press, 2001, pp. 39–50.

[4]   T. Carpentier, N. Barrett, R. Gottfried, and M. Noisternig, "Holophonic sound in IRCAM's concert hall: technological and aesthetical practices," *Computer Music Journal,* vol. 40, no. 4, pp. 14–34, 2016.

[5]   "GRM Acousmonium," GRM. https://inagrm.com/en/showcase/news/202/the-acousmonium.

[6]   "BEAST," BEAST. www.beast.bham.ac.uk/.

[7]   J. Harrison, "Diffusion: theories and practices, with particular reference to the BEAST system," 1999. https://econtact.ca/2_4/Beast.htm.

[8]   S. Wilson and J. Harrison, "Rethinking the BEAST: recent developments in multichannel composition at Birmingham ElectroAcoustic Sound Theatre," *Organised Sound,* vol. 15, no. 3, pp. 239–250, 2010.

[9]   M. Gerzon, "Ambisonics. part two: studio techniques," *Studio Sound,* vol. 17, no. 9, pp. 24–26, 1975.

[10] J. Daniel and S. Moreau, "Further study of sound field coding with higher order Ambisonics," in *Proceedings 116th Audio Engineering Society (AES) Convention*, Berlin, Germany, 8–11 May 2004.

[11] J. Daniel, "Evolving views on HOA: from technological to pragmatic concerns," in *Proceedings of the 1st Ambisonics Symposium*, Graz, Austria, 2009.

[12] S. Bertet, "Investigation on localization accuracy for first and higher order Ambisonics reproduced sound sources," *Acta Acustica united with Acustica,* vol. 99, no. 4, p. 642–657, 2013.

[13] F. Hollerweger, "An Introduction to Higher Order Ambisonics," 2013. https://pdfs.semanticscholar.org/40b6/8e33d74953b9d9fe1b7cf50368db492c898c.pdf.

[14] M. Frank and F. Zotter, "Exploring the perceptual sweet area in Ambisonics," in *Proceedings 142nd AES Convention*, Berlin, Germany, 2017.

[15] M. Noisternig, T. Carpentier, and O. Warusfel, "Espro 2.0: implementation of a surrounding 350-loudspeaker array for sound field reproduction," in *Proceedings of the AES UK Conference: Spatial Audio in Today's 3D World*, York, UK, 2012.

[16] F. Zotter and M. Frank, *Ambisonics A Practical 3D Audio Theory for Recording, Studio Production, Sound Reinforcement, and Virtual Reality*, Springer Open, 2019..

[17] M. Gerzon, "The design of precisely coincident microphone arrays for stereo and surround sound," in *L-20 of Proceedings 50th AES Convention*, London, UK, 1975.

[18] mh acoustics. www.mhacoustics.com/.

[19] S. Berge and N. Barrett, "High angular resolution planewave expansion," in *Proceedings of the International Symposium on Ambisonics and Spherical Acoustics*, 2016.

[20] N. Barrett, "Kernel expansion: a three-dimentional Ambisonics composition addressing technical, practical and aesthetically issues," in *Proceedings of the 2nd International Symposium on Ambisonics and Spherical Acoustics*, Paris, France, 2019.

[21] N. Barrett, Artist, *Urban Melt in Park Palais Meran.* [Art]. 2019.

[22] S. Favrot and J. Buchholz, "Reproduction of nearby sound sources using higher-order Ambisonics with practical loudspeaker arrays," *Acta Acustica united with Acustica,* vol. 98, p. 48–60, 2012.

[23] D. Menzies, "W-Panning and O-Format, tools for object spatialization," in *Proceedings of the AES International Conference: Virtual, Synthetic, and Entertainment Audio*, Espoo, Finland, 2002.

[24] N. Barrett, Composer, *Peat and Polymer (3DB020).* [Sound Recording]. +3DB. 2014.

[25] M. Kronlachner and F. Zotter, "Spatial transformations for the enhancement of Ambisonic recordings," in Proceedings of *2nd International Conf. on Spatial Audio*, Erlangen, Germany, 2014.

[26] T. Lossius and J. Anderson, "ATK Reaper: the Ambisonic toolkit as JSFX plugins," in *Proceedings of the International Computer Music Conference*, Greece, 2014.

[27] S. Jan, H. Sun, P. Svensson, X. Ma, and J. Hovem, "Optimal modal beamforming for spherical microphone arrays," *IEEE Transactions on Audio, Speech, and Language Processing,* vol. 19, no. 2, pp. 361–371, 2010.

[28] A. Favrot and C. Faller, "B-Format decoding based on adaptive beam-forming," in *Proceedings of 146th AES Convention*, Dublin, Ireland, 2019.

[29] F. Zotter, M. Frank, M. Kronlachner, and J.-W. Choi, "Efficient phantom source widening and diffuseness in Ambisonics," in *Proceedings of the EAA Joint Symposium on Auralization and Ambisonics*, Berlin, Germany, 2014.

[30] T. Carpentier, "Ambisonic spatial blur," in *Proceedings of 142nd AES Convention*, Berlin, Germany, 2017.

[31] T. Schmele and U. Sayın, "Systematic evaluation of perceived spatial quality," in *AES Conference: 24th International Conference: Multichannel Audio, The New Reality*, Banff, Canada, 2003.

[32] N. Barrett, "A musical journey towards permanent high-density loudspeaker arrays," *Computer Music Journal,* vol. 40, no. 4, p. 35–46, 2016.

[33] N. Barrett, "Composing images in space: Schaeffer's allure projected in higher-order Ambisonics," in *Proceedings of the International Computer Music Conference*, New York, USA, 2019.

[34] IRCAM. https://forum.ircam.fr/projects/detail/spat/.

[35] N. Barrett, Composer, *Dusk's Gait.* [Sound Recording]. 2018.

[36] J. Berg and F. Rumsey, "Systematic evaluation of perceived spatial quality," in *AES Conference: 24th International Conference: Multichannel Audio, The New Reality*, Banff, Canada, 2003.

[37] S. Emmerson, *Music, Electronic Media and Culture*, Ashgate, 2000, p. 62.

[38] S. Emmerson, *Living Electronic Music*, Ashgate, 2007, p. 161.

[39] H. Lynch and R. Sazdov, "A perceptual investigation into spatialization techniques used in multichannel electroacoustic music for envelopment and engulfment," *Computer Music Journal*, vol. 40, no. 1, pp. 13–33, 2017.

[40] N. Barrett, "Interactive spatial sonification of multidimensional data for composition and auditory display," *Computer Music Journal*, vol. 40, no. 2, pp. 47–69, 2016.

10

Sound and space: learning from artistic practice

Pedro Rebelo

10.1 INTRODUCTION

This chapter makes reference to key historical *soundmarks* in the use of space in music while aiming to distil compositional and performative approaches into a set of practical strategies that can be applied across various aspects of 3D audio production. It will consider the intrinsic relationship between sound design and sound systems, from the Iannis Xenakis designed 1958 Philips Pavilion – incorporating an immersive multisensorial experience with light, projection, and over 400 loudspeakers – to more recent sound art installation practice. The quality of an immersive experience cannot be addressed by separating the often-polarized activities of technological development and content creation. Artistic practice engaged with sound and space has developed to combine the two in reciprocal and interdependent ways, as in David Tudor's 1968 *Rainforest* with numerous sound objects activated with transducers and playing the intrinsic resonant characteristics of each material. The development of loudspeaker orchestras in the acousmatic music tradition since the 1960s represents a symbiotic relationship between the creation of content (i.e. music) and the system used to acoustically diffuse it in space.

Questions arise – around spatial listening and the dichotomy between visually orientated metaphors for placing objects in space and an embodied approach to listening. By learning from the rich history of the relationship between sound and space, a framework can be advanced, one that is much needed for the current developments in immersive sound.

10.2 WHY SOUND AND SPACE?

Discussions on the relationship between sound and space are typically intertwined with the development of technologies that afford some level of control of how sound sources are projected in space through a loudspeaker array [1], [2]. Although it is tempting to correlate the ability to control sound projection with the emergence of space as an intrinsic element in music and sound-design creation, there is an inherent relationship between the sonic and the spatial that goes well beyond a technicist or indeed perceptual understanding. The alignment of various musical traditions with specific types of spaces as discussed, for example, by Blesser and Salter [3], illustrates a symbiotic relationship between sound production and its spatial acoustics manifestation. The directionality of a sound source, its propagation profile, early reflections, and reverberation, among others, all embed sound with spatiality to an extent that it is not possible to think about sound without it existing in space. This relationship is evident in the development of numerous instrumental ensembles and associated performance practices, from the string quartet to orchestras, opera, drum circles, alpine horns, didgeridoos, and so on. The alignment between these practices and *their* spaces comes to the fore when a specific sonic practice is displaced – a didgeridoo in a cathedral?

Musical traditions from across the globe present strategies and solutions to fundamental sound-space concerns, which determine the use of space and the positioning of sound sources. The relevance of sonic fusion in certain traditions (i.e. the ability for multiple sound sources to blend as one, such as in string quartet performance practice) determines how close sound sources need to be to each other, as well as a certain distance for the listening position (this is discussed by Smalley as perspective in the context of electroacoustic music [1]). The sound of Gregorian chant is forever tied into large and reverberant cathedrals, in contrast to the complex polyphonic texture and use of vernacular text that J. S. Bach created for Thomaskirche in Leipzig, which required more intimate acoustics – devoid of the echoes so appreciated in those earlier gothic cathedrals [4].

Jahn [5] identifies the unequivocal connections between the cave spaces chosen by early hunter-gatherers, rock drawings, and specific acoustic resonances. Their acoustic measurements of six Palaeolithic cave configurations across Britain, Ireland, and France clearly shows a common resonance around 110 Hz – which matched the range of the male voice. As discussed by Boren [6], the deliberate choice for spaces of sound production and listening goes back to pre-history and can be tracked steadily through ancient history, early Christian worship, the renaissance, baroque, classical, and romantic periods to the present day.

These intrinsic sound/space relationships go well beyond documented musical practices and apply to our everyday soundscape. The acoustic threshold of a given source determines its spread in space. Referring to Truax's book *Acoustic Communication*, Garrioch [7] describes the geography of soundmarks such as bells and their importance in community building in early modern European towns. "[T]he familiar soundscape helped create a sense of belonging: it was part of the 'feel' of a particular city, town or neighborhood, a key component of people's sense of place" [7]. The sonic spread of each given church bell is the acoustic footprint that determines the size of a given parish, with most important churches typically having the largest bell, and hence a larger acoustic and congregation spread. "The hierarchy of religious authority that allowed each church to possess a particular number and size of bells seems to have been reinforced during the seventeenth century. The respective rights of bishops, chapters of canons, and others, to ring cathedral bells were carefully defined" [7].

These points of reference in a long history of sound space relationships point towards an understanding of sound as "always spatial." Our clear concern with aligning certain sound-production practices with specific spaces for listening then also suggests an understanding of listening as always spatial.

10.3 BEFORE IMMERSIVE

Despite the symbiotic relationship between sound and space being present across numerous listening situations, there is a significant body of creative work that specifically explores the sonic and the spatial by embedding meaning in the way sound sources are arranged in space, and how space itself is designed through sound. In the same way that elements such as spectrum, pitch, rhythm, or amplitude afford various compositional and design strategies, spatiality in sound presents specific conditions and situations as discussed below.

Before the current wave of fascination for immersive experiences, artists and composers have devised ways to expand and at times disrupt conventional listening situations to highlight the spatial characteristics of sound. Although "immersive" can be a somewhat generalized term for certain types of sonic experiences, despite its current ubiquity, it is yet to attain a specific meaning. Examples from artistic practice (addressed below) articulate immersion in the sense that a listener shares an acoustic, aesthetic, and conceptual space with sound sources. As such, immersion is less to do with spatial effects but rather with an unspoken contract between the listener and sound environment: a contract that invites the listener to be a participatory and active part of the environment as opposed to being a passive agent. The embodiment of a listener in such an environment suggests a breakdown of listening "frames" such as those created by the thresholds between stage and audience areas in a conventional concert-hall situation. These thresholds are, needless to say, not only acoustic but cultural, social, and historical, conditioning behavior and hence, they frame their own modalities of listening. This sense of participating in an environment is not only present in art situations, but also in everyday life. Perhaps it is worth considering what makes our everyday listening situations "immersive." Is it the possibility of sound sources arriving at our ears from any direction? The ability for our bodies to move and hence actively navigate through a myriad of sonic phenomena? Is it the sonic and semantic coherence that we embed a given space with? Or is it the irresistible capacity for our own body to inhabit an environment and make its innate sonic presence manifest?

These questions touch disciplines from philosophy to spatial audio, and cognitive science to embodiment. The body of work in music and sound art concerned with space does however provide us with directions of exploration. These will now be distilled in the form of sound-design-orientated language while addressing all aspects of the experiential modes of listening that might be involved in so-called "immersive sound."

10.4 SPATIAL ATTRIBUTES

In order to refine the terminology and address different conditions of sound and space, research in spatial audio and spatialization provides potentially useful concepts. Under the umbrella term of spatial attributes, there lie specific sonic scenarios that determine the relationship between a listener and a given sound field. These include so-called "immersive attributes" such as envelopment, engulfment, presence, and spatial clarity. For the purposes of this discussion, a brief outline of term envelopment is required. Sometimes referred to as ensemble envelopment, the term addresses the experience of being surrounded by sound. Lynch and Sazdov [8] point out that there is a difference between concert hall and reproduced audio when it comes to a sensation of envelopment. Two key concepts are apparent: first, source width – relating to the perception of a sound source as being broader than its visual boundary (a consequence of early acoustic reflections) and second, listener envelopment – relating to late lateral reflections from walls. While pointing out the ongoing debate around a definition of envelopment, their discussion of definitions in the literature demonstrates a variety of views, in particular with regard to the notion of being surrounded by sound, versus that of reverberation arriving at a listener from all directions. An operational definition advanced in the same paper states that "[e]nvelopment is a subjective attribute of audio quality that accounts for the enveloping nature of the sound. A sound is said to be enveloping if it wraps around the listener." Examples from everyday life in which envelopment is apparent tend to include both a large number of sound sources distributed in space and complex reflections: concert hall applause, rain-storm, a busy city centre, a forest…

The related concept of engulfment relates to sonic experiences that include sounds above and below the listener suggesting the notion of covering the listener with sound. A full 3D spatial-scene would therefore combine envelopment and engulfment to create a 360° sound environment. Although this might suggest specific strategies for either sound reproduction or placement of acoustic or electroacoustic sound sources, rather it is the approach to sound design with an understanding of sound – from the semantic to the concrete – that becomes the enabler for such experiences to be effective. The interested reader will find a useful literature review that explores the concept of immersion as a "psychological concept as opposed to being a property of the system or technology that facilitates an experience" in Agrawal *et al.* [9].

10.5 LEARNING FROM ARTISTIC PRACTICE

This section aims to devise a sound-design-focused framework to detail specific strategies that address the intended perceptual and experiential conditions discussed in the previous section. Currently terminology sits either in the field of spatial audio and is derived from an engineering perspective (e.g. [10], [11]) or architectural acoustics [12], or in the field of musical composition, which often requires an understanding of music theory and electroacoustic music [1]. A sound-design-based framework attempts to define these sound and space strategies in a sound-design context, while emphasizing the notion that the success or failure of any given technical approach is dependent both upon context and its design.

The following examples – from practices spanning almost 500 years – point towards specific strategies, which while clearly rooted in particular aesthetics, genres, and performance practices, provide useful templates for an abstraction of sound-space relationships. Key works from the lineage of music making will now be outlined – works that are concerned with space, which develop and implement specific strategies for articulating relationships between sound and space, and which aim to produce specific results for a listener. Although not all were implemented with height, it is important to view them as spatial concepts that can be translated to contemporary 3D systems.

Traditions for multiple instrumental forces within the same family are abundant in the history of music – from the string quartet to the brass band. A composer's motivation for multiplying a given instrument or family of instruments can be spatial. For example, in comparison to a solo string instrument, although a given arrangement for a group of string instruments *may* extend the pitch range and spectrum, it always broadens the spatial image with an ensemble sound. Surrounding the listener with sound sources of equivalent spectral content provides opportunities both for envelopment, call and response, or spatial echoes, and illusion of moving sources. Giovanni Gabrielli's "cori spezzati" technique, following from practices by Willaert and others, developed space musically by distributing choral voices across two, three, or more spaces in St Mark's Basilica in Venice in the late Renaissance period. These were often at different heights, and at times combined with multiple organs. The compositional strategy that proved most striking was not so much a call and response in which liturgical text is divided between two choirs, but rather the rapid interchange of musical material. A reverberation time (RT) of almost 7 seconds (when the building was empty) is an element that brings acoustic fusion or blur, hence allowing for this choral separation. Although long RTs are normally considered as a problem for contrapuntal music, it can be argued that a drive for increasingly complex polyphonic and contrapuntal music led to the exploration of space and architecture as a way of distributing musical materials and achieving a higher degree of polyphony. It is interesting to note that not all were convinced by this innovative spatial music. Contemporary theorist Giovanni Maria Artusi notes: "[n]owadays composers, in the *cantilenae* composed for *concerti*, place the lowest parts (that is, the bass parts of the one and of the other choir) at the interval of a fifth, a third or an octave; almost always, one hears [a] wretched [sound, such as] I cannot describe, which offends the hearing […]" [13].

Multiple organ complexes have been a feature of certain European cathedrals since the 16th century. The Basilica of Mafra in Portugal is unique in that the building was designed with six organs in mind. The organs all have different features and were designed and built by different constructors, although it is clear that the motivation for creating a spatial ensemble was the drive behind the initial design, subsequent reconstructions and modifications to each of the instruments [14]. A publication from 1761 by Santo António indicates that the six organs were used in liturgical festivities [14]. The positioning of the organs clearly highlights a concern for spatial distribution, and an early 19th century mass for three choirs and six organs "reveals a clear intention of exploring the spatial capabilities of the Basilica" [14]. The practice of distributing both choirs and instrumental ensembles to different positions – including at different heights – also points towards 3D spatial exploration as well as a way of articulating musical structure.

The Philips Pavilion was designed and constructed for the Brussels World Fair of 1958. It is arguably still to this day one of the most ambitious immersive experiences, one in which sound and space play a key role in articulating the artistic, musical, and architectural discourse at the genesis of the project. Le Corbusier, the most renowned architect of the time, led a star team, alongside architect and composer Iannis Xenakis, composer Edgard Varèse, plus a team of Philips engineers. This made the project one of the 20th century's most unique and iconic blends of art and technology. The system consisted of multiple projections on walls, plus lights and a loudspeaker array of over 400 loudspeakers, all designed in tandem with the unique architecture of the pavilion. It provided a platform for an immersive journey for 500 spectators at a time. Technological accomplishments aside, the pavilion project provided Xenakis with an opportunity to connect musical and architectural thought: "[…] when the architect Le Corbusier, […] asked me to suggest a design for the architecture of the Philips Pavilion in Brussels, my inspiration was pin-pointed by the experiment with *Metastasis*. Thus I believe that on this occasion music and architecture found an intimate connection. *Metastasis*" [15].

Xenakis would go on to explore these relationships with the *Polytopes* project, but also through re-thinking performance spaces such as La Villette in Paris. "First of all, there will not be a single floor, but rather platforms, perhaps mobilized, arranged in space so that one can look in all directions" [16]. He goes on to describe the notion of the spectator as suspended in space and a configuration of platforms to be used by either performers, spectators, or equipment. Xenakis is quick to point out that this is not a polyvalent extension to concert-hall design, but rather a way of embedding a distribution of sonic elements in 3D space. Instead of a convergence on neutrality (visually and acoustically), Xenakis advocates for strongly characterful spaces: "This space, it would at first be like an envelope which would serve as a sonorous shelter […]. The envelope could have certain sonorous qualities based on the forms and volumes, and not on correcting panels, as usually happens" [16].

In *Terretektorh* (1965–1966), Xenakis dissolves the threshold between the ensemble and audience by placing the audience in a concentric circle with the conductor in the middle and the audience interspersed between the musicians. In a radical break with orchestral tradition and notions of idealized listening position, Xenakis embraces the possibilities of each listener engaging with a complex spatial sound world from different perspectives. Techniques such as the rotation of musical material, particles of short percussive sounds, and strategic distribution of timbral sections around the orchestral circle allow for a truly spatial musical discourse in which each audience member represents a listening point, perceiving trajectories of seemingly moving sound sources and enveloping textures based on multiple sources distributed in space.

Stockhausen's *Gruppen for three orchestras* (1955–1957) surrounds the audience with orchestral forces in a combination of equal, similar, and distinct instruments across the three groups. The combination of instruments across the three orchestras allows for timbral similarity, providing the illusion of moving sources when related musical materials are distributed in space, while distinct instrumental forces suggest location-specific timbres (e.g. one single electric guitar in orchestra 2 in a central position). The most iconic moment in the piece is at times referred to as the brass battle in which a chord is "tossed around" the brass instruments across the three orchestras fully employing the possibilities of timbral similarity to highlight spatial movement. Although this piece operates in horizontal surround, as mentioned, the principles might easily be translated into a scenario that featured elevation.

In counterpoint to Xenakis' work, Stockhausen develops a characteristically more formalized sound environment in the design of the 1970 Osaka Expo Spherical Concert Hall. With the audience sitting on an acoustically transparent grid,[1] 50 groups of loudspeakers provide uniform point-source sound projection around and above the audience – giving a 3D sound field. In works like the 1968 *SPIRAL for a soloist and shortwave receiver*, sound sources are distributed in the hall through spiral movements controlled at the sound desk by a "rotation mill." In *POLE (poles) for 2 players/ singers with 2 shortwave radio receivers* from 1970, the score includes notation for spatial trajectories with, as the title suggests, creating polar opposites between the two performers. In a hemispherical division of the sound space supported by the architecture of the pavilion, Stockhausen created a musical logic of extremes, articulated by his "plus-minus" notation scheme, and allowing for the projection of curved or straight lines and polygons as discussed in detail in [17].

David Tudor's *Rainforest* (there were various versions between 1968–1973) employs "sounds electronically derived from the resonant characteristics of physical materials" [18] to create an active listening environment in which multiple sound objects resonate in space and invite the listener to physically engage with the most minute changes created by the relationships between the objects and one's body. A series of suspended sculptures and found objects become resonant bodies through the use of transducers, imparting each object with its own unique spectral and spatial propagation profile. For the listener, this forest of objects and sound suggests an environment with an immediately understood inner logic. The acoustic characteristics of each object, made sonic through signals played by transducers, situate each component of the piece in space, interacting with the natural acoustics of the site to convey a sense of exploration and sonic immersion. In *Rainforest*, it is precisely the physical "limitations" of each object-turned-loudspeaker that provide an entrancing listening experience, in no small part enhanced by giving the listener agency through movement and exploration of space. Janet Cardiff's *Forty Part Motet* (2001) employs a similar strategy to re-spatialize Thomas Tallis's *Spem in Alium* (1573), in which a recording of the work is played back by placing each of the 40 voices in an individual loudspeaker. An oval, sculptural layout of 40 loudspeakers articulates the polyphony through space in which a listener moves and navigates through the 40 individual voices, hence constantly shifting the listening perspective. In the work of both Cardiff and Tudor, it is the agency given to the listener, manifested in the notion that each listener experiences the work in a unique way, that suggests a sense of shared space and immersion.

The 1998 installation by the author – *Partial Space* – explores an interactive sound world based on the resonant frequencies of the installation site. The measurement of peak frequencies in an impulse response of a space results in a coherent spectrum that when played back into the same space, suggests a sense of acoustic amplification and belonging. The frequencies, synthetized through sine waves and amplitude-modulated sine waves are triggered by the audience's position in space, creating a map and set of relationships between a spectrum and space as manifested by the audience location, number of visitors, and acoustics of the site itself. The close relationship between the synthesized tones and the acoustics of the space suggests that the interaction is not so much to engage the audience in playing a virtual instrument, but rather navigating their own listening through the site as it gets activated: with increased movement and number of visitors, it suggests a sense of social listening. Commenting on the work, Bo Kampmann Walther (in [19]) writes "Partial Space is not only a discrete construction; it is furthermore a participatory experementarium in which interferential fulfilments of physical and scanned space are stretched and 'toned' together."

As in much of Bernard Leitner's work in sound installation, *Tuba Architecture* from 1999 embraces the sculptural elements of sound to create an interplay between sound objects, architecture, and the listener. "A line of sound is produced when sound moves along a series of loudspeakers, from one loudspeaker to another. Space can be defined by lines. Lines of sound can also define space: space-through-moving-sound" [20]. In *Tuba Architecture*, 60 suspended metal sheets form walls articulated sonically through their resonating response to recordings of tuba multi-phonics – with height. "The metal sheets are transformed into vibrating resonance walls, therefore dissolving the boundaries of the narrow passage space, contracting and expanding it." Leitner feeds on the ability of sound to create "aural architecture" [3] often defying visually perceived boundaries and thresholds and generating what he calls "secondary space" [20] and its "continuously changeable" ability.

In Simon Waters's *Line* of 2018, the listener encounters a line of small and delicate sounds facing them, which takes advantage of our perceptual acuity for frontal localization. By employing spectral differentiation in the mid-high-frequency content, Waters establishes an intimate and discrete sound environment that inhabits space through plausibility. In addition to the crafting of sound sources that play off both the 12-speaker linear system and the acoustics of the performance space, the work crucially deals with scale, or to be more precise, the non-scalability of sound. Much talked by sound recordist and artist Chris Watson [21], the notion of scale refers to the need to reproduce sound sources at their inherent "natural" amplitude. Particularly relevant to field recordings, an alignment between naturally occurring sources and their reproduction through recording and playback imbues the listening experience with a sense of realism and shared listening space by reassuring the listener of the dynamic range that is plausible and believable. The apparent simplicity of a work like *Line* is deceiving and creates immersion through the notion of the listener inhabiting the same space as the sound sources as discussed above, situating itself well away from the spatial acrobatics of much "surround sound" rhetoric and indeed much of the electroacoustic use of diffusion.

Arguably, all of the examples discussed above represent a minute fraction of practices in sound and space that fall under Blesser and Salter's *Aural Architecture*. They are deliberate manifestations of the design of spatial and sonic experiences that are in many ways ubiquitous in everyday life – "[a]ural architecture exists regardless of how the acoustic attributes of a space came into existence: naturally, incidentally, unwittingly or intentionally" [3].

10.6 STRATEGIES FOR SOUND AND SPACE

The works addressed above all employ strategies that aim to highlight the spatial perception of sound phenomena. Each work admittedly treats space in a unique and distinct way, in particular, related to location and scope of sound sources, architectural acoustics, and listening position. Although each work employs multiple strategies and is primarily driven by a given musical or artistic discourse, it is possible to identify patterns in how sound and space strategies have developed. I will now summarize these strategies into a framework – from the point of view of sound and space design that can be applied to the creation and production of immersive experiences in a broad sense, emphasizing the shortcomings of treating sound and space in an analogy to an understanding of a 360° visual field.

10.6.1 Spectrum-space continuum

This continuum is rooted in an alignment between spectral content and sound projection. On the one hand, a field of cicadas with thousands of sound sources of closely related spectral content leads to a spatial experience of expansive scatter, determined not only by the sources and their distribution in space, but critically by micro variations in resonance and propagation as determined by the landscape and the relative position of a listener. In the other extreme, a situation in which spectrally different sources are positioned in space can lead to a kind of spatial counterpoint that highlights relational position. Whereas Stockhausen's *Gruppen* brass chord utilizes spectral similarity to convey a sense of movement for a predetermined listening position (central), Xenakis' *Terratektor* explores the spectrum-space continuum by dispersing both listeners and sound sources (performers) to achieve a sense of spatial counterpoint with each listener as a fixed point in ever evolving sonic trajectories.

Normandeau's notion of timbre spatialization [22] reflects this same continuum in relation to the electroacoustic distribution of timbre across points in space so that the entire spectrum of sound is recombined in the listening space itself. Not unlike Gabrielli's approach of splitting bass parts from the rest of the choir, techniques such as "spectral splitting" distribute different frequency regions in space and rely on the acoustics and projection systems to fuse the timbral content of a given signal. These strategies can provide a sense of spatial richness since any minute movement of the listener translates into frequency-domain variations and a sense of being "inside the sound," although needless to say, the relationship between sonic content, the splitting of frequency bands, their placement in space, sound projection, and acoustics is critical for effective results. Barrett discusses such concepts further in Chapter 9.

10.6.2 Sound-scale continuum

The scalability, or rather non-scalability, of sound is highlighted by Waters's *Line* in relation to the plausibility that is perceived when sound sources are reproduced at their own inherent scale. While this relates to the amplitude and distance relation between the source and listener, other considerations include the size and propagation profiles. The sound of an electric plant can be perceived at the same amplitude as the sound of white noise on a small battery-powered radio – depending on the listener position. The sense of scale of the source is however maintained and perceived by the listener. When this relationship is inverted (e.g. playing "small," low volume sounds in a large and loud sound field) the notion of plausibility is compromised. The way in which a given sound source imparts its own scale is also at the core of Tudor's *Rainforest*. It is precisely our ability to understand the sonic potential of each object (and the trust projected by the listener through the sense that an object wouldn't do anything other than what is expected by its very physical and acoustic profile) that allows for an immersive exploration of the work. The sound-scale continuum is also present in electroacoustic diffusion practices in which a stereo or multichannel fixed-media work is projected in real time to a loudspeaker array. Sound diffusion practitioners normally aim to enhance the sonic space and scale embedded in the fixed-media tracks in order to adapt to the concert-hall listening situation as Harrison highlights in the context of dynamic range: "In performance I would, at the very least, advocate enhancing these dynamic strata – making the loud material louder and the quiet material quieter – and thus stretching out the dynamic range to be something nearer what the ear expects in a concert situation" [23].

The notion of the acoustic threshold plays an important role when it comes to scale. The relationship between the scale of a given sound, its movement profile and thresholds of audibility is key, especially for sound sources with inherent movement, for example, a fly or a passing car. The emergence and disappearance of these sources from the listening field is a consequence of both amplitude changes and frequency-domain changes, caused by directionality and acoustics.

10.6.3 Activated-space continuum

This continuum relates to the projection of sound sources, and deliberately activates space acoustically, thus emphasizing the connection between the sound field and the listening space. One only has to observe the architectural shape of the Philips Pavilion to understand that the over 400 loudspeakers are not there to provide uniform or idealized sound distribution, but to activate the space and the architecture as each speaker will have a different resonant response based on its position. Arguably present in any listening situation involving physical acoustics, the notion of activating space can be easily overlooked in virtual environments and often completely forgotten in the rhetoric of generalized uniform 360° sound fields. The virtual creation of a uniform sound sphere with fixed or moving sound sources removes the rich spatial cues provided by the complexities of refraction and reflection embedded in architectural (physical or virtual) ways of thinking about sound and space. Another manifestation of activated space normally left out of discussions around spatial audio refers to a constant source navigating through spaces with different acoustic properties. A radio show in mono in which the presenter talks while moving through different rooms in a house is a spatial listening experience. Although no positional information (other than distance) is captured by the mono recording, with the exception of possible directionality of the voice in relation to microphone position, the listener gets a clear sense of size, shape, and materials associated with each space and perhaps even more noticeably, the physical thresholds between spaces as articulated by doors or windows. Within this continuum, one can also consider the spatialization of microsound [24] where short-duration sound particles are distributed in space governed by specific trajectories, shapes or biologically inspired algorithms [25], [26].[2] These microsound particles function as distributed bursts of sound energy resonating a given space. Alvin Lucier's *I am Sitting in a Room* remains the archetypal example of a work whose process manifests itself in the gradual revealing of the resonances of a space by playing back Lucier's voice into a room repeatedly till the emergence of an "aural palimpsest" [27]. The work moves from a focus on Lucier's voice to the space itself speaking back.

10.6.4 Listening-space continuum

Listening situations with freely moving audiences in which sonic perspectives are constantly shifting are not new, as we've seen from the examples above.

Current virtual, augmented, and mixed reality technologies simply allow for different implementations of sound fields that are interactive, responsive, or adaptive to a listener's presence. The experiences provided by the works by Tudor, Cardiff, and Rebelo discussed above are all designed to convey agency to the listener and create sound worlds that are coherent in themselves, but afford multiple perspectives enacted by the attention and position of the listener. In these works, the listener becomes a "performer" of space [28]. The "hybrid listening" framework developed at the Sonic Arts Research Centre [29] explores the relationship between position and orientation of a listener with open-back headphones and a sound field projected by loudspeakers. The implementation of tracked (position and orientation) open-back headphones allows for the exploration of a dynamic sound field in the headphones while rooting the listener in a plausible sound field projected by the loudspeakers, which themselves activate the space of the installation and therefore provide further spatial cues. As with all experiences that depend on any type of response or interaction from a listener, a critical design concern relates to the *why* of the experience. What is it that entices a listener to actually explore a space with the sense of wonder and discovery one might have during a walk in the countryside or wandering around a city, as so eloquently discussed by de Certeau in his text *Walking the City* [30]. Christina Kubisch's *Electrical Walks* from 2004 invites the listener to explore the city through its electromagnetic signals. The city becomes a sonic space for listening agency and exploration – changing our perceptions of everyday space. The design of these performed spaces needs to attend to spatial logics (placement of sonic events around space taking into account the movement of the listener), thus inviting an exploration of both temporal and spatial narratives and structures.

10.7 CONCLUSION

This chapter attempts to distil numerous approaches to the relationship between sound and space into a framework that I hope is of use for sound designers and aural architects. In contributing to current thinking around immersive sound, the chapter moves away from generalized and uniform approaches for positioning sound sources in virtual space to delineating strategies that have been explored across different creative practices for almost 500 years. It is clear from analyzing these practices that notions of immersion and presence are at play, not because of the use of a specific technology (which this chapter has consciously avoided) or deliberate application of an understanding of spatial auditory perception, but rather through sonic and conceptual frameworks determined by context. This context relates both to the semantics of a given work and to an understanding of sound as a multimodal phenomenon in which interactions between our senses contribute to our own listening attention and capacity. Furthermore, it is a consideration of all aspects of the sound experience that frames the notion of a performed space (physical or virtual) in which one gains agency to actively listen; what Pauline Oliveros called "deep listening" [31].

Of course, the theme of this book is 3D audio, yet this chapter has not always explicitly dwelt upon that concept, preferring to generalize around spatial sound. It is important to understand that this is not an omission, but rather a conscious attempt to illustrate that contemporary 3D sound design has a richer palette of heritage from which to draw than many current practitioners realize. Concepts that might have been originally implemented in horizontal surround can still be deployed periphonically, and the legacy of spatial placement through physical positioning or acousmatic approaches are just as relevant for those implementing binaural synthesis in a game engine or middleware.

This chapter delineates between different strategies for designing a sound-space relationship through four different continua relating to spectrum, scale, activation, and listening. It falls to readers to absorb the concepts and decide how best to apply them in their own unique context. These continua are porous models – ideas that come into play through the notion of liminality. It is often the design of *modes* of listening in the sonic experience that precipitate a higher degree of engaged listening and maximize the plausibility of a sound world.

NOTES

1   The spherical concert hall was one of the main inspirations for the design of the Sonic Arts Research Centre's Sonic Lab, a three-storey rectangular space with an acoustically transparent grid with four vertical layers of loudspeaker arrays.
2   Many of these strategies have been implemented in the commercial software – Sound Particles.

REFERENCES

[1]   D. Smalley, "Space-form and the acousmatic image," *Organised Sound*, vol. 12, no. 1, pp. 35–58, Apr. 2007.
[2]   S. James and C. Hope, "2D and 3D timbral spatialisation: spatial motion, immersiveness, and notions of space," *ECU Publ. 2013*, Jan. 2013. https://ro.ecu.edu.au/ecuworks2013/309.
[3]   B. Blesser and L.-R. Salter, *Spaces Speak, Are You Listening? Experiencing Aural Architecture*. The MIT Press, 2006.
[4]   H. Bagenal, "Bach's music and church acoustics," *Music & Letters*, vol. 11, no. 2, pp. 146–155, 1930.
[5]   R. G. Jahn, P. Devereux, and M. Ibison, "Acoustical resonances of assorted ancient structures," *The Journal of the Acoustical Society of America*, vol. 99, no. 2, pp. 649–658, Feb. 1996.
[6]   B. Boren, "History of 3D sound," in *Immersive Sound: The Art and Science of Binaural and Multi-Channel Audio*, 1st edition. A. Roginska and P. Geluso, Eds. Focal Press, New York; London, 2017.

[7]   D. Garrioch, "Sounds of the city: the soundscape of early modern European towns," *Urban History*, vol. 30, no. 1, pp. 5–25, May 2003.

[8]   H. Lynch and R. Sazdov, "A perceptual investigation into spatialization techniques used in multichannel electroacoustic music for envelopment and engulfment," *Computer Music Journal*, vol. 41, no. 1, pp. 13–33, Mar. 2017.

[9]   S. Agrawal, A. Simon, S. Bech, K. Bæntsen, and S. Forchhammer, "Defining immersion: literature review and implications for research on audiovisual experiences," *Journal of the Audio Engineering Society*, vol. 68, no. 6, pp. 404–417, 2020.

[10] F. Rumsey, *Spatial Audio*, 1st edition. Focal Press, 2001.

[11] A. Roginska and P. Geluso, Eds., *Immersive Sound: The Art and Science of Binaural and Multi-Channel Audio*, 1st edition. Focal Press, New York; London, 2017.

[12] W. J. Cavanaugh and J. A. Wilkes *Architectural Acoustics: Principles and Practice*. Wiley, 1999.

[13] D. Bryant, "The 'cori spezzati' of St Mark's: Myth and Reality," *Early Music History*, vol. 1, pp. 165–186, 1981.

[14] J. Vaz, "The six organs in the Basilica of Mafra: history, restoration and repertoire," *Organ Yearbook*, vol. 44, pp. 83–101, 2015.

[15] I. Xenakis, *Formalized Music*. Indiana University Press, 1971.

[16] I. Xenakis, R. Brown, and J. Rahn, "Xenakis on Xenakis," *Perspectives of New Music*, vol. 25, no. 1/2, pp. 16–63, 1987.

[17] M. Fowler, "The ephemeral architecture of Stockhausen's Pole FüR 2," *Organised Sound*, vol. 15, no. 3, pp. 185–197, Dec. 2010.

[18] D. Tudor, "David Tudor: Rainforest," *David Tudor*, 2001. https://davidtudor.org/Works/rainforest.html (accessed Aug. 03, 2020).

[19] L. Qvortrup, *Virtual Space: Spatiality in Virtual Inhabited 3D Worlds*. Springer Science & Business Media, 2011.

[20] B. Leitner, "Sound – architecture: space created through travelling sound," *Artforum*, Mar. 1971.

[21] C. Watson, "Chris Watson," 2020. https://chriswatson.net/ (accessed Aug. 03, 2020).

[22] R. Normandeau, "Timbre spatialisation: the medium is the space," *Organised Sound*, vol. 14, no. 3, pp. 277–285, 2009.

[23] J. Harrison, "Sound, space, sculpture: some thoughts on the 'what', 'how' and 'why' of sound diffusion," *Organised Sound*, vol. 3, no. 2, pp. 117–127, Aug. 1998.

[24] C. Roads, *Microsound*, New Ed edition. The MIT Press, 2004.

[25] B. Manaris *et al.*, "Zipf's law, music classification, and aesthetics," *Computer Music Journal*, vol. 29, no. 1, pp. 55–69, 2005.

[26] D. Kim-Boyle, "Sound spatialization with particle systems," in *Proceedings of the 8th International Conference of Digital Audio Effects, Madrid, Spain*, 2005, pp. 65–68.

[27] C. Migone, *Sonic Somatic: Performances of the Unsound Body*. Errant Bodies Press, 2012.

[28] "(PDF) Performing space | Pedro Rebelo – Academia.edu." www.academia.edu/1413789/Performing_space (accessed Dec. 14, 2019).

[29] P. Rebelo, "Spaces in between – towards ambiguity in immersive audio experiences," presented at the International Computer Music Conference, New York City, USA, 2019. www.academia.edu/41019446/Spaces_in_Between_-Towards_Ambiguity_in_Immersive_Audio_Experiences (accessed Dec. 13, 2019).

[30] M. de Certeau, *The Practice of Everyday Life*. University of California Press, 2011.

[31] P. Oliveros, *Deep Listening: A Composer's Sound Practice*. iUniverse, 2005.

11

Hearing history: a virtual perspective on music performance

Kenneth B. McAlpine, James Cook, and Rod Selfridge

11.1 INTRODUCTION

Sound, or, to be more specific, the perception and decoding of the spatial characteristics of sound, is one of those things that is fundamental to how we make sense of our position within and relationship to the world around us. We use it for localization and positioning (see, e.g [1]); we use it to judge size and distance (see [2], [3], and [4]), and we use it to get a sense of the solidity and tactility of the things we see around us (see [5], [6], and [7]). Collectively, these component psychoacoustic aspects of sound, and understanding and modelling of the specific psychophysical phenomena that give rise to them, is known as auralization [8].

Together with music, soundscapes and the auralization of the virtual spaces that they inhabit are key – though often neglected – points of interface in video game and virtual reality (VR) experiences. For example, at a conference at the Victoria and Albert Museum in London, which focused on using game concepts, mechanics, and technologies to design new VR experiences and augmented realities, Laura Dilloway, one of the team who created Guerilla Games's *Rigs* [9], noted that the developers had to direct so much of their time and effort to creating a control system that did not actually induce motion sickness in the player, that the sound was little more than an add-on tacked on at the end [10].

This is undoubtedly a missed opportunity. As Collins [11] discusses, the ludic functions of sound are a distinctive feature of video games and – by extension – VR, and they offer enormous potential for the ludic and performative exploration, not just of music, but of acoustic space as a dimension of performance, something that is almost impossible to achieve by any other means given the physical constraints of reconfiguring performance venues, and the social conventions of audience behaviour at concerts.

Looking beyond VR-as-entertainment raises interesting questions about the role of immersive music experiences in other domains, and particularly within a heritage education context. The notion of making playful exploration an end in itself, taps into a very basic human instinct. Play is how we all, as children, begin to make sense of the world around us and our position in it [12], and specific examples of exploring our relationship with music through interactive play can be found throughout computing history (see [13], [14]).

Moreover, the multimodal immersive technologies that comprise VR provide opportunities to embed playful exploration with sound in an artificial, though very convincing, acoustic environment that offers scope to explore the relationship between space and sound in a very systematic way.

That is the principal aim of this chapter. By examining a case-study example, which uses dynamic virtual-acoustic modelling to recreate the sound of two historic auditoria, the chapter will explore how 3D sound and music might be used as the focal points of an interface for an immersive VR experience, and how the technology of VR might be co-opted and used as a platform for heritage organizations to curate the ephemeral elements of music performance that might otherwise be lost, so as to engage visitors experientially not just with a historic performance space, but with the cultural activities that the space represents.

11.2 EXPLORING MUSIC PERFORMANCE THROUGH PLAY

Participation in music is fundamentally linked with the notion of play [15], whether in the ultimate performance of a piece or in the way in which musicians explore the musical territory mapped out by dots on staves and their relationship with these as they learn new pieces and begin to shape what their ultimate realization might be.

This idea connects music directly with video gaming, which can likewise be a playful activity that is mentally and physically stimulating, and that encourages the development of fine motor skills; the agility with which gamers link together complex sequences of keypad combinations to trigger a special move, for example, can be every bit as impressive as the fingerwork required to play a Bach Invention.

Video games such as *Guitar Hero* [16] and *Dance Dance Revolution* [17] have become global phenomena by building on that link, connecting the innate human desire to make and respond to music with that for play, using custom gaming interfaces, guitar-shaped controllers or dance mats, to develop a combination of spatial awareness and a high degree of rhythmic precision in players. While the technical skills developed while "shredding" in *Guitar Hero* might not translate directly to more traditional musical interfaces, that conceptual sense of rhythm and performance most certainly do.

Indeed, it was precisely this relationship that was the primary motivation behind a collaborative project between members of the project team and the National Trust in London, which sought to offer visitors to the Trust new ways to engage with the Trust's musical instrument collections by building an accessible virtualized visitor attraction based on the Benton Fletcher Collection of historic musical instruments [18].

Using a mixed-methods approach, combining interaction design, high-definition sound sampling, and gamification, a series of virtualized software models of the instruments were created and installed in a bespoke hardware interface – a two-manual keyboard using modified MIDI keybeds (the dynamic action of the keyboard), running an embedded software system, and with an integrated amplifier and speaker system that used the natural body resonance of the casing to provide a form of natural amplification. The system was installed in situ at Fenton House in Hampstead, the home of the Benton Fletcher Collection, and this was used as the basis for end-user research.

The research suggested that there was real value in providing tactile, experiential access to historical music resources, and although there were limitations around the tactile response of the keyboard interface, a consequence of there being no commercial keybeds that effectively mimic the feel of a historic keyboard, the public expressed a strong appetite for tactile play with the collection so as to better understand the relationship between the physical form of the instruments and their voice (see [13], [18]). Indeed, in use, the installation increased the Trust's key performance indicator for visitor engagement, the number of visitor hours heard, more than tenfold, and the instrument itself featured as a case study in a subsequent review of the role that immersive virtualized technologies might play in heritage strategy [19].

This project established two key features. First, it demonstrated the principle that virtualization and interactive play were effective modalities to provide meaningful and authentic – as judged by visitors to the collection – access to heritage objects, and second, it established a framework for the analysis and technical production of the sound qualities and sonic fidelity of those models, with a focus on sound sources – the characteristic qualities of the musical instruments.

Musical performance, however, has always been as much about people and places and experiences as it is about sound, as important as that is. Much of the experiential aspect of a performance consists of the interpersonal relationships between performers and the shared experience of their audience. This is arguably thrown into sharper relief in the case of historical performances, which bring with them their own rituals quite apart from the post-19th-century concert-hall rituals we experience today. Whether this is the boisterous social occasion that was the 18th-century concert event [20], or the equally social but intensely ritualized 15th-century church service, social interaction is key to live performance, and it was one aspect that the Benton Fletcher project did not explicitly address.

Space and place too, however, are important aspects of performance.

The acoustics of a space are used by performers for tuning and intonation (see, e.g. [21]), and play an important role in animating a performance and in shaping the resultant sound. The vocal polyphony of the High Middle Ages was designed for the particular acoustics of the church space, something which is replicated by few modern concert halls.

Indeed, 15th-century liturgical music took place as a multimedia spectacle [22]; as the audience moved through the sacred space unencumbered by pews, they would have been able to explore a richly decorated world of wall paintings and sculptures: decorative, instructive, and devotional.

As many contemporary accounts attest, these sculptures, lit by the play of flickering candlelight, would be transformed into living stone: the very saints similarly enlivened in their polyphonic decoration seeming to become animate. Sight and sound here work reciprocally to full effect, and even the faculty of scent would have been brought in to play by censing with a thurible.

Considered as such, early music performance was a fully immersive, perhaps even overwhelming, multimodal sensory experience. Audiences – if, indeed, audiences is the correct term in this very different listening context – were not bound by the social conventions of music performance that we adhere to today, and would have had much more agency to move around the space and experience the shifting qualities of sound in a changing acoustic.

How then can modern performance seek to recapture these aspects? Certainly, it is possible live, although even here we can never recapture the absolute "authenticity" of a performance since, even if the space, place, performers, and music are entirely accurate, the audience members are still aware that they are watching a re-enactment. It is something that is far harder to achieve in recorded performance; while historically informed performance practice may capture a sense of the sound of the past on (say) CD, it will necessarily stop short of giving a sense of presence, participation, multimodal immersion, and the broader aspects of space and place.

This is something that VR can address, at least in part. By drawing on the abilities of VR to embody music performance in a virtual environment that encourages a sense of presence and agency in the listener, we contend that it is possible to bring together the experiential aspects of live performance with the reproducibility and accessibility of a recording. In the same way that the architectural and acoustic spaces of historic performances were mechanical structures for creating immersive sensory experiences, modern VR technology provides a digital mechanism to do likewise.

11.3 SPACE, PLACE, SOUND, AND MEMORY

This is the idea that underpins a new project titled "Space, Place, Sound, and Memory," which focuses on the replication of the acoustics of heritage performance spaces, and thus the authentic timbre of the music that might have been originally performed within them. Two historic venues were selected: St. Cecilia's Hall, which now forms part of the estate of the University of Edinburgh and houses the Russell Collection of Historic Keyboard Instruments, and the Chapel at Linlithgow Palace, a site that is of international archaeological interest, and is managed and maintained by Historic Environment Scotland (HES). There was a clear contrast between a pristine, contemporary working space that was built to closely replicate the original layout of a historic venue and an architectural ruin.

These sites were selected in part because of their historical significance – St. Cecilia's Hall was designed to be Scotland's first purpose-built concert hall [23], while we know from surviving records that King James IV had "chapele geir" and "organis" in the royal chapel at Linlithgow [24] – and in part because they represent spaces for which there exist detailed historical records and archaeological information about the construction and nature of the spaces as they were in the past, providing a mechanism by which we might be able to digitally reconstruct the historic environments from accurate models of the spaces as they currently appear: a modern physical reconstruction of the historic space in the case of St. Cecilia's Hall, and little more than a stone shell in the case of the Chapel at Linlithgow Palace (see Figure 11.1).

Originally built in 1763 and named after the patron saint of music, St. Cecilia's Hall is the oldest purpose-built concert hall in Scotland, and one of the oldest in Europe. At its heart is its concert room, which hosted regular concert meetings for the Edinburgh Music Society until 1798. Following the disbandment of the Society in 1801, the hall was sold to a Baptist congregation, and was later used as a Freemasons' lodge, a warehouse, a school, and a ballroom. As its function changed, the interior was drastically remodelled, and when the building was purchased by the University of Edinburgh in 1959, the elliptical concert room bore little resemblance to its original appearance [23].

In 2016, St. Cecilia's Hall underwent a £6.5m restoration and renovation in order to improve the concert hall and return it to its original condition, or as close to its original condition as contemporary public-access building regulations allow (see Figure 11.2).

The spatialization was recreated using current VR 3D audio plugins, Steam Audio [25] and Google Resonance [26]. The primary benefit of real-time VR plugins is that they integrate seamlessly into the VR development

Figure 11.1 The chapel at Linlithgow Palace is currently a stone shell. There is no roof and no interior fixtures or fittings survive.

Figure 11.2 The £6.5m reconstruction of St. Cecilia's Hall.

pipeline via a game engine and allow users to be immersed in the scene, to move around, and to experience the music in an interactive way.

However, in order to achieve a high degree of interactivity, these packages have to perform acoustic calculations in real time, and have to employ computational shortcuts in order to achieve this. This type of auralization is more typically performed by calculating a static impulse response (IR), either measured within the space or based upon a model within a software package. This can be accurate for a fixed source and fixed listener, but it requires re-calculation when either moves relative to the other, and so is not suitable for interactive VR. Six-degree-of-freedom movement is discussed in detail by Llewellyn and Paterson in Chapter 3.

This immediately raises the following question: can real-time acoustic modelling provide a sufficient level of accuracy to create an immersive experience that is suitable for use in a heritage context?

11.4 CAPTURING AND EVALUATING ACOUSTIC SPACES

The initial stage of the project involved a site survey to create an auditory point of reference for the physical acoustics of the spaces. Using a binaural microphone and the sine sweep method (see [27]), detailed binaural room IRs (BRIRs) were recorded at both sites and from a variety of sound source and listener locations. The measured BRIRs were obtained using the MikIRAM MATLAB® App for recording and analysing IRs [28]. A 15-second sine sweep was output from a Yamaha 600i PA loudspeaker and recorded with a Soundman Dummy Head using DPA 4060 Stereo microphones.

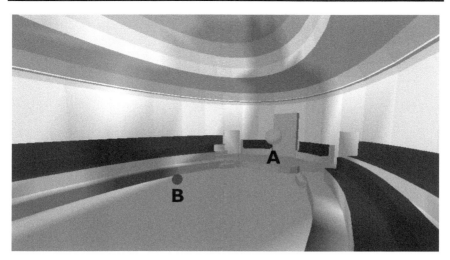

Figure 11.3 The positions of the sound source, marked as A in the image, and the listener, marked as B, in St. Cecilia's Hall when recording the BRIR. These positions represent the physical locations of the sound source (Yamaha 600i speaker) and listener (Soundman binaural microphone) in the hall, and these were subsequently translated to the 3D models to provide an equivalent BRIR.

The intention here was to provide a technical record of the room response to serve as a direct point of comparison with the modelled acoustics, both quantitatively, by carrying out systematic statistical analyses of recorded audio signals, and qualitatively, by carrying out perceptual listening tests. In the case of Linlithgow Palace, the lack of roof and any interior fixtures or fittings in the physical space precluded any useful recordings being made; however, St. Cecilia's Hall provided a very interesting point of reference.

For the purposes of recording, the binaural microphone was placed in the middle of the hall, and the sound source was placed on the stage area at a height of 1.96 m from floor level, as shown in Figure 11.3. This position was subsequently used in all the virtual reconstructions.

In addition to this sampling of the acoustic spaces, we employed a second approach to auralization – modelling the sound using acoustic ray-tracing – to provide a further analytical basis for our work. We then had a measurable control case, the sampled acoustic of St. Cecilia's Hall, against which a modelled acoustic of the same space could be compared and judged. In principle, this provided some procedural validation for the notion that we might then remodel the spaces as they were according to architectural records and have some confidence that the modelled spaces provided an accurate representation of their acoustics. This aligns with Mills's suggestion of new approaches to the study of archaeological spaces that consider "sound as a medium of past social interaction" so that the acoustic properties of historic spaces should "be acknowledged and fully integrated into strategies for their management and preservation" [29, pp. 72–74].

In terms of acoustic modelling, our primary intention was to investigate the sonic fidelity, that is, how accurately different aspects of the acoustics

Figure 11.4 Scanning the Chapel at Linlithgow Palace created a detailed point cloud, a dense cluster of spatially indexed data points that captures digitally, and with phenomenal accuracy, the dimensions and layouts of the physical site.

were modelled. For comparison, we selected one off-line package, Odeon Auditorium [30], which uses prediction algorithms (image-source method and ray-tracing) to calculate fixed binaural and multichannel room IRs, and two real-time packages, Steam Audio and Google Resonance.

In order to create these acoustic models, we started by creating accurate digital *architectural* models of both St. Cecilia's Hall and Linlithgow Palace, as they currently appear. Working with key partners, NC-Tech and HES, both experts in the field of visual scanning of complex physical locations using light detection and ranging (LIDAR; a type of laser telemetry that is used for accurate distance measurement and surveying of physical spaces), we created detailed closely architectural scans of the two sites to create two very high-resolution point clouds that represented the spaces as they were at that time (see Figure 11.4).

Although incredibly accurate, a point cloud is unsuitable for use in a VR engine. While point clouds can be directly rendered and inspected (see [31], [32]), most VR middleware and acoustic ray-tracing software is designed to efficiently manipulate and display low-polygon mesh or triangle mesh models. Therefore, the next step in our process was to take the point clouds and convert them into low-polygon textured models to be imported into VR middleware for use as the basis for both offline and real-time modelling and auralization (see Figure 11.5).

By converting the point cloud to an accurate low-polygon model, we were able to create a virtual architectural framework that allowed us to strip back the modern elements of construction – or, in the case of Linlithgow Palace, those elements of its deterioration – and using detailed architectural records, recreate the original performance spaces *in virtuo*.

In both cases, it was necessary to modify the point-cloud-derived models to digitally reconstruct both the structural elements and interiors of the historic buildings, a process that was carried out with reference to original

Figure 11.5 Low-polygon model of St. Cecilia's Hall derived from a high-resolution point-cloud LIDAR scan.

Figure 11.6 Virtual reconstruction of the interior of the Chapel at Linlithgow Palace in panoramic view. The layout and the qualities of the materials were selected in consultation with architectural historians from HES.

architectural drawings in the case of St. Cecilia's Hall, and in the case of Linlithgow Palace, in consultation with HES's architectural historians, who were able to advise on the design and material qualities of the interior fixtures and fittings, including the virtual plasterwork, wood panelling, and drapes and hangings (see Figure 11.6), all of which have a material impact on the quality of the acoustic and the liveness of the space.

The BRIRs were then simulated in three different ways. The acoustic simulation software Odeon Auditorium was used to provide an architectural-standard acoustic model to be used as a direct comparison to the real-time VR acoustic modelling plugins, Steam Audio and Google Resonance.

Odeon states that between 100 and 2,000 surfaces are normally sufficient to model a space, and recommends a simplified boxed area to represent the audience and seating. Initially, simplified models of St. Cecilia's Hall from the present day and from 1769, to represent its historical heyday, were imported to Odeon for simulation. These models were then imported into the game engine Unity to use as a platform for evaluating the two VR plugins.

An initial evaluation of the 1769 model suggested that the simplified audience area recommended by Odeon was unsuitable for Google Resonance. The simplified area behaved like a sound box in Google Resonance and increased the reverb time to the extent that the acoustic sounded unnaturally large and the intelligibility of sound and music in the space was very badly affected. The boxed area for the audience was removed and a more detailed model was substituted. This had the effect of increasing render times in Odeon, but it did not significantly alter its BRIR output.

Once completed, the present-day 3D model of the hall had 1,165 surfaces and the historic version 7,730. Tables 11.1 and 11.2 show the acoustic properties used for the simulations for Odeon.

This was replicated as closely as possible in Steam (Tables 11.3 and 11.4), where the absorption coefficients for 500 Hz in Odeon were used for 400 Hz in Steam. The values for 2 kHz in Odeon were used for 2.5 kHz in Steam, and the values for 8 kHz used for the values for 15 kHz in Steam.

Google Resonance does not give the user the ability to customize the settings, instead it offers a choice from 23 preset options. These were

Table 11.1 Absorption and scatter coefficients for St. Cecilia's Hall modelled in Odeon (present-day model).

Surface	63 Hz	125 Hz	250 Hz	500 Hz	1 KHz	2 KHz	4 KHz	8 KHz	Scatter
Floor	0.2	0.2	0.15	0.1	0.1	0.05	0.1	0.1	0.05
Walls	0.14	0.14	0.1	0.06	0.04	0.04	0.03	0.03	0.1
Roof	0.14	0.14	0.1	0.06	0.04	0.04	0.03	0.03	0.1
Cupola	0.18	0.18	0.06	0.04	0.03	0.02	0.02	0.02	0.05
Seat platforms	0.4	0.4	0.5	0.58	0.61	0.58	0.5	0.5	0.7
Doors	0.14	0.14	0.1	0.06	0.08	0.1	0.1	0.1	0.2
Organ	0.28	0.28	0.22	0.17	0.09	0.1	0.11	0.11	0.6
Orchestra rail	0.19	0.19	0.14	0.09	0.06	0.06	0.05	0.05	0.7
Pilasters	0.19	0.19	0.14	0.09	0.06	0.06	0.05	0.05	0.8
Rug	0.02	0.02	0.06	0.14	0.37	0.6	0.65	0.65	0.1

Table 11.6 Absorption and scatter coefficients for St. Cecilia's Hall modelled in Resonance (historical model).

Method	Period	T30 (s)	C80 (dB)	EDT (s)
Measured	Present-day	0.6816	6.2655	0.7382
Odeon	Present-day	0.8921	-12.8400	0.9619
	Historic	0.6730	1.5863	0.7503
Steam	Present-day	0.5516	0.4274	0.7763
	Historic	0.5125	1.3949	0.7551
Resonance	Present-day	1.3598	6.5814	0.1045
	Historic	1.7390	-2.6986	0.9275

Table 11.7 Statistical analysis of the room impulse responses.

Method	Period	T30 (s)	C80 (dB)	EDT (s)
Measured	Present-day	0.6816	6.2655	0.7382
Odeon	Present-day	0.8921	-12.8400	0.9619
	Historic	0.6730	1.5863	0.7503
Steam	Present-day	0.5516	0.4274	0.7763
	Historic	0.5125	1.3949	0.7551
Resonance	Present-day	1.3598	6.5814	0.1045
	Historic	1.7390	-2.6986	0.9275

The results of the statistical calculations for all the IRs is given in Table 11.7, averaged between the left and right binaural channels.

Examining the T30 values we can see that there is a difference between the present-day virtual model in Odeon and the measured response in the physical hall of 200 ms. This is significant and could be due to an error in the measurement due to noise or discrepancies in the material choices used in Odeon. The T30 value in Odeon clearly decreases from the present-day model to the historic model. This is expected due to the different seating, pilasterwork around the walls, and use of wall hangings to dampen the room response. Although the T30 values obtained from Steam are shorter than both the measured values and those simulated in Odeon, they do show a slight decrease in T30 from the present to the past. By contrast, in Google Resonance, the T30 times obtained increase from the present to the past. It is believed this is due to the lack of flexibility Google Resonance offers when assigning acoustic properties to materials.

The clarity values are given in decibels and a positive value is obtained from the measured IR. Positive values imply that speech might be clearer compared to negative values, although negative may be preferred for some types of music performance. The values obtained from Odeon for C80 show a large change from higher energy in the late sound in the present day to higher energy in the early sound in the historical model. In Steam, the C80 values indicate higher energy in the early sound for both the present day and historic models, whereas Google Resonance again gives opposite results from Odeon, with the early sound containing higher energy in the present-day model and the late sound in the historic model. The C80 values are compared based on the just-noticeable difference (JND). In [35], the value of JND for C80 values is 1 dB. We can see from Table 11.7 that there is a large JND (14.4) between Odeon for the present day and historic models. This is much smaller for Steam, with approximately 1 JND. Google Resonance has a relatively high value of 9.3 JND.

The EDT values show a similar trend to the T30 results, with Odeon's present-day model giving a higher value than that measured. The JND is given as 5% in [35]. Therefore, between the present-day model and historical model in Odeon, the EDT has a JND value of 4.4. In Steam, this decreases to 0.5 JND, but in Google Resonance, it is a substantial 157.5 JND.

All of this suggests that none of the acoustic modelling packages truly capture the detailed acoustic qualities of the contemporary space, although interestingly, Steam gets closest in terms of the T30 and EDT measurements, and behaves as predicted when moving from the present day model to the past, and so may represent the best overall method for simulating a 3D acoustic.

11.6 QUALITATIVE ANALYSIS OF THE MODELS

A listening test was used to simply indicate whether participants could hear the difference between the different auralization methods used to model St. Cecilia's Hall. This venue was selected because we had a recording of the physical space that we could use as a reference.

The listening test was hosted using the Web Audio Evaluation Toolkit [37] and took the form of an A-B listening test where participants were invited to listen to a reference audio clip then select the identical one from two choices presented. In total, there were 25 participants with ages ranging between 17 and 62 with a median of 31. There were 15 males and 10 females, all of normal hearing. Six had experience in evaluating room acoustics. Results are listed associated with nominal dates – "then and now" (1769 and 2018).

The reference audio clips and test clips were taken from music material that formed part of the wider project as discussed in the next section. Each rendering method was compared against each other, as well as the same method over the two different models (present-day and historic), with the null hypothesis being that listeners would be able to perceive no difference in the acoustic qualities of any of the rendering methods used. Each test was carried out twice, with a different piece of music for the second trial.

The results of the listening test are presented in Figure 11.7. In the majority of the results, a high percentage of participants were able to tell the difference between each auralization method, with the greatest challenge being in distinguishing between the present and past using the same method. The

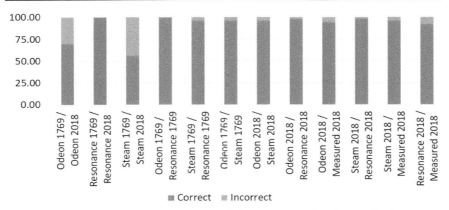

Figure 11.7 Participants' ability to identify between auralization methods.

option participants found that the most difficult to distinguish was between Steam in 2018 and Steam in 1769. This is understandable by examining the quantitative results in Table 11.7. Here, the statistics for both these options are relatively close. There is a greater difference between the past and present in Odeon, and even more with Google Resonance.

11.7 PROGRAMMING AND REPERTOIRE

Having constructed two virtual auditoria, our intention was to build a VR listening experience around them, using the virtual acoustic spaces as platforms for historically informed performance. This required careful consideration of three aspects: programming, in terms of the selection of repertoire, performance, and recording and production.

Programming for St. Cecilia's Hall was relatively straightforward. Edinburgh Music Society kept very detailed programme notes for all of the concerts that it held, and, with the help of Dr Jenny Nex, Curator of Musical Instruments Collections at Edinburgh University, we were able to select a representative concert programme featuring music that had been performed in the hall in 1769, including an ensemble piece, Thomas Erskine's *Overture XIII*, and two pieces for voice and harpsichord, J. C. Bach's *Blest with Thee, My Soul's Dear Treasure*, and the Scottish traditional song *The Lass of Peaty's Mill*.

Programming for the Chapel of Linlithgow Palace was more challenging. The historic model relates to the building as it was under the rule of James IV of Scotland, this being the period for which the architectural historians have the most detailed information. Unfortunately, specific knowledge of the repertory performed for James IV, and indeed more widely within his country at this period, is scant. We do know, however, from surviving records that the king spent many important occasions at Linlithgow, particularly at Easter, by far the most important religious occasion for the medieval Christian. He spent Paschaltide at Linlithgow as a 16-year-old in 1489, and again heard Easter Mass there in the 1490s, most probably for the first official use of the new chapel.

For our reconstruction, however, we focused on a third Easter visit in 1512. It was during this visit that his son, the future James V, was born on Easter Saturday. On the following morning, after James IV had heard Mass, James V was baptized in the chapel. While we will never know what music was performed on this occasion, we do know that the Scottish Chapel Royal followed the *Sarum Use*, originating at Salisbury Cathedral [38], so accordingly, we constructed a programme around Easter chants, some of them augmented by means of the improvised contemporary procedure – which we know was employed in Scotland – of faburden [39].

Alongside these plainsong items we programmed two groupings of works that, whether or not known to James IV, certainly could have been. The contents of the Scottish Carver Choirbook, owned by the Chapel Royal and named for probable canon of the chapel, Robert Carver, who may have been its scribe and who composed seven works for the manuscript, testify to a wide diversity of a kind clearly suited to the chapel of a cosmopolitan royal house that patronized musicians, some of whom (Robert Carver possibly included) received musical training in the Low Countries. This source mixes Scottish, English, and continental music for a court clearly interested in sampling the best music that Europe had to offer.

These include works dedicated to St Katherine – Byttering's motet *En Katerina Solennia* and *Virgo Flagellatur*, the foundation of Walter Frye's Mass – *Nobilis et Pulcra* presented in the contemporary improvised style of faburden – and a mass cycle, *Rex Virginum*, from the Carver Choirbook, the oldest in the source, and a setting that represents the repertoire of the Chapel Royal in the last quarter of the 15th century. It could quite possibly have been the foundation of any Marian celebration in Linlithgow Palace during this period.

With the repertoire in place, two ensembles, each of which specializes in historically informed performance, were contracted to perform the work for recordings. Because the recordings were intended to be heard in a VR environment with an artificial acoustic applied, they had to be made in an environment that had no natural acoustic of its own, and to that end, all production work took place at the anechoic chamber at the University of Edinburgh.

From a music-production perspective, this presented a number of significant challenges that disrupted the traditional music-production pipeline. First, the chamber itself, although large enough to accommodate multiple musicians and their instruments, was designed as an acoustically isolated environment without windows or ventilation, which made it an uncomfortable environment to work in for extended periods (see Figure 11.8), and so sessions had to be broken down into short sessions with recovery breaks in between.

Second, the lack of any natural acoustic meant that many of the performance cues that the musicians normally used for balance, intonation, tuning, and timing, particularly of entries and exits, were missing, and this impacted significantly on the quality of performance: one criticism commonly directed at live performances captured anechoically is that they sound "flat and lifeless," and this was a situation that we were actively striving to avoid.

Figure 11.8 The anechoic chamber at the University of Edinburgh was large enough for performance and provided a controlled and neutral dry acoustic, but it was an uncomfortable performance space for performers, who required frequent breaks, disrupting the recording process.

To try and recreate some of the "liveness" of performance, we had initially planned to feed all performers a headphone mix with artificial reverberation applied, but the singers, who were used to singing in an open space with a large, natural acoustic and without headphones, found this distracting and it impacted upon the quality of performance. Working with the musicians and the sound engineer, it became clear that the most effective way of capturing performance was to allow the musicians time to acclimatize to the sound of the chamber and recalibrate the acoustic cues that they used for performance validation. We then devoted time to rehearsing each section in situ, before attempting to record a take.

Finally, given that ultimately the anechoic recordings were intended to have a natural-sounding acoustic applied artificially, and that the effect of that acoustic is broadly to soften the harder elements of performance, particularly sibilants, monitoring the multitrack recordings proved to be challenging. On the one hand, the signal had to be monitored dry to ensure that it was clean, but also had to be evaluated in terms of its final context, which required an artificial acoustic to be applied.

Ultimately, the decision was made that our sound engineer would monitor the signal dry while recording, while the conductor would monitor via a headphone mix with the VR acoustic applied to it so that he would be able to judge the quality of performance. After each take, all of the musicians and the production team would then review the take with the acoustic applied, where necessary bypassing the effect if there arose performance issues that needed to be addressed in a retake.

Although none of the production challenges were insurmountable, all of them introduced additional workflow elements, which meant that the session time was increased, on average by a factor of three. Clearly, were this work being carried out in a commercial context, this would have a considerable impact on production costs. This approach also required the musicians to perform at high level in less-than-ideal conditions, and in particular, were it not for their professionalism, and willingness and ability to engage with new performance contexts that were completely outside of the norm, we would have struggled to create high-quality and engaging musical content for the virtual auditoria.

11.8 CONCLUSION AND FUTURE WORK

Alongside the controlled listening test and statistical analysis detailed above, the project team conducted end-user testing of the VR experience and music with 243 people at Manchester Science Festival in 2018. After experiencing the VR acoustic models, users were invited to complete a questionnaire based on their experience. Participants overwhelmingly expressed satisfaction with their VR experience and an appetite to have more cultural and heritage VR experiences. Users also reported a strong sense of presence within the immersive space, and, in general, a preference for the sound of the historical spaces, which, in the case of St. Cecilia's Hall, was described as "warmer" and "better suited to the music."

As a result, HES and Edinburgh University have both committed to installing VR visitor installations at each of the sites. Visitors will be able to don VR headsets and experience music in the spaces as they currently are, before stepping back into the past to see and hear the acoustics of the historic environments. This should ensure that the work can be experienced by as broad a range of international visitors as is possible. In 2018, HES reported visitor numbers at Linlithgow Palace of 94,718 [40], and while similar numbers are not yet available for St. Cecilia's Hall, we anticipate footfall in the tens of thousands.

HES is also keen to develop similar VR visitor experiences at three other major sites: St Andrews Cathedral and Abbey, Stirling Palace Chapel, and Melrose Abbey. This provides a wonderful opportunity to further refine the acoustic modelling process, and for the development of new, bespoke simulations derived from the underlying game engines. It also provides a unique platform for exploring our shared cultural heritage in new ways, and to experientially understand the architectural design of some of Scotland's most significant historic buildings, as well as offering new perspectives on historical repertoire and its performance: the VR acoustic models we have developed offer genuinely new methods for interrogating and exploring history and historical musicology in a systematic way.

The project team has been contracted by the independent classical label Hyperion to produce a full CD of early Scottish music – *Music for the King of Scots: The Pleasure Palace of James IV* – using post-production tools derived from the VR acoustic models described above. The CD repertoire was collated and edited by Dr James Cook and Professor Andrew Kirkman, and performed anechoically by the Binchois Consort, with Hyperion carrying out the engineering and post-production.

In addition to a CD release, a companion app, again derived from the original VR models, will allow consumers to download additional recorded content and remix it at different points in time and space. We plan to investigate the impact of this as an approach to developing audiences and new modes of listening; of monetizing existing back catalogue, and of maximizing the commercial potential of newly recorded content.

Finally, that notion of new and extended modes of listening raises interesting questions about the representation and embodiment of performance in VR. Our approach thus far has been to provide a single, first-person perspective on action, with static avatars representing performers. There are many unanswered questions around the most effective way of achieving a sense of user-engagement, presence and agency within a virtual auditorium: Should we represent other audients within the space, and if so how is this best done? How should we represent performers and performance? What are the performance cues that lead to a similar sense of cognitive and emotional engagement with performance in a virtual auditorium as is experienced in a physical one?

It is through that confluence of technical development, performance practice, and interactive play that we hope, not just to better understand the role of 3D audio on immersive user-experiences in VR, but to provide a powerful and systematic way of understanding its role in performance of the past.

ACKNOWLEDGEMENTS

This project would not have been possible without the generous support of the Arts and Humanities Research Council. The project team would also like to extend their thanks to Professor Andrew Kirkman, Dr Jenny Nex, the Binchois Consort, and Soluis Group.

REFERENCES

[1]   A. S. Bregman, "Auditory scene analysis: hearing in complex environments," in *Thinking in Sound: The Cognitive Psychology of Human Audition*, S. McAdams and E. Bigand, Eds. Oxford University Press, Oxford, 1993, pp.10–36.

[2]   A. Tajadura-Jiménez, P. Larsson, A. Väljamäe, D. Västfjäll, and M. Kleiner, "When room size matters: acoustic influences on emotional responses to sounds," *Emotion*, vol. 10, no. 3, pp. 416–422, 2010.

[3]   S. Hameed, J. Pakarinen, K. Valde, and V. Pulkki, "Psychoacoustic cues in room size perception," in *Proceedings of 116th Audio Engineering Society (AES) Convention*, 2004, Paper 6084.

[4]   C. Arias and O. Ramos, "Psychoacoustic tests for the study of human echolocation ability," *Applied Acoustics*, vol. 51, pp. 399–419, 1997.

[5]   E. Kirkland, "Resident evil's typewriter: survival horror and its remediations," *Games and Culture*, vol. 4, no. 2, pp. 115–126, 2009.

[6]   S. Delle Monache, P. Polotti, and D. Rocchesso, "A toolkit for explorations in sonic interaction design," in *Proceedings of the 5th Audio Mostly conference (AM '10)*, 2010, pp. 1–7.

[7]   G. Dubus and R. Bresin, "A systematic review of mapping strategies for the sonification of physical quantities," *PLOS One*, vol. 8, no. 12, e82491, 2013.

[8]   M. Kleiner, B. Dalenbäck, and P. Svensson, "Auralization: an overview," *Journal of the AES*, vol. 41, no. 11, pp. 861–875, 1993.

[9]   Guerilla Games, "Rigs [PlayStation 4]," 2016.

[10]   L. Dilloway, "The realities of virtual reality." Parallel Worlds: Designing Alternative Realities in Videogames, V&A Museum, London: UK, 2016.

[11]   K. Collins, *Game Sound: An Introduction to the History, Theory, and Practice of Video Game Music and Sound Design*. The MIT Press, 2008.

[12]   L. Davis, E. Larkin, and S. B. Graves, "Intergenerational learning through play," *International Journal of Early Childhood*, vol. 34, no. 2, pp. 42–49, 2002.

[13]   K. McAlpine, "Shake and create: re-appropriating video game technologies for the enactive learning of music," in *Serious Games and Edutainment Applications*, M. Ma and A. Oikonomou, Eds. Springer International Publishing AG, New York, 2017, pp. 77–97.

[14]   K. McAlpine, *Bits and Pieces: A History of Chiptunes*, Oxford University Press, 2018.

[15]   J. Richmond, "Introduction," in *The New Handbook of Research on Music Teaching and Learning*, R. Colwell and C. Richardson, Eds. Oxford University Press, New York, 2002, p. 3.

[16]   Red Octane, "Guitar Hero [Playstation 2]," 2005.

[17]   Konami, "Dance Dance Revolution [Arcade Cabinet]," 1998.

[18]   K. McAlpine, "Sampling the past: a tactile approach to interactive musical instrument exhibits in the heritage sector," in *Innovation in Music 2013*, R. Hepworth-Sawyer, J. Hodgson, R. Toulson, and J. Paterson, Eds. Future Technology Press, Shoreham-by-Sea, 2014, pp. 110–125.

[19]   M. Waitzman, "The Challenges of Caring for a Playing Collection," in *Perspectives on Early Keyboard Music and Revival in the Twentieth Century*, R. Taylor and H. Knox, Eds. Routledge, Abingdon, 2017, pp. 86–97.

[20]   W. Weber, "Did people listen in the 18th century?" *Early Music*, vol. 25, no. 4, pp. 678–691, 1997.

[21]   S. Ternström, "Physical and acoustic factors that interact with the singer to produce the choral sound," *Journal of Voice*, vol. 5, no. 2, pp. 128–143, 1991.

[22]   A. Kirkman and P. Weller, "Music and image/image and music: the creation and meaning of visual-aural force fields in the later Middle Ages," *Early Music*, vol. xlv, no. 1, pp. 55–75, 2017.

[23]   J. Blackie, "A new musick room: a history of St. Cecilia's Hall". Friends of St. Cecilia's Hall and the Russell Collection of Early Keyboard Instruments, Edinburgh, 2002.

[24]   T. Dickson and J. B. Paul, Accounts of The Lord High Treasurer of Scotland vol. IV, H.M. General Register House, Edinburgh, 1902, p. 347.

[25]   A. Valve, Benchmark in Immersive Audio Solutions for Games and VR. https://valvesoftware.github.io/steam-audio/.

[26] Google, Resonance Audio. https://resonance-audio.github.io/resonance-audio/.

[27] D. Frey, V. Coelho, and R. M. Rangayyan, *Acoustical Impulse Response Functions of Music Performance Halls*. Morgan & Claypool Publishers, 2013.

[28] M. Newton, Mike's Impulse Response App for Matlab [MikIRAM] – Public version (GitHub). https://github.com/acoustimike/MikIRAM_GitHub.

[29] S. Mills, *Auditory Archaeology: Understanding Sound and Hearing in the Past*. Routledge, 2016.

[30] Odeon. Odeon A/S: Room Acoustic Software. http://odeon.dk.

[31] M. Levoy and T. Whitted, "The use of points as a display primitive," Technical Report 85–022, Computer Science Department, University of North Carolina at Chapel Hill, 1985.

[32] S. Rusinkiewicz and M. Levoy, "QSplat: a multiresolution point rendering system for large meshes," Siggraph, 2000, pp. 343–352.

[33] C. Hummersone, IoSR Matlab Toolbox. https://github.com/IoSR-Surrey/MatlabToolbox.

[34] ISO/TC 43/SC 2 Building Acoustics, " ISO 3382-1:2009 Acoustics – measurement of room acoustic parameters – part 1: performance spaces," International Organization for Standardization, Brussels, Belgium, Jun. 2009. www.iso.org/cms/render/live/en/sites/isoorg/contents/data/standard/04/09/40979.html, (accessed: Mar. 25, 2021).

[35] H. Massey, *Behind the Glass Volume 1: Top Record Producers Tell How They Craft the Hits*. Rowman & Littlefield, 2000.

[36] D. Murphy, S. Shelley, A. Foteinou, J. Brereton, and H. Daffern, "Acoustic heritage and audio creativity: the creative application of sound in the representation, understanding and experience of past environments," *Internet Archaeology*, vol. 44, 2017. https://intarch.ac.uk/journal/issue44/12/index.html.

[37] N. Jillings, B. De Man, D. Moffat, and J. Reiss, "Web audio evaluation tool: a browser-based listening test environment," Presented at *12th Sound and Music Computing Conference*, 2015.

[38] J. R. Baxter, "Music, ecclesiastical", in The Oxford Companion to Scottish History M. Lynch, Ed. Oxford University Press, Oxford, 2001, pp. 431–432.

[39] J. Aplin, "The fourth kind of faburden: the identity of an English four-part style," *Music & Letters*, vol. 61, no. 3/4, pp. 245–265, 1980.

[40] K. Smith, "Heritage visitors' £620 million tourism boost," Scottish Field, 2019. www.scottishfield.co.uk/travel/scotland-travel/heritage-visitors-620-million-tourism-boost/.

12

3D acoustic recording

Paul Geluso

12.1 INTRODUCTION

The everyday sounds that enter our auditory system contain complex spatial information. As such sounds travel from their sources to our ears, the acoustic signals gather this spatial information as it meets room reflections, objects around us, and our bodies – and our ability to hear this spatial quality is extraordinary. For example, even in early mono recordings, we can hear which instruments or voices were closer to the microphone than others, and we glean a sense of the recording space. Soon after the invention of the telephone, Alexander Graham Bell began experiments around transmitting pairs of signals for spatial effect, and this was successfully and commercially implemented by Clement Aders with the introduction of the "Theatrophone" – perhaps the first binaural system. Early experiments with microphones and loudspeaker line arrays were taking place at this time as well [1]. Alan Blumlein's historic patent from 1933 defined a two-channel recording and playback systems grounded in physics and psycho-acoustics. It was intended to give the listener a true *directional impression* that relied on the intensity differences picked up by a pair of directional microphones, thus laying the groundwork for Ambisonic technology [2]. Although he is best known for his contributions to two-channel recording and playback systems, as well as the horizontal, Blumlein also suggested the capture of vertical spatial information. This concept of 3D sound capture was formalized decades later with the introduction of Ambisonics in the 1970s. Modern day listeners now have options beyond stereo and surround sound through more recent 3D consumer formats developed by researchers at Auro Technologies, Dolby Labs, DTS, the MPEG group, and others. With the recent proliferation and development of immersive sound used in gaming, extended reality (XR), podcasting, and music production, 3D production sound is finally more accessible and can be enjoyed by many more people.

This chapter is focused on practical recording techniques to capture acoustic sounds in 3D using binaural, Ambisonic, and microphone-array techniques. It is meant to be a guide that enables recordists tasked with capturing 3D sound to both use their own creativity and harness available technologies. Some historical and theoretical context is provided here, but in addition, 3D sound recording and reproduction technologies are also discussed in detail with theoretical depth throughout this book in other chapters.

12.2 BINAURAL RECORDING

Our auditory system has the ability to locate sounds in 3D space with amazing accuracy. As the position of a sound changes relative to us, so does the relationship between the acoustic signals reaching the ears. The term *binaural recording* implies a two-channel system that captures the time, intensity, and spectral cues that are recognized by the human auditory system. These cues can either be encoded naturally in the recording process with the use of physical ears – prosthetic or human, or synthesized electronically [3]. Left ear to right ear interaural differences can be described in terms of two primary auditory cues: interaural time difference (ITD) and interaural level difference (ILD). These are the primary cues that aid our auditory system in determining the location of sounds in the horizontal plane. Sound localization in both the vertical field and front-to-back is largely determined by the effect of the geometry of our pinnae on the incoming acoustic signals. As acoustic signals enter the ear, the superposition of reflections from the pinnae cause complex spectral filtering [4]. The resulting frequency response of the ear-input signal is called head-related transfer function (HRTF), and it is unique for both an individual's morphology and also each direction of sonic arrival. Both Pike and Lord offer specific contexts for binaural issues in Chapters 1 and 13, respectively, and much more detail is provided by Sunder in Chapter 7.

12.2.1 Binaural microphones

In the early 1930s, Harvey Fletcher's research at Bell Labs led to the development of a mannequin named "Oscar" with microphones built into its *dummy head* at each ear location [1]. Since then, binaural microphones resembling a human head – with microphone capsules built into the ears – have continued to evolve (see Figure 12.1). Alternatively, placing miniature microphones in our own ears forms a compact and unobtrusive system, and in this way, a personalized binaural recording can be captured with one's own personal HRTF superimposed. Research has shown that personalized HRTFs improve the sense of externalization, reduce front-back confusion, and enhance the perception of elevation [5].

12.2.2 Binaural recording techniques

Binaural recording can deliver a "you are there" experience as opposed to a "they are here" experience associated with stereo and other conventional channel-based systems [3]. Since a binaural system captures sounds from all directions, there is no perceived *zooming* in due to the lack of off-axis sound rejection normally associated with conventional directional

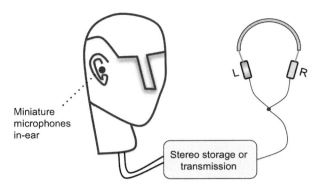

Figure 12.1 Binaural recording system.

microphones. Therefore, the microphone location and orientation at the time of recording is of paramount importance because a complete 3D audio scene is captured all at once – a streamlined approach. At the same time, a potential drawback to using a static binaural system (i.e. one without head or motion tracking) is that location and distance cues are fused together in the recording and cannot be undone or even modified in post-production. Yet another benefit to binaural recording is that it can potentially deliver a very *natural* 3D experience accessible to the headphone listener, although the extent of this is heavily dependent upon the compatibility of an individual's HRTF to that of the capture arrangement. Applications include radio drama, podcasts, sports broadcast, music production, gaming, augmented tours, theatric sound design, and human auditory-system research. Binaural recording techniques are also widely used to make acoustic measurements and for collecting sounds in challenging environments.

As mentioned above, the binaural microphone-system's position and orientation relative to the sound sources should be considered carefully and creatively. If a sound source is positioned very close to the microphones, as illustrated in the first zone by example in location zone A in Figure 12.2, a heightened and intimate listening experience can be achieved. In this zone, direct signals overpower reflections and the effect of acoustic head shadowing is heightened. Also, sound levels change dramatically as a function of distance as sound sources move closer and farther from the binaural recording system in this zone (a function of the inverse square law). On listening back, the effect is like someone whispering right into your ear or a bee flying close to your head. Therefore, close-up binaural recording has become popular with those seeking to create an autonomous sensory meridian response (ASMR) or seeking dramatic effects for radio plays, podcasts, and live theatre experiences that are enhanced with head-phone monitoring. Placing the sound source farther away from the bin-aural recording system in the second zone, as illustrated by example in location zone B in Figure 12.2, acoustic reflections from close boundaries and objects, such as the floor, walls, ceiling, or furniture in some cases, are fused with the direct signal thus causing some location-specific spectral colouration. Sound levels change less rapidly as a function of distance in

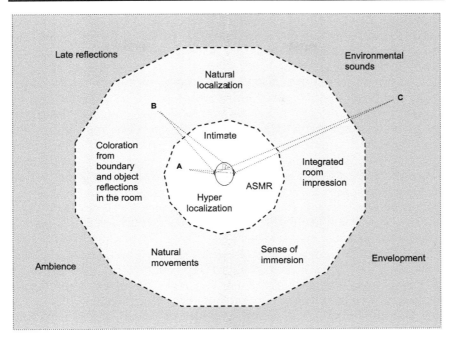

Figure 12.2 Recording attributes illustrated as a function of distance from a binaural recording system.

this zone, and further, later room-reflections arriving from all directions are more balanced with the direct sounds than in the first zone, thus giving the listener a better sense of external space. For more distant sounds in the third zone, as illustrated by example in location zone C in Figure 12.2, the direct signal may be hard to distinguish from its many reflections arriving from all directions to the binaural recording system, which will in turn envelop the listener in a very natural way. Lastly, sound-level shifts relative to changes in distance are less distinguishable. Therefore, with a single binaural recording system, a wide range of close, medium, and distant sounds can be captured and these coexist naturally in an integrated 3D audio scene.

12.2.3 Delivery over headphones and stereo compatibility

Static binaural recordings are delivered via a two-channel stream just like stereo programmes and (unlike many 3D formats) are compatible with stereo broadcast and recording equipment, but for a full 3D experience, they must be monitored with headphones or with crosstalk cancellation over speakers, as detailed by Sunder in Section 7.2.3. It should be noted that some tonal colouration due to HRTF-related spectral filtering imposed on binaural recordings can be audible over conventional loudspeakers. When monitoring a conventional stereo programme over headphones, sound images are largely perceived as being somewhere between our ears, thus "*in* our head." Compared to conventional stereo recordings, with binaural recordings, the auditory localization cues such as ITD and ILD are encoded more accurately for both direct and reflected sounds. The sound images are therefore perceived more externally – both surrounding and "outside" the head – with both less "in the head" effect as described above, or as can be the case when listening over speakers, arriving at our ears or "on the nose."

The combination of stereo and binaural sources can be used creatively for effect over both headphones and stereo speaker systems. The author has found that they are certainly not mutually exclusive and can complement each other within a mix. For example, a vocal mixed in as a dual-mono stereo signal intended to be perceived *in the head* can coexist with field recordings captured using a binaural recording technique that will be perceived externally in 3D with a feeling of *outside the head*.

Conventional mono and stereo recordings can be up-mixed using binaural synthesis methods. The downside of using a generic HRTF to up-mix is that front-back and up-down localization might be compromised [6] and it has different degrees of effectiveness for different individuals, although many are based on statistical averages that will be fairly effective for a large proportion of listeners. Personalized HRTFs are perhaps ideal, but are costly and still generally complex and time-consuming to measure. Further, they can also introduce a certain level of complexity to the workflow. Working towards simplifying this process, the spatially oriented format for acoustics (SOFA) [7] defines a standard way for storing HRTF and other spatially related acoustic data that has been already adopted by academic institutions and software developers. There are increasing numbers of commercial plugins that facilitate binaural synthesis and panning, sometimes augmented with synthetic spatial information which can enhance externalization.

12.3 AMBISONICS

12.3.1 Ambisonics and sound-field recording technology

At present, Ambisonics is perhaps the best known and most widely used technology for 3D recording. Michael Gerzon and Peter Fellgett introduced Ambisonics to the broadcast, studio, and home audio world in the early 1970s. The intention was to extend monitoring from a 60° stereo perspective to a full 360° surround sound experience, and additionally, with 180° elevation coverage. The reproduction techniques that Gerzon termed as "periphonic" (which came to refer to adding elevated reproduction), treat horizontal and vertical sounds equally and thus the system is designed not to favour any particular recording direction or loudspeaker playback position. Fellgett and Gerzon had an important design goal of creating a non-*speaker-centric* system to provide a natural listening experience [8], [9]. As opposed to stereo, quadraphonic, and other channel-based systems, Ambisonics provides a recording and playback method for which all directions are treated equally and different numbers of loudspeaker signals can be derived to serve different configurations. Fellgett and Gerzon's design intent has clearly succeeded today; in the age of the proliferation of immersive experiences, Ambisonic technology has been rediscovered by a new generation of sound designers and rescued from near obscurity to support interactive and immersive listening experiences. Ambisonics can support static and dynamic binaural systems as well as a wide array of multichannel speaker systems. A comprehensive tutorial is offered by Armstrong and Kearney in Chapter 6, but here, a simplified necessary context will be established in order to help the reader to understand practicalities of such recording.

"The term *sound field* refers to the capture, reproduction and description of sound waves" [10], and to make Ambisonic-compatible acoustic recordings, sound-field recording technology can be used. The concept of a sound field is based on the idea that a sound wave can be defined by a sum of directional (pressure-gradient) components. This concept resonates with and is compatible with Blumlein's early work with coincident X-Y and mid-sides (M-S) recording techniques. 3D sound-field recording systems are designed to capture such directional components of sound at a single point or with receptors evenly spaced along the surface of a rigid sphere. To obtain a fully 3D (so called "first order") recording, either a combination of omnidirectional and figure-of-eight microphones, or a minimum of four tetrahedrally positioned microphones must be used, and these capture what is termed as an A-format Ambisonic recording. This recording can be processed – encoded – into four components: a spherical omnidirectional (pressure) component, which represents the necessary zeroth order (W), and three bi-directional (pressure-gradient) components (X, Y, and Z), which define the first order. This set of four components, defined as W, X, Y, and Z are termed spherical harmonics, and together can make up what is known as first-order Ambisonic B-format (see Figure 12.3). The maximum number of spatial components that define a sound field is proportional to the number of system channels, and also help determine the "order number" of the system [10]. As the number of channels increases, so does the spatial resolution of the system – and of course its complexity.

12.3.2 Tetrahedral microphones

Conceivably, the Ambisonic B-format W component can be captured natively with an omnidirectional microphone, whereas components

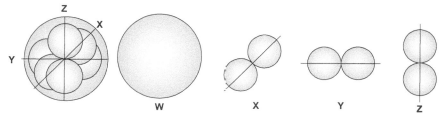

Figure 12.3 Ambisonic B-format components; shown superimposed (far left) and individual (middle to right) W, X, Y, and Z.

Table 12.1 Tetrahedral microphone output channel numbering conventions.

Channel	Orientation	Abbreviation
1	Left front up	LFU
2	Right front down	RFD
3	Left back down	LBD
4	Right back up	RBU

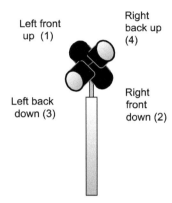

Figure 12.4 An example of a tetrahedral microphone system.

X, Y, and Z can be captured natively with three figure-of-eight studio microphones. Although it is theoretically possible to build such an array with studio-grade microphones, it is challenging to configure due to physical positioning restrictions caused by the microphone bodies – for this superposition to function as intended, the sound field should be captured at a precise single point in space. A far more eloquent and practical solution was offered by Craven and Gerzon [11]; they proposed a compact first-order microphone array based on a tetrahedral configuration. A typical tetrahedral microphone system consists of four cardioid capsules situated on a tetrahedral frame, with each capsule equally spaced and angled 109.5° from its neighbouring capsule in 3D space. The angle between capsules measured along the 2D horizontal or vertical plane is 90° (see Figure 12.4).

The multichannel output of a tetrahedral microphone is quad-buss compatible and is described in Table 12.1 [12]. The microphone delivers four directional signals with capsules one (LFU) and four (RBU) orientated upward by 45°, and capsules two (RFD) and three (LBD) orientated downward by 45°. As above, the four discrete outputs of a tetrahedral microphone make up the four components of Ambisonic A-format.

When recording, matched microphone preamps and A-to-D converters should be used and when possible, with the preamps linked to electronically provide equal gain. Setting microphone pre-amp levels independently is difficult because the four signals may not match and may even greatly vary in some recording situations. Maintaining equal pre-amp gain for all four channels during recording is vital to later accurately decode the spatial information in post-production.

Typically, at some point in the capture, post-production, and broadcast signal flow, A-format signals are converted into B-format with appropriate weighting and summing, before finally "matrixing" (decoding – again with weighting and summing – into discrete audio-channel streams) the signals into D-format (a little-used term) to drive loudspeakers and so on, but the author has found that native A-format microphone signals can be used creatively for many applications as well – without the need for matrix processing. For example, to create an immersive sound installation using only four speakers, A-format signals can directly feed a 3D 2.2 cube speaker system with channel one feeding an upper-left speaker, channel two feeding a lower-right speaker, channel three feeding a lower-left surround speaker, and channel four feeding an upper-right surround speaker. In another native A-format application, one capsule of a tetrahedral microphone can be orientated directly at a source to be used either as a spot microphone or as the A (or B) microphone of an A-B or spaced-omni pair. Since the A-B technique is reliant upon identical omnidirectional pairs, it is ideal to have a second tetrahedral microphone available to devote one of its capsules to being the second microphone. At least on one tetrahedron, while capturing the *main* signal directly on one capsule, the remaining other three capsules will capture room reflections, reverberation, and other indirect sounds that can be used creatively in post-production. The direct output from multiple tetrahedral microphones can be used effectively to form a surround spaced array with height as well [13]. The advantage of using the raw A-format signals is that no encoding or subsequent matrixing is applied since that could add unwanted colouration or spatial aliasing.

12.3.2.1 B-format conversion

As mentioned, B-format can be easily derived from A-format for conventional stereo and surround applications. B-format is supported by commercial surround decoders, immersive sound software, as well as streaming and broadcasting services including YouTube, Facebook, and broadcasting formats including MPEG-H. B-format Ambisonic signals are supported by both game-creation platforms such as Unity and other audio middleware that might be employed for XR application development to facilitate the creation of 3D-sound experiences that respond sonically to the user's virtual navigation and/or head rotations.

Encoding from A-format to B-format is accomplished using the following summation matrix (here, seen unweighted):

$$W = LFU + RFD + LBD + RBU$$

$$X = (LFU + RFD) - (LBD + RBU)$$

$$Y = (LFU + LBD) - (RFD + RBU)$$

$$Z = (LFU + RBU) - (RFD + LBD).$$

Since W is the summation of all four cardioid capsules using a tetrahedral microphone system, there will be some pick-up pattern overlap
causing greater on-axis gain for the W component than the figure-of-eight
components. There is also pick-up pattern overlap when creating the X,
Y, and Z signals – forming *wider* figure-of-eight pick-up patterns than are
ideal. Also, due to the spacing of the microphone capsules, high-frequency
compensation may be needed above frequencies with wavelengths smaller
than the space between the capsules. With proper weighting, cardioid directional signals can be easily derived from B-format components using
familiar M-S-matrixing techniques. Once opposing cardioid signals are
derived, additional pick-up patterns can further be created as discussed
below. Sound-field microphones typically come with their own matrixing
software or hardware from the manufacturer that provides precise
weighting and equalization specific to a given microphone, to create both
"virtual microphones" (e.g. with remotely modifiable directivity) and B-
format signals. That said, one can always experiment with matrixing manually for custom applications or to better understand the inner workings of
Ambisonics.

12.3.2.2 Deriving M-S and stereo signals from B-format

Virtual M-S microphone pairs can be easily derived along the X, Y, and Z
axes. For example, to create virtual M-S signals along the X and Y axes,
the following weighting and summing can be applied to the B-format
components:

$$M = 0.707W + X$$

$$S = Y.$$

The 0.707 weighting compensates for pick-up-pattern overlaps when
adding the four tetrahedral microphone capsule outputs to create the W
component. It translates to a gain reduction of about 3 dB (such a use of
scalars will be applied in the remainder of this chapter). From a virtual
M-S pair, a compatible virtual stereo-pair of cardioid left (L) and right
(R) signals can be obtained with matrixing and weighting as follows:

$$L = (M + S)/1.21$$

$$R = (M - S)/1.21.$$

This time, the resulting (M + S) and (M −S) expressions have been
reduced by about 1.7 dB to maintain the same on-axis sensitivity as the M
signal [14]. Similar M-S matrixing can be applied to other pairs of signals.
Once a full set of 180° opposing cardioid patterns are derived, a full range
of directional virtual microphone patterns can be obtained through additional summation and weighting. Expanded M-S and X-Y matrixing
concepts will be explored later as we discuss native B-format recording

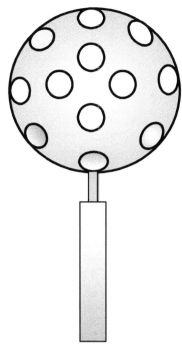

Figure 12.5 Side view of an HOA sound-field microphone. The small circles represent multiple capsules on a spherical head.

techniques, including double M-S + Z. X-Y + Z, and the triple dual-capsule coincident (3DCC) systems [14], [15].

12.3.3 Higher-order Ambisonics

Higher-order Ambisonic (HOA) systems are considered to be those that have more than four spherical harmonics. The order (N) is captured/described by $(N + 1)^2$ capsules/channels. A wide range of commercially produced higher-order microphone systems are currently available in models with 8, 16, 32, and even 64 capsules (see Figure 12.5). These microphones are typically paired with proprietary hardware and software that allow the user to form remarkably directional sound-capture "beams" in 3D space (Figure 12.6), either during the recording or in post-production. As will be discussed, as the order number increases, the directional beams become narrower, thus better isolating sound sources in post-production or broadcast.

12.3.4 Sound-field recording techniques

A sound-field microphone system can be used to capture ambiences or a complex acoustic scene, such as crowd noise or room reverberation, or when placed closer to the sound sources, be used in place of one or multiple spot microphones to capture directional acoustic signals for subsequent

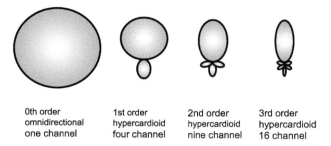

0th order	1st order	2nd order	3rd order
omnidirectional	hypercardioid	hypercardioid	hypercardioid
one channel	four channel	nine channel	16 channel

Figure 12.6 Maximum directivity-factor patterns based on the order number and number of Ambisonic channels [16].

balancing, equalization, or re-spatialization. For playing back such ambient recordings, the derived loudspeaker-channel signals should be equal in coverage angle and loudness in all directions to achieve a natural "you are there" listening experience. This is appropriate for capturing both sounds of nature or city soundscapes alike. The microphone should be placed on a stand in a fixed position for the entirety of the recording, and longer takes are recommended so that unwanted sounds can be removed and reduce the need for looping sound in post-production. Like using any microphone outdoors, a good wind protection device is needed. The author has used a boom pole to get close enough to capture close-up nature sounds with ambience both by the sea and in woods – with great success.

To use in place of a main stereo microphone system, placing the sound-field microphone just inside the reverberation radius (i.e. critical distance) is advised. Again, keep in mind that the sound-field microphone behaves similarly to an omnidirectional microphone when using it traditionally. Although highly directional beams can be formed using HOA microphones, a natural sounding playback environment where all directions are treated equally will result in a perceived directivity factor about equal to that of an omnidirectional microphone. Care should be taken so that the microphone is not placed too distant, or too close, from the sound stage. If headphone monitoring on location, it is a good practice to sum the signals in mono or monitor a folded stereo or binaural signal to help determine the optimum microphone placement.

A sound-field microphone can be used in place of multiple spot microphones in certain applications, and when doing this, an HOA microphone is recommended. The creation of virtual directional microphones, also known as beamforming, is accomplished by weighting and summing the raw microphone capsule signals to create hybrid spherical-harmonic responses that can emulate virtual microphone signals with a variety of directional properties. As mentioned above, the spatial resolution of the system is governed by the number of capsules which in turn determines the order number of the system. For most applications, a first-order microphone does not deliver a narrow enough beam to take to the place of multiple spot microphones whereas an HOA microphone can deliver impressive directional signals.

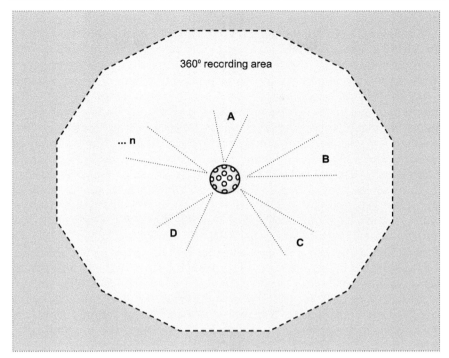

Figure 12.7 Sound-field recording technique with sound sources surrounding an HOA microphone.

For example, musicians can be placed around an HOA sound-field microphone as shown in Figure 12.7 at positions A to E. To achieve a good balance, moving instrumentalists and singers closer or further away from the HOA microphone is effective and most likely necessary. This balancing technique can be used just as it was in the early days of monaural wax cylinders and live disk-cutting recording sessions! In this way, the sound-field microphone can be conceived as a *pseudo-omni* microphone whose signal can later be divided into spatial areas in 3D space, like slices of a pie, in post-production. By focusing the virtual microphone beams in post-production on sound sources, additional balancing and equalization can be achieved but keep in mind only so much can be done; therefore, getting the *pseudo-mono* balance and sound colour near perfect during the live recording is vital.

12.3.5 Workflow and the delivery of a sound-field recording

Once A-format signals from a sound-field microphone are converted to B-format, familiar M-S matrixing methods can be used to derive a directional signal in virtually any direction. Note that the original aim of Ambisonic decoding is to physically reconstruct the sound field as accurately as possible in reproduction, so that it can produce ILD and ITD cues that closely match the real listening conditions. This requires strict mathematical calculations of weighting factors for spherical harmonics. Again, the reader is directed to the theoretical approaches of Ambisonic decoding as provided by Armstrong and Kearney in Chapter 6. In the present section, we discuss flexible ways of decoding B-format signals using virtual microphone techniques and how they can be used in *practical* 3D reproduction.

As mentioned, any number of virtual microphone signals, also known as *beams*, can be derived from a multichannel Ambisonic signal stream during post-production – likely using aforementioned software provided by the microphone manufacturer. Sound-field microphone decoders typically allow dynamic and continuous control of the virtual microphone's pick-up pattern from omnidirectional to subcardioid, cardioid, hypercardioid, figure-of-eight, and so on. In addition, the pick-up angle can be dynamically controlled from a 0° to 360° azimuth and a 0° to 180° elevation. Once the directional signal beams are formed, additional 3D panning can be performed using XYZ vector-based controls for VR applications or by using a surround-plus-height approach for theatrical and music-listening applications. The resulting mix can then be formatted into a channel-based mix, an object-based format such as Dolby Atmos® or MPEG-H, or back into an Ambisonic format for physical-installation-based or XR applications. Binaural synthesis can be applied to any of these formats during production to create an immersive mix for headphone monitoring. Mixing over headphones is incredibly useful when a multi-speaker monitoring environment is not available. This workflow is illustrated in Figure 12.8.

12.4 3D MICROPHONE ARRAYS BASED ON M-S AND X-Y TECHNIQUES

12.4.1 Double M-S + Z

The M-S stereo recording technique can be expanded to capture both horizontal surround and height information with the addition of a rear-facing cardioid M' microphone and a vertically orientated figure-of-eight Z microphone [15]. Traditional M-S consists of a forward-facing directional M microphone, or any other polar pattern including omni, and a sideward-facing bi-polar S microphone. When recording in 3D, it is important to maintain equal real and virtual microphone sensitivity in all directions. Therefore, the left channel (L), right channel (R), and (in this case, using a cardioid) the middle channel (M) must all maintain equal gain for an on-axis signal. To achieve this, assuming that the figure-of-eight side (S) signal is created using two opposing cardioid signals with the same sensitivity as the M cardioid signal, the S figure-of-eight signal is first attenuated by 6dB and then the resulting sum or difference, (M + S) or (M − S), are attenuated by 1.66 dB as expressed below:

$$L = (M + 0.5S)/1.21$$

$$R = (M - 0.5S)/1.21.$$

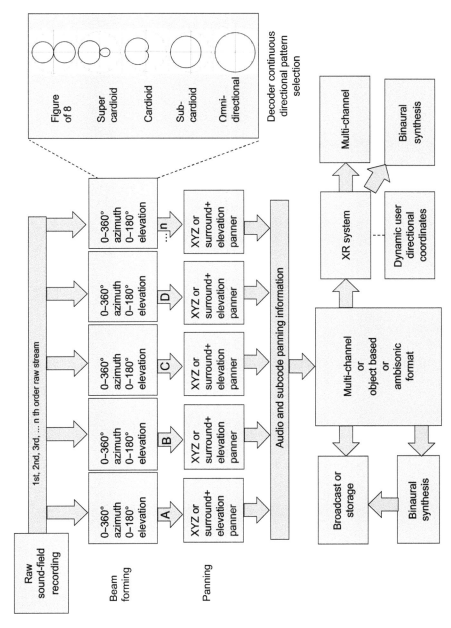

Figure 12.8 Post-production workflow for sound-field recording.

Note that the resulting L and R pick-up patterns are directional, but not as directional as a pure cardioid. Once a 180° opposing cardioid signal is derived, the pick-up pattern for any signal can be tuned electronically and varied dynamically from omnidirectional to cardioid to figure-of-eight.

By adding a rear-facing M' cardioid microphone to capture a surround signal, we can expand the system to a double M-S (i.e. MSM) 2D surround system. We can then derive left surround (L_s) and right surround (R_s) polar patterns in a similar way:

$$L_s = (M' + 0.5S)/1.21$$

$$R_s = (M' - 0.5S)/1.21.$$

To expand the system to 3D, we can add a vertically orientated bipolar Z microphone to the system. Pairing the vertically orientated figure-of-eight Z signal with any of the horizontally orientated signals creates a vertical M-S pair, therefore further M-S matrixing yields vertically orientated ± 45° signals. For example, to obtain quadraphonic height (h) channels, the following summation and weighting expressions will apply:

$$L_h = (L + 0.5Z)/1.21$$

$$R_h = (R + 0.5Z)/1.21$$

$$L_{hs} = (L_s + 0.5Z)/1.21$$

$$R_{hs} = (R_s + 0.5Z)/1.21.$$

Similarly, downward-facing signals (d) will be:

$$L_d = (L - 0.5Z)/1.21$$

$$R_d = (R - 0.5Z)/1.21$$

$$L_{sd} = (L_s - 0.5Z)/1.21$$

$$R_{sd} = (R_s - 0.5Z)/1.21.$$

This arrangement is illustrated in the upper area of Figure 12.9.

12.4.2 Double X-Y + Z

Similarly, the X-Y recording technique can be expanded to capture 3D. Summing a pair of cardioid microphone signals, orientated at −45° and +45° respectively, creates a virtual centre signal (C) at 0°. Again, weighting must be applied to maintain uniform sensitivity to all directions and so C will need to be attenuated by 4.66 dB to compensate for the overlap of the left (L) and right (R) cardioid pick-up patterns. To expand the system to capture a full 360° on a 2D plane, a rear-facing pair of cardioid receptors, X' and Y', can be matrixed in a similar way to create left surround (L_s),

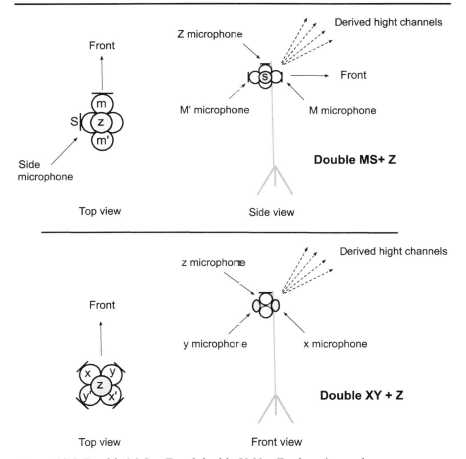

Figure 12.9 Double M-S + Z and double X-Y + Z microphone placement.

right surround (R_s), and centre surround (C_s) signals. By adding the L to the L_s signal and attenuating the result by 4.66 dB, we obtain a left-side signal (L_{ss}) at −90°. In a similar way, we can derive a right-side signal (R_{ss}) at +90° adding the R to the R_s signal and attenuate it by 4.66 dB. To expand the system to 3D, a figure-of-eight Z microphone signal can be combined with any horizontal-plane signals to steer the signals upward or downward. This setup can be seen in the lower area of Figure 12.9.

For example, the channel output for a 7 + 4 playback system can be obtained with the below expressions (four height channels are computed), and these 11 signals routed to a 7.0.4 loudspeaker array for playback:

$$L = X$$

$$R = Y$$

$$C = (L + R)/1.71$$

$$L_s = Y'$$

$$R_s = X'$$

$$C_s = (L_s + R_s)/1.71$$

$$L_{ss} = (L + L_s)/1.71$$

$$R_{ss} = (R + R_s)/1.71$$

$$L_h = (L + 0.5Z)/1.21$$

$$R_h = (R + 0.5Z)/1.21$$

$$L_{sh} = (L_s + 0.5Z)/1.21$$

$$R_{sh} = (R_s + 0.5Z)/1.21.$$

12.4.3 3DCC recording technique

For purely coincident recording, dual-capsule microphones provide an excellent tool for sound-field capture. Based on back-to-back cardioid receptors and discrete microphone outputs for each, the front (f) and back (b) signals can be used independently or combined to create a range of polar patterns from omni, to cardioid, to figure-of-eight. Table 12.2 shows the summing and weighting expressions to derive various patterns from a dual-capsule microphone termed X.

The advantage of the dual-capsule microphone system is that the polar pattern of each signal can be adjusted dynamically during recording or in post-production. By doing so, the weightings for the secondary signals will also have to be adjusted accordingly. See Zhang and Geluso's AES paper [14] for an expanded table of weightings and polar equations. The system can be used in double M-S + Z and double X-Y + Z mode as well – using the respective matrixing described above. Using three-dual capsule coincident microphones as shown in Figure 12.10, with each microphone orientated along the X, Y, and Z axes, respectively, a native B-format signal is easily obtained following the expressions below:

Table 12.2 Dual-capsule microphone: capsule weightings and resulting polar patterns.

Signal combinations	Polar pattern
x(f) + x(b)	Omni
x(f) + 0.5x(b)	Subcardioid
x(f)	Cardioid
x(f) − 0.5 x(b)	Hypercardioid
x(f) − x(b)	Figure-of-eight

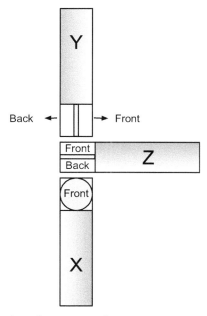

Figure 12.10 3DCC microphone array placement.

$$X = x(f) - x(b)$$

$$Y = y(f) - y(b)$$

$$Z = z(f) - z(b)$$

$$W = [x(f) + x(b) + y(f) + y(b) + z(f) + z(b)]/3.$$

12.5 SURROUND PLUS HEIGHT MICROPHONE-ARRAY RECORDING

Although compact coincident systems introduced in the previous section have the potential to produce a precise image when the listener sits at the sweet spot during loudspeaker reproduction, they cannot capture interchannel time differences, which are important for producing spaciousness and increasing the size of the listening area [17]. Such limitations can be overcome to varying degrees by using spaced arrays – albeit at the expense of channel count and complexity. Typically, a 2D 5.0 spaced array consists of five spaced microphones to capture multichannel signals compatible with the speaker plan of a surround sound playback system with a channel that is often paired with a film screen in a cinema, or a visual display in a home theatre. Height channels can be added to 2D surround sound systems to add a vertical sonic dimension for music, sound effects, soundscapes, and ambiences – thus delivering 3D. This section now explores some recording techniques developed for surround playback systems with height channels.

12.5.1 Height-layer recording techniques

For example, a 2D 5.0 surround system using five main-layer microphones augmented with four height microphones becomes an effective 3D-capture system that can be denoted as 5.0.4. As illustrated in Figure 12.11, spaced surround-microphone arrays with height channels can be classified by the vertical spacing of the height layer and the shape of the array. The height layer can be positioned close to the main layer using coincident or near-coincident directional microphones spaced anywhere from 0 to 30 cm. Alternatively, the height layer can be spaced further from the main layer, from 100 to 200 cm, thus forming a cuboid and is then more compatible with the use of spaced omnidirectional height-microphones.

For acoustic music applications, a spaced surround microphone array with a height layer can be placed so that the front of the array is facing the sound stage just inside the reverberation radius so that the surround and height microphones are focused on collecting more ambient sounds. If there is an elevated sound source – like a choir in a balcony or a large pipe organ – these microphones will pick up direct sound as well. Even though microphones have directional characteristics, all microphones will "hear" all sound sources in the acoustic space, thus a cohesive 3D sound scene

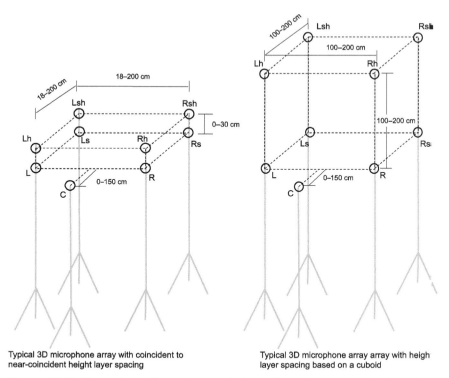

Typical 3D microphone array with coincident to near-coincident height layer spacing

Typical 3D microphone array array with height layer spacing based on a cuboid

Figure 12.11 General configurations and dimensions for surround-microphone arrays with height layers: coincident/near-coincident systems (left) and spaced cuboid systems (right) are illustrated.

can be captured with a 3D microphone-array. The amount of interchannel crosstalk can be controlled through microphone placement and microphone polar-pattern choices. At one extreme, a lack of correlation may result in perceived *holes* between speakers upon playback, whereas too much crosstalk may result in a loss of localization accuracy in playback. Herein lies the art; it is determining both the optimum microphone type and location – to achieve the desired immersive experience for the listener – based upon the balance between the acoustic signals received at the microphones.

A basic surround-array setup can be expanded by adding additional microphones. Ambience and spot microphones can be added to the recording plan to give the creative team more balancing, panning, ambient, and spectral control during post-production or broadcast. Ambience microphones can be placed to reject direct sound from the sound stage, whereas spot microphones can be placed close to, and directed at the individual sound sources. Using the philosophy of "one main microphone per playback channel," two additional directional sideward-facing microphones can be added to a 5.0 array to create a 7.0 main array. Additional directional height-microphones can be added as needed to support playback environments with additional height speakers (see Figure 12.12).

Acoustic events captured with surround-plus-height systems can be integrated into a channel-based mixing environment using a

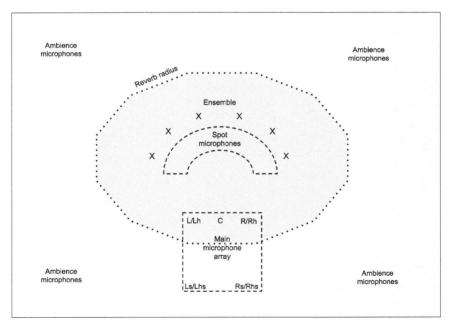

Figure 12.12 Suggested placement of a 5.0.4 main microphone array, spot microphones, and ambience microphones in the relation to the room's reverberation radius (also known as critical distance where direct and reflected sound levels are about equal).

microphone-to-speaker approach. Many recording engineers prefer this approach without any electronically introduced crosstalk between channels, and minimum signal processing. With the adoption of object-based systems, channels can still be confined to single speakers or virtually placed across several speakers with the spread determined at the time of playback by a local renderer.

12.5.2 Surround-plus-height microphone arrays

12.5.2.1 PCMA-3D

The perspective control microphone array (PCMA) is a horizontal-surround-capture system proposed by Hyunkook Lee [18]. It is a microphone array comprised of five coincident pairs, for which adjusting the mixing ratio could instead give the impression of virtual microphones with different polar patterns and directions. The PCMA-3D array [19], as shown in Figure 12.13, extended this work to capture elevation in preference to synthesizing virtual microphones. This two-layer array has a lower layer based upon the original PCMA layout, but uses only single cardioid microphones instead of pairs. The centre microphone is placed 25 cm forward from the baseline between the front-left and front-right microphones, which are spaced 100 cm apart angled towards the sound sources and around 30°-45° outward. Left and right surround microphones are spaced 100–200 cm apart and placed 100–200 cm directly behind the left and right channels that are orientated to the rear. For the height layer, which was based upon [20], four supercardioids – all pointing vertically upwards – were positioned directly above the front-left, front-right, rear-left, and rear-right microphones (see Section 12.5.2.6). This "horizontally spaced and vertically spaced" design was based on the findings by Lee and Gribben [19] showing that changing the vertical spacing of the height layer did not have a significant effect on the perceived spaciousness of the system and test listeners marginally preferred both layers to be at the same vertical height.

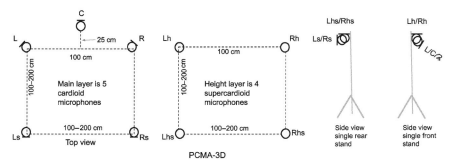

Figure 12.13 PCMA-3D microphone array configuration.

12.5.2.2 ORTF-3D

The main layer of the ORTF-3D system [17] consists of four supercardioid microphones spaced on an 18 cm square base (a much smaller footprint than the above) angled 45° off centre and 45° downward as shown in Figure 12.14. The height layer is coincident to the main layer as in the PCMA-3D array based on the findings in [19], consisting here of four further supercardioid microphones angled 45° upward. Since the front, back, and side acoustic signals are captured equally, this system is well suited for recording of 3D ambience, particularly for binaural VR contexts.

12.5.2.3 Bowles array

The Bowles array [21] is another two-layer approach where the main (lower) layer consists of four omnidirectional microphones spaced on a square and angled outwards. Both the width of the square (*d*) and the microphone angles can be adjusted to best suit the size of the ensemble being recorded. A directional centre channel is placed midway on the front baseline of the main layer. The height layer consists of four supercardioid microphones placed at a height of d/2 above the main layer (see Figure 12.15). The vertical angle of the height microphone is determined by directing the null of

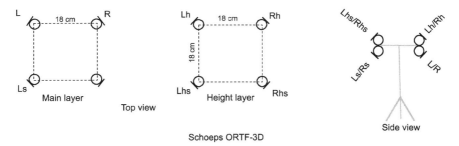

Figure 12.14 The indoor version of an ORTF 3D microphone array configuration. The outdoor version has a 20 cm width and 10 cm depth.

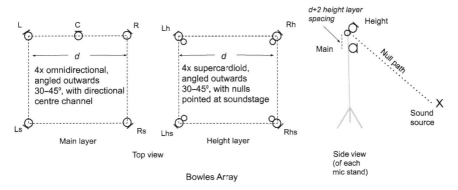

Figure 12.15 Bowles microphone array configuration.

the supercardioid, at about 126° to the sound stage for maximum rejection of direct signals. This is designed to provide very stable main-layer localization, and Bowles commented that his clients had remarked on the "realness" and "spaciousness" of the recordings, and that when played back on a 9.1 system, there was presence of height, but that the focus remained on the main microphones.

12.5.2.4 2L Cube

The 2L Cube array developed by Morten Lindberg [22] was largely based on the Auro-3D® speaker configuration. It consists of nine matched omnidirectional microphones in a 5.0.4 configuration. The main layer can also be expanded to accommodate a 7.0.4 configuration – with the addition of two more omnidirectional microphones. The cube dimensions can vary and can be sized from 40 cm for chamber music up to 120 cm to accommodate an orchestra (see Figure 12.16). Lindberg has recorded with the players completely surrounding the microphone array and on occasion will use acoustic pressure equalizer (APE) spheres to enhance the directionality of the microphones. The microphones can be angled as needed to maximize localization. Although this array appears optimized for channel-based Auro-3D® reproduction, its recordings play back well on Dolby Atmos® and DTS:X as well.

12.5.2.5 AMBEO Square and Cube

The AMBEO Square and Cube systems were developed by Gregor Zielinsky at Sennheiser [23]. They are inspired by A-B recording techniques and the Auro-3D® (5.0.4) speaker configuration. The use of multi-output dual-cardioid microphones make for another versatile system with dynamic pick-up patterns. Each microphone has two isolated outputs, front and rear, and these can either be used independently or combined to create variable polar patterns (see Table 12.2) – in either direction – on the fly, or in postproduction. In the AMBEO Square configuration, five microphones can

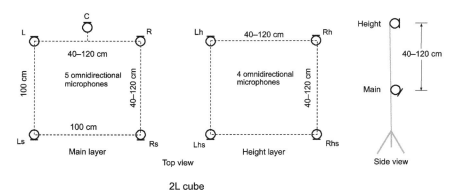

Figure 12.16 2L Cube microphone array configuration.

fit onto three stands in a row. The left microphone can provide signals for the left and left-surround channels, and the right microphone can provide signals for the right and right-surround channels. The left-height microphone can provide signals for the left-height and left-surround height channels, and conversely for the right-height microphone. The system still provides a spacious surround signal and optimizes the delay time between the direct signal and reflections arriving from the rear of the room.

The AMBEO Cube system has nine microphones (see Figure 12.17) thus delivering front-to-back time-of-arrival information and this also increases the front-to-rear decorrelation of the system, thus delivering greater perceived spaciousness on playback than the AMBEO Square system. The microphones can be angled as needed to maximize localization.

12.5.2.6 OCT-3D

Yet another two-layer configuration is the optimized cardioid triangle (OCT-3D) system, which was developed by Theile and Wittek [20]. The lower layer consists of two supercardioid microphones opposed at 180°, and spaced 40–90 cm apart with a centre cardioid placed 8 cm in front. Changing the spacing can alter the image; for the following supercardioid spacings, the associated stereo recording angles apply [24]:

40 cm: 160°
50 cm: 140°
60 cm: 120°
70 cm: 110°
80 cm: 100°
90 cm: 90°

The surround channel microphones are spaced by up to 100 cm using cardioids facing away from the sound stage. The system can be expanded to 3D by adding height channels on a height layer that consists of four

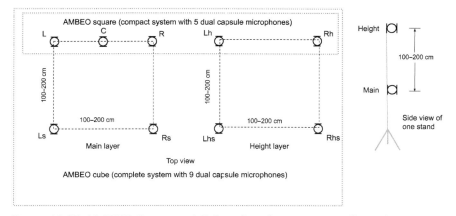

Figure 12.17 AMBEO Square and Cube microphone-array configurations.

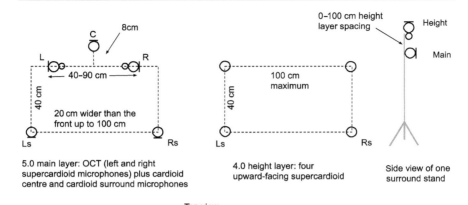

Top view

Figure 12.18 OCT-3D microphone-array configuration.

upward-facing cardioids, positioned up to 100 cm above the main array as shown in Figure 12.18. Compared to other systems, the OCT-3D delivers optimal frontal localization by minimizing the acoustic crosstalk between the front microphones. Since all the microphones are directional, two flanked omnidirectional microphones can be added to extend the low-frequency response of the system.

12.5.2.7 Decca tree hybrid

The Decca tree stereo system typically consists of three omnidirectional microphones with acoustic spheres to enhance directionality. The system has been widely used to capture film orchestras – historically using Neumann M50 microphones that feature an omni capsules mounted on the internal surface of an acoustic mounting-sphere (the capsule faces an open port on the opposing surface of the sphere); this provides enhanced directionality between 1 kHz and 4 kHz [25]. The Decca system can be adapted for surround sound by adding rear-facing cardioid microphones – placed to reject direct signals from the sound stage. By way of example, the system in Figure 12.19 is shown here as being expanded further to pick up height sound with the addition of a Hamasaki Square as a height layer, consisting of four figure-of-eight microphones orientated vertically. The L and R microphones can be angled in to adjust the frontal image-width and localization.

12.6 CONCLUSION

In this chapter, examples of microphone systems designed to capture soundscapes and musical performances in 3D were discussed. Needless to say, there are many more excellent 3D capture systems designed by recording engineers and researchers that were not included due to the scope and size of this chapter. That said, most 3D recording methods can be classified as binaural, sound-field, spaced arrays with height, or mixed

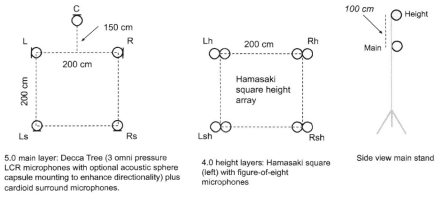

Figure 12.19 Decca tree hybrid configuration.

recording techniques. Binaural recording offers a compact system to capture, store, and reproduce a fully integrated immersive sound scene. One of the limitations of the system is that headphones or a crosstalk cancellation speaker system are needed on playback for full effect.

Ambisonic recording techniques treat all directions the same and thus support full 3D-sound environments yet are still compatible with mono, stereo, surround sound, binaural (using binaural synthesis), and XR systems. Unlike static binaural and channel-based recording techniques, the listener's perspective is not fixed and can be adapted in real time. Coincident microphone arrays from which a native B-format signal can be derived share the same flexibility thus are fully compatible with Ambisonic systems.

Surround-with-height microphone arrays allow recordists to work in 3D using familiar spaced microphones techniques using studio-grade microphones. These systems may not be as compact as sound-field systems, but meet the needs of the music and film sound industries.

REFERENCES

[1] B. Boren, "History of 3D sound," in *Immersive Sound: The Art and Science of Binaural and Multi-Channel Audio*, 1st edition, A. Roginska and P. Geluso, Eds. Focal Press, New York; London, 2017.

[2] A. D. Blumlein, "British Patent Specification 394,325 (improvements in and relating to sound-transmission, sound-recording and sound-reproducing systems)," *Journal of the Audio Engineering Society*, vol. 6, no. 2, pp. 91–98, 130, 1958.

[3] A. Roginska, "Binaural audio through headphones," in *Immersive Sound: The Art and Science of Binaural and Multi-Channel Audio*, 1st edition, A. Roginska and P. Geluso, Eds. Focal Press, New York; London, 2017.

[4] E. M. Wenzel, D. R. Begault, and M. Godfroy-Cooper, "Perception of spatial sound," in *Immersive Sound: The Art and Science of Binaural and*

Multi-Channel Audio, 1st edition, A. Roginska and P. Geluso, Eds. Focal Press, New York; London, 2017.

[5]   F. L. Wightman and D. J. Kistler, "Headphone simulation of free-field listening. II: psychophysical validation," *J. Acous. Soc. Am.*, vol. 85, no. 2, pp. 868–878, 1989.

[6]   E. M. Wenzel, M. Arruda, D. J. Kistler, and F. L. Wightman, "Localization using nonindividualized head-related transfer functions," *J. Acous. Soc. Am.*, vol. 94, no. 1, pp. 111–123, 1993.

[7]   P. Majdak *et al.*, "Spatially oriented format for acoustics: a data exchange format representing head-related transfer functions," in *Proc. 134th Audio Engineering Society (AES) Convention*, Rome, Italy, May 2013. www.aes.org/e-lib/browse.cfm?elib=16781.

[8]   P. Fellgett, "Ambisonics. Part one: general system description," *Studio Sound*, vol. 17, no. 40, pp. 20–22, Aug. 1975.

[9]   M. Gerzon, "Ambisonics. Part two: studio techniques," *Studio Sound*, vol. 17, no. 40, pp. 24–26, Aug. 1975.

[10]   R. Nicol, "Sound field," in *Immersive Sound: The Art and Science of Binaural and Multi-Channel Audio*, 1st edition. A. Roginska and P. Geluso, Eds. Routledge, New York; London, 2017.

[11]   P. G. Craven and M. A. Gerzon, "Coincident microphone simulation covering three dimensional space and yielding various directional outputs," US4042779A, Aug. 16, 1977.

[12]   "Core Sound — TetraMic." www.core-sound.com/TetraMic/2.php (accessed Dec. 11, 2020).

[13]   W. Woszczyk and P. Geluso, "Streamlined 3D sound design: the capture and composition of a sound field," presented at the 145th Audio Engineering Society Convention, Oct. 2018. www.aes.org/e-lib/browse.cfm?elib=19724.

[14]   K. (Y.-Y.) Zhang and P. Geluso, "The 3DCC microphone technique: a native b-format approach to recording musical performance," in *Proc. 147th AES Convention*, New York City, USA, Oct. 2019. www.aes.org/e-lib/browse.cfm?elib=20668.

[15]   P. Geluso, "Capturing height: the addition of z microphones to stereo and surround microphone arrays," in *Proc. 132nd AES Convention*, Budapest, Hungary, Apr. 2012. www.aes.org/e-lib/browse.cfm?elib=16233.

[16]   M. A. Gerzon, "Periphony: with-height sound reproduction," *J. of the AES*, vol. 21, no. 1, pp. 2–10, 1973.

[17]   H. Wittek and G. Theile, "Development and application of a stereophonic multichannel recording technique for 3D audio and VR," in *Proc. 143rd AES Convention*, New York City, USA, Oct. 2017. www.aes.org/e-lib/browse.cfm?elib=19266.

[18]   H. Lee, "A new multichannel microphone technique for effective perspective control," in *Proc. 130th AES Convention*, London, UK, May 2011. www.aes.org/e-lib/browse.cfm?elib=15804.

[19]   H. Lee and C. Gribben, "Effect of vertical microphone layer spacing for a 3D microphone array," *J. of the AES*, vol. 62, no. 12, pp. 870–884, 2015.

[20]   G. Theile and H. Wittek, "Principles in surround recordings with height," in *Proc. 130th AES Convention*, London, UK, May 2011. www.aes.org/e-lib/browse.cfm?elib=15870.

[21] D. Bowles, "A microphone array for recording music in surround-sound with height channels," in *Proc. 139th AES Convention*, New York City, USA, Oct. 2015. www.aes.org/e-lib/browse.cfm?elib=17986.

[22] M. Lindberg, *3D Recording with "2L-Cube."* Oslo, Norway: VDT, 2016.

[23] Personal communication with Gregor Zielinsky.

[24] SCHOEPS Mikrofone, "OCT sets," 2020. https://schoeps.de/en/products/surround-3d/oct-surround/oct-sets.html (accessed Dec. 11, 2020).

[25] SCHOEPS Mikrofone, "Decca tree set," 2020. https://schoeps.de/en/products/stereo/sets/decca-tree-set.html (accessed Dec. 11, 2020).

13

Redefining the spatial stage: non-front-orientated approaches to periphonic sound staging for binaural reproduction

Jo Lord

13.1 INTRODUCTION

Multichannel surround sound has existed in varying formats for three quarters of a century and for most of this time with undulating salience and favour. Through all of its developments, surround sound has managed to survive this vacillation of popularity versus purpose, and today is implemented in theatre, cinema, live performance, DVD recordings, and gaming audio, among other sound-to-picture applications. However, despite the achievements of researchers and the success garnered across the audio-visual industries, two-channel stereo continues to be the medium of choice for the delivery of recorded music [1]–[4].

The lack of spatial-audio assimilation within record production is due to perceptually and technologically limited creative practice, format wars, and the difficulties in consumption [5]. It could be said that even the most creative recording producers were never fully able to exploit the surround medium, locking themselves into a rigid four-corner reproduction and a front-reliant sound stage. Although the technology and delivery methods have evolved, the creative limitations within practice remain the same today.

With integration of periphony (surround sound with captured or artificial height) there is a requisite change in the way we actively listen to recorded audio and this change subsequently facilitates new approaches to sound staging and record production. This chapter does not specifically deal with loudspeaker configurations or Ambisonic technology, but does adopt the term periphony – as was first introduced by Gerzon in the 1970s to describe the phenomena of both vertical and horizontal sound reproduction around a listener [6]. The etymology of periphony comes from the Greek words "peri" – meaning "around or about" - and "phone" – meaning voice or sound. Periphony can therefore be defined as literally meaning "sounds from around" and will be used here to refer to surround *or* binaural audio with height. Given the widespread popularity of personal headphone listening, and with the democratisation of spatial audio technologies, there has never been a better time to explore new ways of composing records and utilizing space within production.

In a contemporary music-production context, sound staging can be defined as the organization of sound sources within the perceived boundaries around a schematized performance environment [7]–[11]. This chapter seeks to answer the question: "How do we approach the structuring of a periphonic sound stage for music?" In doing so, it redefines our approach to spatial staging through research and practical exploration, proposes key considerations, and offers a suggested approach to non-front-orientated periphonic staging – for binaural record production.

A headphone-based periphonic sound stage can provide more creative agency than that offered through a stereophonic or single-tier surround sound production system, while providing a democratic delivery method well suited to current consumer listening behaviour. This chapter explores how such agency can be exploited by using proxemics [12], sonic embodiment, and metaphor [13] as vehicles to enhance the representation and expression of the musical concepts presented through the staging.

The techniques discussed within this chapter focus on exploiting binaural perceptual phenomena as an aesthetic enhancer in music production, although not all of the techniques defined herein are strictly reliant on a specific binaural-synthesis tool (such as Dear VR Pro, used within this project) as a means of generating this perception, although it is acknowledged that the use of generic head-related transfer functions (HRTF) can compromise the perception of elevation. In many instances the same or a closely similar experience can be garnered through our "natural" spatial perception when listening via a domed, tiered multi-speaker system (such as Auro-3D® 13.1), thus these techniques are considered *transferable* – defined in this instance as being system-agnostic and therefore not reliant upon or governed by any specific tool or playback system. There are, however, some specific techniques that are designed to employ proxemic reaction through binaural sensation [12], and thus are implicitly reliant on the binaural phenomena as presented over headphones in order to convey a more embodied experience, especially where the staging explores intimacy and in/externalization. Figure 13.1 shows a hierarchy of production considerations that will be broken down in the subsequent text. The reader is also referred to the engineering treatment of binaural audio given by Sunder in Chapter 7 and Pike's discussion of its broadcast applications in Chapter 1.

13.2 SONIC CONTEXT

"Context determines creativity" [14] – in his book *How Music Works*, David Byrne put forward that it is the context behind a composition and relevant adaptations to new technologies that subsequently determine the output of creativity. In the case of this study, the same can be argued. Creative agency is reliant on not just the technology that is used, but also the context of the

Sonic context: performance music vs. production music		Sonic content: instrumentation, spatial, spectral, and lyrical content
Periphonic staging: non-front-orientated, focused, and unfocused		
Vocal staging: pitch-height periphonic placement, omnimonophonia, polyperiphonia or, hybrid combination	Instrumental staging: pitch-height placement, quadrants, hybrid mono/ stereo/periphonic sound field	Acoustic effects: non-traditional use of reverb, delays, Doppler, and Haas.
Binaural physiological phenomena: equalization, depth, intensity, and position as function of panning, frequency mapping, and externalization		
Proxemics and embodiment: sonic cartoons and metaphor		
Dynamic staging: perceived movement using static and kinetic sound sources		Rhythm and melodic contour: spectromorphological staging approach, utilizing temporal and tonal shape, and movement
Perceived space and depth: conceptual blending, internalization and externalization, and acoustic effects		

Figure 13.1 Periphonic production framework: outlines a hierarchy of production considerations when approaching periphonic sound staging.

musical production being worked on. By exploring periphonic approaches to music production, we think beyond the confines of the stereo "(*sound*) box" and the front-respecting front-projecting triangular sound field that we are now so conditioned to, as defined by Dockwray and Moore [10]. The proper holistic utilization of the periphonic sound field, as governed by a non-fronted approach, presents new opportunities to creative production that could not have been achieved through the traditional front-orientated approaches as employed in stereo and surround sound productions. Importantly, through the current democratization of spatial audio production tools, this opportunity for heightened creativity can now be explored, and consumed, by anybody with an interest and a set of headphones. This can be contrasted with "traditional" spatial audio practices of the not-so-bygone days that required specialist technology, complex workflows, high-end specialist studios and expert knowledge (mostly only attainable by industry and academic elite).

The sonic context of a production plays an important role in informing the creative practice and the level to which the periphonic sound field can be utilized. Two conceptual approaches – performance versus production – narrow down the appropriate creative response through the contextual requirements relating to the respective sonic content and schema. The characteristics of the two approaches can be defined as follows:

- *Performance music* represents the capture and reproduction of a performance in a given space and time; however, Zagorski-Thomas states that in such music, the boundaries between creation, performance and staging can become blurred through the creative and collaborative mediation processes [13, p. 76]. In an attempt to clarify such distinctions, Novotny states that the undertaking of performance music often involves minimal technological intervention and often does not involve virtual staging [15].
- Whereas, *production music* allows for the creative interpretation of a theoretical performance in imagined space(s), often undertaken through "technological means such as performance overdubbing, hyper-real microphone placement, MIDI & synthesizers, editing and processing, click tracks or prepared loops resulting in exaggerated sound and virtual soundscape" [15].

A study was conducted whereby the creative agency pertaining to each category was explored through practice. The limitations and affordances were defined, collated, and demonstrated via a two-production case study. Joey Clarkson's *Sort Yourself Out* [16] and Beautiful Thing's *Waiting* [17] were recorded as pilots for a separate research project whereby unusual spaces and locations were utilized for a live audio-video performance series (see [16] and [17] for links to audio). The performances took place at the Brentford Water and Steam Museum, a former Victorian steam-works in London. The artists were recorded live in differing acoustic spaces using common stereo approaches to live-performance recording: a combination of direct sound and acoustic capture using close and ambient microphone techniques.

Moylan articulates the relationship between staging and performance across various texts [7], [8], [11]. The *sonic content* of these productions intrinsically had a space imprinted within them and so too an environment. The nature of live-performance capture also presents a staging relationship within that environment. The ambient microphones captured the physical live stage at a fixed position along with the reverb generated by and containing the sonic properties of all the acoustic instruments and voices. The close microphone techniques also captured an amount of the natural reverb and spill from other instruments nearby. This limited the available variation in their placements due to their relationships with one another [8, pp. 164–167]. In Joey Clarkson's *Sort Yourself Out* [16] the trumpet spills across most of the microphone channels; it was loud and the space was particularly responsive to its energy. This presents much less autonomy over the placement of the trumpet within a virtualized stage, as the other tracks contain varying degrees of trumpet spill, with the violin presenting the most. In order to achieve both level balance and the appropriate perceived position of a given instrument in the mix, it requires a balance between the direct sound capture, the spill and the captured ambience.

If (when later mixing) it were desired to elevate the perceived source of the trumpet through panning, it could not be perceived as being high overhead if it could also be heard in the violin channel front right, unless it was much louder or arrives much sooner, as governed by the principles of the Haas effect. However, making the trumpet either louder or arrive sooner are often not viable options when working with captures of physical live stages. Increasing loudness affects our perception of sound source distance – it makes it feel closer, thus affecting perception of the intended position and the overall mix balance. "Increasing" the time of arrival of the trumpet – by delaying the other tracks that contained trumpet spill – could compromise the tone and would render the mix messy and the performance incoherent – care must be taken with time delays to maintain phase coherence, transient preservation, and musical timing. It is truly what it is – a performance captured in space and time – and this presents staging limitations pertaining to the instrument sources and acoustics that one has to work with. It is possible that capturing the performance through a 3D microphone array may have given a more coherent spatial aesthetic. However, there would still be limitations pertaining to the retrospective creative auralization of a sound stage given the physical limitations in the physical positioning of instruments, and the nature of ambient recordings of a performance dictating and fixing the stage position within the capture.

This case study evidences that both the sonic context (live-performance capture in a particular space) and the content (ambient recording, spill, and acoustic imprint) limit the creative agency when periphonically staging these pieces. Therefore, a more hyper-realistic production method was sought and a lightly enhanced spatial reproduction of the physical live stage was deemed the best approach to present both the most coherent sound stage, while also affording some means of spatial enhancement. Production-music pieces, such as Hidden Behind Static's *Monomorphic* and *Far From Here* [18], [19], and Jerome Thomas's *Late Nights* [20] present more flexibility and opportunity for conceptualizing meta-realistic virtual-staging concepts and constructing a more creative, surrealist auralization that makes full use of the periphonic sound field (see the audio links in the references for examples). This is not only due to the instrumentation and more complex textural layering that featured throughout these studio productions, but fundamentally also due to the absence of any particular perceived environment or space dictating the staging relationship and the staging freedom that drier, more direct sound affords.

13.3 SONIC CONTENT

As touched upon in the previous section, the "sonic content" also plays an important role in defining the spatial production.

By utilizing the lyrical, melodic, and temporal content as a production narrative, the musical and lyrical content can be reinforced within the spatial production staging, and vice versa. The melodic and temporal techniques are featured and continually developed across several spatial productions and presented through both instrument and vocal staging arrangements. The lyrical production technique can be demonstrated to good effect throughout Hidden Behind Static's *Far From Here* [19] and is discussed in detail in the vocal staging case study to follow.

Research states that vertical perception is influenced by spectral cues residing in higher frequency bands, with frequencies near 6–8 kHz being of particular importance for elevation decoding [21]–[24]. The "pitch-height effect" is a potent example of spectral governance over vertical perception; to some extent, the higher the frequency, the higher the perceived height of the point of origin relative to a point-source loudspeaker [22], [23], [25]. Therefore, it seemed a logical approach to consider sound sources with higher-frequency content as being particularly suited for a periphonic height placement, and therefore counter-defining sources with more low-frequency content as being best placed in the lower layers of periphony, or presented in mono or stereo. In Chapter 5, Lee provides a more detailed discussion of vertical auditory localization.

Having defined the differences in approaching a production based on the "sonic context" and "sonic content," there are two further categories to inform the production approach to non-front-orientated periphonic sound staging. These staging categories are presented as *focused* and *unfocused*:

- A focused stage can be defined as having a main focal point situated within a given area – typically the phantom centre of stereo playback, or perhaps the median plane for surround with height
- Unfocused staging can be defined as the holistic sound stage having no particular fixed focal position, even though the mini-stages that make up this holistic stage may be of a fixed-focus, or the point of focus may shift across different areas (such as above, behind, to the sides, or front)

To offer examples of each of these, we could say that a stereo or a traditional surround sound production has a typically front-*focused* stage, whereas Hidden Behind Static's *Far From Here* [19] presents a holistically *unfocused* stage throughout (see the audio link in the reference), with the change in the focal area being dependent upon the spectromorphology of the composition and the lyrical narrative. Spectromorphology, as defined by Smalley, refers to how sound spectra are shaped and change through time [26].

13.4 PERIPHONIC VOCAL STAGING

One of the most effective areas of periphonic production practice is vocal staging and this is due to several factors. The first is that the voice is naturally easier for human beings to localize as we are physiologically attuned to it. It has a familiar and recognizable sonic quality, and a higher frequency range to which elevation perception can readily respond. Second, in popular-music production, the voice is generally the principal point of attention in any stage, making it a particularly suitable candidate for staging experimentation [27]. Due to our familiarity with vocal placement most often being "upfront-centre" in the stereo sound field, this presents an expectation, due to a lifetime of stereo listening, that the vocal should be front-centre. Therefore, the voice requires careful consideration when placing within a non-front-orientated periphonic sound stage.

A concern of a non-fronted lead vocal is that the mix would sound unbalanced, unusual or incoherent if the vocal was placed off to one side or less defined and impactful if placed to the rear of the listener. Although "true" stereo recording and reproduction prevailed from the outset of classical-music stereo reproduction, one can often find examples in early stereo popular-music recordings where (say) the entire drum kit is in the left channel, the vocals in the right, or vice versa, and the rest of the instruments are similarly hard-panned – a two-channel-in-preference-to-stereo approach. Although this was more mono compatible than some contemporary stereo mix approaches, the disjointed hard-panned placement presents a less coherent sound stage with no interlinked stereo phantom imaging. As research, and familiarity surrounding the two-channel system progressed, techniques were defined that presented a relationship between the two channels – stereo. This relationship better reflected our binaural means of listening (and Blumlein's intensions) and subsequently real-life staging concepts were more commonly accommodated within the schema of the stereo sound field. These staging concepts use phantom imaging techniques generated through the physical position and signal distribution of sound sources in relation to the two speakers. However, when looking at a periphonic-binaural sound field, the entire production could be perceived as a "phantom image" that the user sits centrally within, as opposed to the stereo phantom image that is projected in front of the listener. This paradoxically presented creative opportunity in terms of staging practice, while also at the same time posing the problem of where to situate the lead vocal.

This case study investigates the theory that the localization and balance issue pertaining to non-fronted vocal staging could be rectified if the vocal cannot be localized as being in any one particular place but instead being perceived as emanating from everywhere; thus appearing as if the listener was inside the voice (although using binaural externalization to consciously avoid the "in the head" phantom-centre experience of normal stereo-headphone listening). Hidden Behind Static's *Monomorphic* [18] seeks to address the aforementioned issue of vocal placement by implementing an "omnimonophonic" aesthetic to the vocal stage. "Omnimonophonic" is a term coined through this research practice to describe the phenomenon of the voice being perceived as "one voice from everywhere." It can be considered as synonymous to "omnipresent," but specifically relating to audio perception and record production. The etymology can be broken down as "omni" meaning "in all places," "mono" meaning "one," and "phonic" meaning "voices or sounds."

Prior to establishing this, the study had investigated an alternative approach to implementing such an omnimonophonic image. This involved positioning the lead vocal directly overhead and filling the lower layers with a cascade of vocal reverb as a means for creating the surround. Although this did not provide the desired omnimonophonic effect, it did however, provide a basis for the vocal-staging experiment that will be discussed in the following section, whereby pitch-height staging and the omnimonophonic vocal placement were successfully addressed.

13.5 PITCH-HEIGHT VOCAL STAGING STUDY – *MONOMORPHIC* (2016)

Monomorphic can be described as an electroacoustic piece that experiments with vocal layering and rhythms [18], the salient aspects of which are summarized in Table 13.1.

There is considerable understanding of the perceptual and cognitive boundaries that underpin multi-voice composition. Individual voices can be heard singularly as melodic lines, whereas when in combination they might create a harmonic effect that must be balanced with melody, and although mutually intertwined, melody and harmony do not always work together in the same manner [28]. Such understanding of the different psychological effects that melody and harmony can have on musical cognition helped to further inspire the concept underpinning the following vocal technique.

The concept behind the *Monomorphic* "Vocal Tree" was to experiment with creating an immersive listener-centric vocal stage using an ecological approach in the construction of a choral sonic cartoon [29]. The *sonic context* and *sonic content* of the vocal suggested an interpretation whereby the harmonic grouping of voices with a very similar timbre was theorized as a possible means to meld the voices into either a "single" omnimonophonic voice or ensemble. The vocal staging was defined based on phrase-matched pairs of single voices, vertically positioned to exploit pitch-height effects,

Table 13.1 "Omnimonophonic Vocal Tree" staging matrix. This table presents a summary of the staging decisions as outlined within the text below.

Single-tracked voice split into parts	5 x vocal sound sources
Pitch and content – determines vertical position	Positioned in wide opposing pairs, lower-pitching pair on the lower layers, a mid-pitching pair in height, highest-pitching single voice top-centre
Result: "omnimonophonic" – a sonic cartoon based on a choral configuration	Height separation can be perceived, plenty of upward movement in melodic contour
	No specific single-voice localization, in unison they act as one voice surrounding the listener

and utilizing the spectromorphology of the melodic contour as a further means to create upward movement and exaggerated height. A single voice recording was used to attain a user-centralized phantom image of the combined vocals in harmony, reinforcing the perception of one-voice ensemble.

The stage consisted of a single-tracked vocal split into three parts over five vocal lines, delivered by the same vocalist and recorded on a single track in one take. Lower-layer voices were positioned on the listeners' shoulders, slightly to the rear and panned left-right, and voices in a height layer were positioned front-back. This made the vocals feel more cohesive and the front-back height positioning was an aid to localizing the voice at the rear.

This configuration of an immersive choir was based upon an ecological and schematic representation of how we might perceive the structure and auditory content of a choir in a performance space. Further, the voices were arranged by pitch – a very common approach to the physical positioning of a choir's parts – soprano, alto, and so on. However, a choir comprising many singers with the same voice is not a real-world phenomenon and therefore could be thought of as sonic cartoonism – a metaphorical representation of real-world sonic structures [30].

Since the vocal parts all came from the same vocalist, they had an almost identical timbre, and this was most beneficial when the voices sang both in unison and in single layers. When singing in single-layer pairs, the wide pairing acted to define and extend the stereo image, and the similar timbre reinforced both unison and harmony between the voices, allowing the listener to perceive the layered pairings as each being one voice. The opposed pairing and group positioning in this fashion also provided vocal coverage around all areas of the listener's head and this was deemed an important consideration for an enveloping and immersive binaural experience. The approach was further supported by matching vocal register to elevation; the lowest register was positioned lowest, the middle register was positioned centrally, and the highest register positioned at the top (see Figure 13.2). The voices were also grouped by phrasing on the X-Y plane in order to create a "call and response" scenario and a wider image via the spaced pairing as indicated in Figures 13.2 and 13.3.

Figure 13.2 A basic representation of the vertical-placement layers defined through pitch-height relationship.

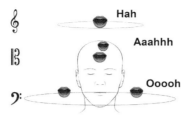

Figure 13.3 A basic representation of the omnimonophonic vocal-staging structure as presented within *Monomorphic*. The vocal height-positioning is defined by phrase, and register/pitch.

As the melodic contour ascends, the voices perform from spatial bottom to top creating a further implied upward movement in the melody. However, when all of the voice layers perform simultaneously, the voices and melodies meld into a unified harmonic structure, where the movement in the layers is identifiable – but the individual voices and melodies are not. This technique utilizes the phenomenon of inattentional deafness, whereby the harmonic vocal stream presents auditory-scene overload, and little differentiation can be made between the familiar timbres arriving simultaneously at the ears of the listener [31]; thus creating a perceivable effect of one voice from everywhere, or omnimonophony. This harmonic melding of periphonically arranged voices places the listener in the centre of a conically shaped phantom image, resulting in a fully encompassing, immersive sonic cartoon based upon a metaphorical choral configuration [18, Sec. 02':19"] (see audio link). This staging structure exists as the first known successful and coherent application of a non-fronted periphonic vocal arrangement that presented a single vocalist emanating from "everywhere." HRTF-matching permitting, there is clearly a perceivable upward vertical movement supported by the pitch-height based vocal placement and spectromorphology of the melodic contour.

13.6 POLYPERIPHONIC VOCAL STAGING AND DEPTH STUDY – *FAR FROM HERE* (2017)

The concept behind the track *Far From Here* reflects mental anguish and escapism. The spatial environment was constructed to represent a "void" or "nothingness." Proxemic theory [12] was exploited via binaural synthesis to explore the possibilities relating to perceived depth and intimacy in the staging [8]. This piece also uses a combination of externalization and internalization phenomena, achieved by combining stereo and periphonic sound fields, to imply a space existing outside of another space (see the "*Time*" cue in the lyric-based production section below).

The sociologist Edward T. Hall introduced the discipline of proxemics as being "the interrelated observations and theories of man's use of space as a specialized elaboration of culture" [12]. Hall is responsible for the notion of so-called "personal space" – the invisible force field that most people

Table 13.2 Polyperiphonic staging matrix. This table presents a summary of the staging decisions as outlined within the text below.

Multi-tracked "gang style" vocal technique	(Female voice) – not layered, multitracks periphonically placed.	Single-tracked vocal (male voice)
	Female lead rear LR (on shoulders)	Positioned top-front
	Depth is dependent on lyrical context	Under-mixed
	Rear left/right call and repeat interaction	Lo-fi, telephone distortion effect
	Contrasting frontal accent	
	Mostly dry with wet ride cymbals for distance enhancement	Drives and retains lyrical intelligibility
	Chorus female vocal top centre	
	Wet vocal swells	
	Upward melodic contour	

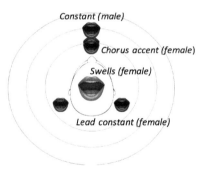

Figure 13.4 Far From Here polyperiphonic vocal-stage topography.

ensconce themselves in while moving through public places. A breach of implied boundaries (Hall suggested that the human ego extends about a foot and a half outside the body) is neither welcome nor tolerated [32].

To extend the periphonic vocal work developed in *Monomorphic* [18], a multitracked lead vocal with an unfocused (and coining the term) *polyperiphonic* stage placement is investigated to describe the attributes of the next development of the Vocal Tree. From its etymology, "polyperiphonic" can be defined as meaning "many voices from everywhere." The associated staging decisions are summarized in Table 13.2, and the associated topology in Figure 13.4.

Far, far from *this*,
go to *anywhere*, than *here*.
Dream, dreams you lost,
slip away from you.
Now you're *lost*.

Time, time won't heal.
Give for anything, not to feel.

Everything you wanted, everything you needed.
Gone.
You're wounded, drag yourself through every day,
always numb.

Ohhhh Ohh Ahhhh Ohhhhh
Never can be found, when you're hiding in the darkness.
Stuck between four walls, built by **you**.

Ohhhh Ohh Ahhhh Ohhhhh
Ahuahuohhhh wahhh ohh ahhh ohhhh
Stuck between four walls, built by **you**.

Figure 13.5 Colour-coded lyric cue-sheet presenting the lyrics and vocal staging cue-points relative to Figure 13.6.

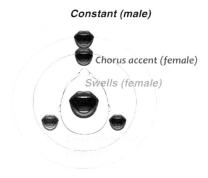

Figure 13.6 Far From Here colour-coded polyperiphonic vocal-stage topography as related to the colour-coded lyric cue-sheet.

This technique of multitracking the lead vocal provides the engineer with the opportunity to split phrases into spatial locations based upon timbral nuance, pitch/frequency content and lyrical content (see Figures 13.5 and 13.6). This technique utilizes small changes in timbre and the perform- ance interactions between voices to create the illusion of many voices from one voice. Typically, in stereo production the multitracked voices would be layered in the same position to create a thicker texture, or panned left, centre, right (LCR) to enhance the width. This is commonly used in "gang" style production and is often present through choruses. However, in this application the multiple voices were spaced, and presented similar, but fundamentally more distinct timbres. The voices interacted as one lead voice throughout the piece, and the stage focus and vocal tone changed depending on the spectromorphology of the musical constructs (verse, chorus, etc.).

Lyric-based production cues also informed the development of the vocal stage, and offered the opportunity to experiment with both depth and intimacy (again, see Figures 13.5 and 13.6). However, the topography of the vocal stage remains focused on utilizing the pitch-height phenomena as a means for defining the tonal structure of the voices, with the exception of the spoken male voice, which was positioned high-front centre as a constant lyrical reinforcement.

The first cue point, "*Far,*" starts the female vocal in the mid-field – this vocal is made up of a double tracked pair of voices binaurally positioned directly to the left and right of the listener, in line with the ears. "*This*" then cues the automation to bring the voices into the closer mid-field. "*Go to anywhere*" takes the voices back to the farther mid-field and then further perceptual contrast is utilized when "*here*" draws them in very close to the listener whereby they can be perceived as almost being upon the listener's shoulders. This feeling of intimacy is further enhanced using a multi-band compressor to apply compression to high frequencies (a familiar stereo technique) creating a sonic cartoon of closeness. This compression technique controls the dynamics of the high-frequency content while allowing the lower spectrum to remain naturally dynamic, resulting in a heightened, hyper-real, breathiness to the vocal.

The next phrase begins with the lead pair of voices again in the mid-field. The cue "*slip away*" automates the voices out to the far field and "*lost*" takes the female voices far enough out that they are almost inaudible, effectively lost in the far reaches of the sound field. The position of the male voice is consistently high-front centre. It is under-mixed throughout (sitting low in the vocal blend) so that it acts as a supporting layer rather than a featured one. However, when the female voices are "*lost,*" the male voice then becomes more of a feature, helping to retain intelligibility of the lyrics when the lead voices are drawn away into the distance. To accentuate the feeling of the voices being lost, the reverb send was switched to pre-fader, and the tail and output level increased. This made the voices appear as if they moved outside of their critical distance, whereby the reflections became far greater and more prominent in perception than their direct sound [33].

"*Time*" is an interesting cue that is unrelated to the vocal stage. It is a cue for exploring the conceptual blending of stereo and periphonic-binaural spaces in order to enhance the void/vacuum aesthetic that contextualizes the production. The two bass-synth pads had a similar tonality and timbre, such that when panned and layered together they created a thicker texture and a slight perception of musical movement through spectral and temporal flux. However, when experimenting with their purpose in the spatial sound stage, it was discovered that panning them in the stereo domain caused the perceived similarities and difference between the two sounds to come alive with what appeared to be interference patterns. This created not only width but a directional and fronted sense of movement between the interactions of the two panned channels. When combined with the binauralized periphonic sound field, this action presented a very interesting spatial duality that lent a new aesthetic to the production. This combination of stereo and periphonic spaces presents a noticeable change in the character of the perceived sound field(s). The stereo sound field appears internalized, as it tends to be on headphones, whereas the periphonic-binaural sound field externalizes due to the effect of the HRTF [34] and the early reflections offered by the DearVR panner. It is this juxtaposition between the two spatial perceptions, along with the synth timbre and spectromorphology that exaggerates the sense of "void" or "vacuum." This adds to the perceived perceptual contrast within the staging schema, creating interest, intrigue and movement within the music, while presenting a new experience that is metaphorically representative of the sonic content and concepts.

Following this, the female voices remain in the near field, but not overly close, or as intimate as the "*hear*" cue was presented. They remain in this position until the cue "*gone*" triggers their cut with the male voice again taking attention and retaining the intelligibility and meaning of the lyrical content.

The "*oooooooh*" pre-chorus swells were positioned high overhead, and present movement through melodic contour and provide a larger sense of space – in contrast to the verses. The swells act to introduce the chorus, which utilize the rear lead-vocals in a close call-and-repeat fashion relative to the lyrical phrasing.

The lyrics "*stuck between four walls*" cue a ping-pong call-and-repeat interaction between the voices, whereas "*built by you*" follows this same pattern, but "*you*" then cues both voices to sing in unison – which doesn't quite present the same definitive frontal phantom image we see with stereo, but the similar experience conveys a more fronted intimacy.

13.7 VARIATION OF (RE)DEVELOPMENT AND (RE)APPLICATION

The techniques were (re)developed and (re)applied across the various pieces within the larger research project.

Jerome Thomas's *Late Nights* [20] (see audio link) presents an amalgam of the omnimonophonic and polyperiphonic versions of the Vocal Tree, redeveloped and reapplied to *Late Nights*. The combination of techniques suited the vocal in this piece particularly well. Characteristic of the production practice associated with the work and style of Jerome Thomas, there are numerous vocal stems, each containing a specific harmony, backing vocal, or lead vocal phrasing layer. In a typical stereo production, these vocals would be layered together "gang style" to create texture and tonal harmony between the voices, and gently panned across the stereo image. In this instance, the periphonic sound field offers the opportunity to completely spread them out and make use of them as individual sound sources in 3D. Accordingly, the backing vocals and vocal harmonies in this production were periphonically organized around the listener and arranged within a way that presented balance between the quadrants of the periphonic sound field.

The lead vocals were arranged in a paired call and repeat fashion in the frontal left-right domain. One backing vocal was automated to sweep rear left-right on the backing-vocal-repeat of the lyrical cue "*Am I too late on arrival?*" [20, Sec. 0':47"–0':53"] as a means of creating a movement-themed sonic cartoon influenced by the movement implied within the lyrical content (see audio link for example). The backing-vocal placements work particularly well in the rear, as they are not a prominent feature in the mix and therefore the filtering and lack of presence is not a concern. The way they respond individually aids to define clarity to their placement, create movement through the vocal interactions and reinforce the immersion generated through their staging. This technique was an adaptation of the polyperiphonic technique, whereby the same voice was used throughout the recordings with different timbres and phrases in order to construct the "many voices from everywhere" aesthetic. Although there are points in this piece where all voices sing together, a different result was achieved from that in *Monomorphic* – the effect still lends itself to presenting an amount of perceived omnimonophony through an immersive and enveloping vocal harmonization.

Although the staging practice discussed throughout this chapter can be varied, redeveloped, and reapplied in a manner dependent on the sonic content of a given production, the sound-stage boundaries remain consistent throughout any variation of concept. It is interesting that upon reflective analysis, it could be observed that periphonic sound staging often presents a 3D cone shape around the listener (although clearly a function of playback medium and HRTF compatibility), defining these as boundaries of the perceived performance environment (see Figure 13.7). This could be considered an upright, multidimensional variation on the *triangular* and *diagonal* mixing taxonomies originally proposed by Moore and Dockwray [10, pp. 185–187], albeit with a new and contemporary framework defining the sound source organization therein.

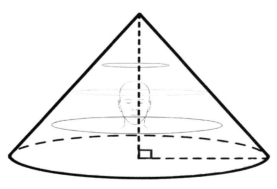

Figure 13.7 Periphonic cone staging taxonomy.

13.8 CONCLUSION

The techniques constructed within the scope of this project address key issues pertaining to periphonic sound staging and offer a new and contemporary approach to music production. This can be seen more specifically through the design and implementation of the vocal-staging practice created within this project, as well as via the philosophical, theoretical, and practical implementation of staging constructs throughout the associated AES practical work. In taking an ecological approach to staging through an applied understanding of embodiment theory (which was largely beyond the scope of this chapter), the study evidenced that non-front-orientated techniques can be creatively utilized to reinforce the sonic narrative, content, and metaphor of a production – beyond that which more traditional production approaches can usually afford. This increase in creative agency is demonstrated through the construction and redevelopment of periphonic sonic cartoons that comprise the surreal, non-frontal sound stages. When combined with the binaural phenomena and the integration of proxemics as a production theory, periphonic spatialization allows for further creative agency to convey emotiveness, depth, and intimacy, although further work should evaluate this effectiveness for a range of listeners with differing HRTF compatibility. The inclusion of head-tracked binaural panning would also help to both resolve potential front-back confusion and enhance the perception of elevation.

The omnimonophonic staging technique helps to resolve the placement and perceptual issues pertaining to non-front-orientated periphonic vocal staging without compromising balance or the listener's experience, offering an enhancement to the musical production that cannot be achieved through traditional stereo or surround sound practice. Thus, the study presents a new perspective for approaching spatial music production, offering a framework of technique that provides pathways for implementation across other musical contexts.

REFERENCES

[1]   E. Torick, "Highlights in the history of multichannel sound," *Journal of the Audio Engineering Society*, vol. 46, no. 1/2, pp. 27–31, Feb. 1998.

[2]   A. Farina, R. Glasgal, E. Armelloni, and A. Torger, "Ambiophonic principles for the recording and reproduction of surround sound for music," presented at the *Audio Engineering Society (AES) Conference: 19th International Conference: Surround Sound – Techniques, Technology, and Perception*, Jun. 2001. www.aes.org/e-lib/browse.cfm?elib=10114 (accessed Nov. 17, 2020).

[3]   J. Wuttke, "Surround recording of music: problems and solutions," presented at the *AES Convention* New York City, USA, Oct. 2005. www.aes.org/e-lib/browse.cfm?elib=13356 (accessed Nov. 17, 2020).

[4]   T. Holman, *Surround Sound: Up and running*. Taylor & Francis Group, 2007.

[5]   S. Guttenberg, "Whatever happened to 5.1-channel music?" *Stereophile. com*, Jul. 07, 2009. www.stereophile.com/asweseeit/whatever_happened_ to_51-channel_music/index.html (accessed Nov. 17, 2020).

[6]   M. A. Gerzon, "Periphony: with-height sound reproduction," *Journal of the Audio Engineering Society*, vol. 21, no. 1, pp. 2–10, Feb. 1973.

[7]   W. Moylan, *The Art of Recording: The Creative Resources of Music Production and Audio*. Van Nostrand Reinhold, 1992.

[8]   W. Moylan, "Considering space in recorded music," in *The Art of Record Production*, S. Zagorski-Thomas and S. Frith, Eds. Routledge, United Kingdom, 2012.

[9]   S. Lacasse, "Persona, emotions and technology: the phonographic staging of the popular music voice," 2005. http://charm.cchcdn.net/redist/pdf/ s2Lacasse.pdf (accessed Nov. 17, 2020).

[10] R. Dockwray and A. F. Moore, "Configuring the sound-box 1965–1972," *Pop. Music*, vol. 29, no. 2, May 2010, doi: 10.1017/S0261143010000024.

[11] W. Moylan, *Recording Analysis: How the Record Shapes the Song*. Taylor & Francis Group, 2020.

[12] E. T. Hall, *The Hidden Dimension*. Bantam Doubleday Dell Publishing Group, 1966.

[13] S. Zagorski-Thomas, *The Musicology of Record Production*. Cambridge University Press, 2014.

[14] D. Byrne, *How Music Works*. Canongate Books, 2012.

[15] P. Novotny, "Performance music or production music?" presented at the 14th Art of Record Production Conference 2019, Boston, MA. United States, May 2019. https://arp19.sched.com/event/L4Dc/performance-music-or-production-music (accessed Nov. 17, 2020).

[16] Joey Clarkson, *Sort Yourself Out* (Live At Brentford Steam Works) [Periphonic Mix]," *Soundcloud*, Nov. 17, 2020. https://bit.ly/30EX6HQ (accessed Nov. 17, 2020).

[17] Beautiful Thing, *Waiting* (Live At London Water Steam Museum) [Periphonic Mix]," Nov. 17, 2020. https://bit.ly/2MZpcWu (accessed Nov. 17, 2020).

[18] Hidden Behind Static, *Monomorphic*, Nov. 17, 2020. https://bit.ly/37sEhct (accessed Nov. 17, 2020).

[19] Hidden Behind Static, *Far From Here*, Nov. 17, 2020. https://bit.ly/3fugtI5 (accessed Nov. 17, 2020).

[20] Jerome Thomas, *Late Nights* [Periphonic Mix]," Nov. 17, 2020. https://bit. ly/3e1okfV (accessed Nov. 17, 2020).

[21] D. Gibson, *The Art of Mixing: A Visual Guide to Recording, Engineering, and Production*. Taylor & Francis Group, 2018.

[22] H. Lee, "Sound source and loudspeaker base angle dependency of phantom image elevation effect," *Journal of the AES*, vol. 65, no. 9, Sep. 2017, doi: 10.17743/jaes.2017.0028.

[23] J. Paterson and G. Llwellyn, "Producing 3-D audio," in *Producing Music*, M. Marrington, R. Hepworth-Sawyer, and J. Hodgson, Eds. Focal Press, New York, 2019, pp. 169–170.

[24] C. I. Cheng and G. H. Wakefield, "Introduction to head-related transfer functions (HRTFs): representations of HRTFs in time, frequency, and space," *Journal of the AES*, vol. 49, no. 4, pp. 231–249, Apr. 2001.

[25] D. Cabrera and S. Tilley, "Parameters for Auditory Display of Height and Size," in *Proceedings of the 9th International Conference on Auditory Display,* Boston, MA, USA, Jul. 2003, p. 4.

[26] D. Smalley, "Spectromorphology and structuring processes," in *The Language of Electroacoustic Music*, S. Emmerson, Ed. Macmillan, London, 1986, pp. 61–93.

[27] S. Lacasse, "'Listen to my voice': the evocative power of vocal staging in recorded rock music and other forms of vocal expression," University of Liverpool, United Kingdom, 2000.

[28] W. F. Thompson, *Music, Thought, and Feeling: Understanding the Psychology of Music*, 2nd edition. Oxford University Press, 2015.

[29] E. F. Clarke, *Ways of Listening: An Ecological Approach to the Perception of Musical Meaning*. Oxford University Press, 2005.

[30] S. Zagorski-Thomas, "The spectromorphology of recorded popular music: the shaping of sonic cartoons through record production," in *The Relentless Pursuit of Tone: Timbre in Popular Music*, R. Fink, M. Latour, and Z. Wallmark, Eds. Oxford University Press USA - OSO, Oxford, United States, 2018.

[31] S. Koreimann, B. Gula, and O. Vitouch, "Inattentional deafness in music," *Psychol. Res.*, Mar. 2014, doi: 10.1007/s00426-014-0552-x.

[32] A. Petrusich, "Headphones everywhere," *The New Yorker*, Nov. 17, 2020. www.newyorker.com/culture/cultural-comment/headphones-everywhere (accessed Nov. 17, 2020).

[33] G. D. White and G. J. Louie, *The Audio Dictionary: Third Edition, Revised and Expanded*, REV-Revised, 3. University of Washington Press, 2005.

[34] B. Xie, *Head-Related Transfer Function and Virtual Auditory Display: Second Edition*. J. Ross Publishing, 2013.

14

Mixing 3D audio for film

Kevin Bolen

14.1 INTRODUCTION

"I've always thought of multichannel film-sound as 3D," says re-recording mixer Juan Peralta [1], whose work includes dozens of immersive film mixes including *Avengers: Endgame* [2], *Ant-Man and the Wasp* [3], *Jurassic World* [4], and *Tomorrowland* [5]. "Whether 5.1, 7.1, or Dolby Digital EX, any sounds that we pan beyond the screen channels exist in the 3D space of the movie theater." Unless heard only through headphones, a film's soundtrack is never experienced in only two dimensions. Even without the overhead speakers and object-based panning now common in modern immersive film-sound formats, the height, width, depth, and acoustics of the cinema have always informed re-recording mixers' creative and technical processes. Film-sound formats like Dolby Atmos®, Barco Auro-3D®, Xperi DTS:X, and IMAX 12.0 empower re-recording mixers to manipulate 3D soundscapes with greater flexibility and precision than multichannel sound, but the creative techniques they employ remain focused on the listeners' perception of the story, whether on or off screen.

This chapter presents a collation of interviews with Hollywood industry professionals who have been at the evolution and cutting edge of 3D audio in movies, and have shaped its current deployment the world over.

14.2 BACKGROUND

I too am an industry professional, and I first offer the reader an overview of my work. My first experience mixing 3D audio for film was assisting Peralta and another re-recording mixer, Gary Rizzo, on Joseph Kosinski's 2010 movie *Oblivion* [6] at Skywalker Sound, north of San Francisco. One of the first films to be mixed natively in Dolby Atmos®, the immersive mix for *Oblivion* extended Kosinski's striking post-apocalyptic vision beyond the confines of the movie screen. As a novice assistant re-recording mixer, I found object-based audio for cinema novel in its creative execution but familiar in its technical implementation because it paralleled video game audio, where I had started my audio career. After *Oblivion*, I had the great fortune of contributing to many immersive 3D audio film mixes, including

The Croods [7], *Epic* [8], *Thor: The Dark World* [9], *How to Train Your Dragon 2* [10], *Maleficent* [11], *Home* [12], *Pixels* [13], *The Jungle Book* [14], and *Ice Age: Collision Course* [15].

14.3 FROM HORIZONTAL SURROUND TO 3D

Dolby Atmos® is one of the most familiar 3D immersive-sound formats for film, although Skywalker Sound's team is well versed with Barco's Auro 3D, Xperi's DTS:X, IMAX 12.0, and a variety of other formats destined for cinemas or theme parks.

Re-recording mixer Will Files (*Stranger Things* [16], *War for the Planet of the Apes* [17], and *Star Trek Into Darkness* [18]) was involved in early beta testing for Dolby Atmos®, which is now one of the most common theatrical 3D film-sound formats. He describes the experience of trying out the new technology: "We took a few scenes from recent movies like *The Incredibles* [19] and *Mission Impossible: Ghost Protocol* [20] and tried re-mixing them into Atmos. Then I had the opportunity to mix the first two Atmos trailers, 'Unfold' and 'Leaf,' as well as the Atmos re-mix of the very first Atmos feature film release ever, Pixar's *Brave* [21]." Recognizing the potential impact of Dolby Atmos® on cinema audio, he recalls, "It was a fun and exhilarating experience!" [22].

Brave may have been the first feature film to be released theatrically in Dolby Atmos®, but it was not natively mixed in that way. It was originally mixed in 7.1 by Gary Rydstrom and Tom Johnson, and up-mixed into the new immersive audio format by Files. Before the Dolby Atmos® mix could begin, assistant re-recording mixer Tony Sereno had to meticulously separate the 7.1 mix into "pre-dub" elements suitable for re-panning as sound objects. Over the course of four evenings during the 7.1 mix, Sereno re-recorded nine discrete 7.1 premixes of sound effects, creatures, ambience, Foley footsteps, Foley cloth, dialogue, loop group, sound effects reverb, and Foley reverb for Files to remix in the Stag Theater, the first room at Skywalker Ranch to be able to play Dolby Atmos®. "This gave Will a starting point," Sereno recalls, "to have the Atmos final mix similar to what Gary and Tom had for the 7.1, yet giving him the flexibility to separate any pre-dub elements into an Atmos environment. Will had everything he needed to remix the movie in Atmos. He had the challenge; I was able to possibly give him a starting point" [23].

14.3.1 Opportunities

While full-frequency sound is often constrained to three full-range loudspeakers behind the screen, immersive audio formats ensure that the impact is not lost when sounds travel around the auditorium. On *Oblivion*, re-recording mixers Rizzo and Peralta were able to take advantage of mixing in 3D from the beginning of the mixing process. As Peralta recalls, "Joe Kosinski, the director, loved mixing in 3D. The sci-fi-ness and technology in his film made it easy to use a lot of sound objects to tell the story." Dialogue, music, and sound effects were all given more latitude to convey emotion and information in concert than traditional multichannel audio formats usually allowed.

Each immersive 3D audio format for film brings unique opportunities into the theatre. In addition to adding speakers for height on the walls or ceiling above traditional multichannel surround sound arrays, immersive audio formats may include expanded frequency ranges in the surround speakers, including bass management, so that a sound can be panned from the screen, around the room, and back, without losing any of its acoustic energy or emotional impact. Whether mixing for the character Dr. Ryan Stone in *Gravity* [24] or the object Thor's hammer Mjölnir in *Thor: The Dark World*, 3D mixing allows a character or object to rotate around the theatre, even through individual speakers, expanding the narrative reach of the storyteller in every direction. "One of my favorite sonic effects offered by the new immersive formats is the increased panning resolution across the screen," says Files, "especially if I can access the two speakers located just immediately off the screen. It's very cool to be able to pan someone talking across the screen and then *off* the screen and continuing into the auditorium. If it's done right, it just feels completely natural." In this way, the storyteller's vision persists, even when it is heard, but cannot be seen. The power of 3D audio is not a new discovery, but the widespread adoption of immersive theatrical formats means that the potential of Walt Disney's Fantasound (the playback system devised in the late 1930s by Walt Disney Studios for the movie, *Fantasia* [25] – the first commercial film release with stereophonic sound) can now be brought to life in theatres worldwide.

Disney's *The Jungle Book* [14] is another film that utilizes 3D audio to the fullest extent. From the beginning, director Jon Favreau wanted to recreate Walt Disney's Fantasound, using 3D audio positioning to tell the story with not only positional dialogue and sound effects, but also with 3D positioned music. The snake character Kaa's looming menace and King Louie's imposing stature are underscored by spreading their dialogue, Foley, and thematic music vertically, making the audience feel as small as Mowgli does in comparison to their epic scale. "Whether slowly panning the flute during the opening sequence, or the voice of Kaa the snake," Lora Hirschberg [26] recalls, "we were able to use Dolby Atmos to apply the Fantasound approach Jon was looking for."

14.3.2 Challenges

Mixing 3D audio for film is not without its challenges. Due to the complexity of rendering additional channels of audio, and recording positional metadata, large theatrical films now re-record up to 128 channels of audio or more, depending on the number of multichannel beds and objects utilized during the final mix. "Disney Stage A was the first Dolby Atmos stage built from scratch as the format was coming into use," remembers re-recording mixer David Fluhr (*Frozen* [27], *Frozen II* [28], *Moana* [29], and *Zootopia* [30]). "I was tasked with innovating a workflow that utilized Atmos in ways that would not add significant time to mix schedules, and would still provide a way to create the traditional, more prevalent 7.1 and 5.1 theatrical formats. I spent three months alone on the stage to create the workflow and test it. Our first use was on *Planes* [31], and then *Frozen*" [32].

Besides the increased complexity of the re-recording process, the ability to individually address up to 64 speakers can distract the audience if not used with intent. In visual effects (VFX)-heavy films, a single sequence may cost millions of dollars to shoot, composite with VFX from multiple studios, and colour correct. The film's soundscape must be mindful not to break the audience's immersion and draw their attention away from the screen just as millions of dollars of work flashes by in a few seconds. "We use surround speakers to help tell the story on screen without getting in the way," Peralta says. "Everything you mix in 3D, you have to mix down into the other sound formats, and the more panning you do in 3D, the trickier it is to fit in the 5.1 or 7.1 box. We have to keep in mind that we want everyone to experience the film in the same way."

Even with the luxury of time and experimentation, sometimes the most effective 3D film mixes require bespoke technology specifically built for the content of the show. According to Rydstrom, "Theatrical 3D can best be achieved, in my experience, in specialty venues, such as dedicated theme-park theaters, in which non-standard speaker placements can pinpoint sound anywhere in the theater itself. Disneyland's *Muppet*Vision 3D* had speakers inside balconies [characters: Statler and Waldorf], and the projection window [character: Swedish Chef]. *Honey I Shrunk the Audience* included floor speakers that could literally pull sounds (rat squeals!) through a specific path in the theater" [33]. The latter was a "4D" film – a 3D-visuals film with physical effects, for example, shaking auditorium – only shown in Disney theme parks, and the former, an "attraction" located in Disney's Hollywood studios. However, Rydstrom is quick to point out that the perception of 3D audio mixing in film sound is sometimes perceived more dramatically than intended. "Disneyland's *Captain EO* [another theme-park attraction 3D film] utilized site-specific speakers to achieve 3D audio effects, though we learned about the psychology of 3D sound and image on that show. We were often astounded at how a great visual 3D effect would trick the mind into hearing the accompanying sound in 3D, both by setting the sound back, seemingly behind the screen, and floating the sound out from the screen, into the audience. The lesson learned was that sometimes with only the slightest nudge, 3D audio was successful when the 3D visuals were."

14.3.3 Panning dialogue in 3D

The same risk of distraction applies for critical narrative dialogue when mixers are tempted to pan characters' dialogue behind the camera, especially during perspective changes during conversations. As Hirschberg has learned, for the most part, no one should be the camera; the camera doesn't exist in the film. As technologies evolve, we are sometimes tempted to try new and novel solutions to familiar challenges. However, novel approaches to panning and positioning often draw attention to themselves. Hirschberg says, "the sound of a film takes a different trajectory than the picture cut, often designed to take a longer arc in order to join more frequent picture cuts. The mix indicates the appropriate beginning and end of movement through a scene." Noticeably, changing the perspective of the mix or rapidly panning sounds can draw unwanted attention to picture cuts, unless that is the director's intent. Lora points out that films like Kathryn Bigelow's *Strange Days* [34] – which was mixed in 5.1 – would benefit from realistic perspective shifts using modern 3D film-mixing tools, as the first person point of view was established as a narrative tool. "For standard format movies, it can be effective to pull sounds into the surrounds, simulating a 3D effect," agrees re-recording mixer Rydstrom, "Katherine Bigelow's *Strange Days* was full of dramatic point of view moments, and we aggressively pulled sounds into the surrounds; most interestingly, we pulled the vocals of the character whose POV we were watching into the surrounds, seeming to put the vocals into the heads of the audience."

The techniques applied in *Strange Days* were pushed even further in Alfonso Cuarón's *Gravity* [24] as the film's mixers were able to effectively pan dialogue because often the camera was locked into a long shot with a relatively stable perspective as the chaos of the scene swirls around the camera. Hirschberg points out that the film's visual and audio aesthetic provided the mixers with an ideal opportunity for 3D panning on dialogue: "they were able to move the dialog through the room because it moved slowly and locked into the visual perspective correctly. It was also band-passed and futzed like a radio transmission, which stands out to the listener's ear, and also helps hide the sonic shift as the sound moves between different speakers that might have a different timbre due to speaker size or tuning."

14.3.4 Timbre shift

Not all speakers in a cinema auditorium are the same type or the same size. Even though 3D audio formats like Dolby Atmos® dictate the use of full-range surround speakers closer in specification to the screen channel speakers, subtle difference between speakers, placement, and room acoustics means that the same sound may be perceived differently between two different speakers. Files reminds us that "[t]he human ear tends to forgive sonic and timbre changes in moving sound effects, but our ears are so closely attuned to the frequencies in dialog that panning often exposes the timbre shift between speakers; being between speakers is a dangerous place for dialog. Sometimes we try it and then undo it. It doesn't get better or more exciting, it just gets different and weird [26]."

Files combines several techniques to keep elements sounding natural as they pan around the room: "I'm a big believer that sounds should change in some way as they leave the screen and venture into the auditorium. For example, when panning dialog, Foley or hard sound effects, I tend to use reverb and EQ to approximate the feeling of the sound going 'off mic' a bit because it just seems to sound more natural that way. If you hear a bone-dry sound panning around the room, it often feels fake or contrived. Though sometimes it can be used to great effect – for instance, in *Gravity*

[24]. Sometimes I'll use a trick of panning a reverb return or printed mono reverb behind the dry object and add a bit of size/diffusion to the reverb so it creates a 'halo' around the dry sound as it moves around the room. That seems to help it sound more natural."

14.3.5 The sweet spot

Many film directors still feel that 5.1 multichannel film mixes are "3D enough" for their tastes. The sweet spot, where the sound mix sounds best to the greatest number of people seated near the middle of the auditorium, is wide due to the diffuse nature of the side and/or rear surround speaker arrays. As Hirschberg says, "you're much less likely to perceive one sound as just being heard by one ear; you're not going to get stuck in a seat next to a cricket." She feels complex 3D film mixes are best for the largest of cinema auditoriums, so that sounds have room to travel before hitting the listeners. However, she feels surround-heavy mixes can be oppressive in stadium seating environments, especially when mixed in 3D: "I've been dialing back the amount of static content in my mixes, because if you have a lot of static objects in specific surround or overhead speakers, the audience in the back of the auditorium is going to be overwhelmed."

14.3.6 Panning music in 3D

Hirschberg feels that putting individual instruments into different speakers often causes the music mix "to fall apart." Fluhr agrees: "I've seen and heard many examples of music in particular ending up 'disembodied' into wide, immersive sound-fields, where it doesn't sound like an orchestra any longer [...] instruments flying around the room in peculiar ways is distracting." Instead of creating many dynamic objects, Fluhr often prefers to create what he calls "custom arrays" of static objects spread above and to the rear of the theatre. "It requires a lot of planning in advance [...] and some experimentation – using objects, delays, verbs, EQ – all the tools [...] in the right combinations. I usually keep music wide – and size plays a big role to keep things from falling apart. It also gives elements on screen more room."

14.3.7 Depth

The earliest forms of multichannel surround sound allowed re-recording mixers to enhance the depth of the soundscape. However, traditional multichannel surround-speakers output at lower volume and lacked bass management, with limited frequency response compared to the primary screen-channel speakers, meaning that any sounds panned off screen "obviously lose fidelity and power," says Peralta. "More impressive than just sound-field placement are the full-range surrounds" common to many 3D film-sound formats. "With 3D mixing, we can bring sounds closer to the people in the theater, but we have to be careful," warns Peralta: "a sound panned into multiple speakers can create the illusion that the sound is coming from everywhere."

In addition to panning sounds back into the room, many mixers have developed more sophisticated 3D mixing techniques utilizing traditional mix tools like reverb, as described by Files: "my main tool for adding depth is the same as what I would use in a traditional mix as well – reverb. But specifically, layers of reverb with different amounts of pre-delay and varied lengths. It's very seldom that we encounter spaces in the real world that only have one reverb characteristic. Especially if characters and the camera are moving around in a space, the character of the reverb will change a lot through the course of the shot and scene. Layering multiple types of reverb is a nice way to add a pleasant complexity to the sonic space. One of my favorite simple tricks is to feed a single reverb send into two or three stereo reverb returns, which I can set with slightly different size and length and pre-delay, so I can stack them in the room with the larger/longer ones in the back of the room. It has the effect of lengthening the apparent auditorium depth, and it just sounds cool."

14.3.8 Width

There are multiple approaches to adding width to a film mix, particularly with regards to music. "Some mixers like to pan the music super-wide into the proscenium," often utilizing additional side surround speakers near the screen as available in Dolby Atmos®, "but you think to yourself – I hope every theater actually has those side speakers," muses Peralta. When spreading sounds like music with 3D panning, Hirschberg cautions "taking the music off screen doesn't necessarily help the movie. I prefer to keep the music mostly on-screen and carefully mixed under the sound effects and dialog in the sound-field."

However, "I love spatial storytelling with sound effects," says Peralta. "I look at the characters' eyes. Anytime they look offscreen, or an object leaves the screen, I try to put an appropriate sound there," to help support the visuals.

14.3.9 Height

"I love the height" in 3D mixing, says Peralta, especially when justified by "anything that leaves the top edge of the screen." However, "panning into the height also changes the width" warns Peralta. "What used to be a mono sound in the center screen channel might then play out of stereo overheads unless you can make it an object and pan it to specific speakers. With discrete left, center, and right overhead channels I could better place things above me." Adding height can also add an unintended sense of depth depending on speaker placements. When mixing DreamWorks Animation's *Home* [12] in Auro 3D, a particularly amusing scene with a rising platform motivated dialogue from Steve Martin's Captain Smek to be panned off the top edge of the screen. However, we found during playbacks in Skywalker's Stag Theater that for the audience members sitting in the front of the auditorium, the overhead speakers are actually above and even behind the front rows of seats. What works in one 3D sound format might not translate as intended to a different speaker configuration without a custom mix.

"I use static overheads very rarely in music," says Hirschberg. When necessary, she uses height objects off to the side or rear of the auditorium to spread the music, or add echo to dialogue. "I don't put ambiences or air in the overhead, just humming along muddying up the mix," agrees Peralta. Using overhead speakers to enhance height is best used when the music or sound effects need to "bloom" to support the story. Hirschberg advises "turn them on only when you need them. When used sparingly, height panning is more obviously noticed when it is used."

Overhead panning is only part of the approach when enhancing the perception of height in the soundscape: "A lot of our experience of height has to do with high frequencies. So sometimes, I will add a little high shelving or some bright reverb to those sounds to give them a bit more 'air' and help the sonic illusion. It also helps to avoid 'noisy' sounds in favor of more clear sounds. For example, it's better to layer a few sounds of raindrops to create a "canopy" of rain rather than a recording that has a lot of white-noise SHHHH kind of sound – it just turns to mud on the ceiling."

14.4 CONCLUSION

Such are the 3D audio and spatial perspectives of many of the top professionals in the world today, often working with the biggest budgets, the best tools, and the most exciting material – and the greatest pressure. They have presented a range of the best working practice and given indications of potential folly, and the reader is invited to reflect upon these not just as isolated tips, but as indicative of the ethos of a much greater best-practice approach that might be applied to numerous other situations.

The power we wield as mixers of 3D sound is to balance the storyteller's vision with the audience's freedom of exploration, while being mindful of the range of venues that might play back our work. Our responsibility is to support the story, and almost always to draw attention towards the screen more often than to draw attention away from the screen. However, when given the opportunity to immerse the audience inside our stories, we should use every tool at our disposal with care and intent, to craft truly transformative entertainment experiences.

REFERENCES

[1]   Interview with J. Peralta, Jan. 15, 2020.
[2]   A. Russo and J. Russo, *Avengers: Endgame*. Marvel Studios, Walt Disney Pictures, 2019.
[3]   P. Reed, *Ant-Man and the Wasp*. Marvel Studios, Canadian Film or Video Production Tax Credit (CPTC), The Province of Ontario, 2018.
[4]   C. Trevorrow, *Jurassic World*. Universal Pictures, Amblin Entertainment, Legendary Entertainment, 2015.
[5]   B. Bird, *Tomorrowland*. Walt Disney Pictures, A113, Babieka, 2015.
[6]   J. Kosinski, *Oblivion*. Universal Pictures, Relativity Media, Monolith Pictures (III), 2013.
[7]   K. DeMicco and C. Sanders, *The Croods*. DreamWorks Animation, 2013.

[8]   C. Wedge, *Epic*. Twentieth Century Fox Animation, Blue Sky Studios, House of Cool Studios, 2013.

[9]   A. Taylor, *Thor: The Dark World*. Marvel Studios, Walt Disney Pictures, 2013.

[10] D. DeBlois, *How to Train Your Dragon 2*. DreamWorks Animation, Mad Hatter Entertainment, Vertigo Entertainment, 2014.

[11] R. Stromberg, *Maleficent*. Jolie Pas, Roth Films, Walt Disney Pictures, 2014.

[12] T. Johnson, *Home*. DreamWorks Animation, 2015.

[13] C. Columbus, *Pixels*. Columbia Pictures, LStar Capital, China Film Group Corporation (CFGC), 2015.

[14] J. Favreau, *The Jungle Book*. Fairview Entertainment, Moving Picture Company (MPC), Prime Focus, 2016.

[15] M. Thurmeier and G. T. Chu, *Ice Age: Collision Course*. Blue Sky Studios, FortyFour Studios, Twentieth Century Fox Animation, 2016.

[16] *Stranger Things*. 21 Laps Entertainment, Monkey Massacre, Netflix, 2016.

[17] M. Reeves, *War for the Planet of the Apes*. Twentieth Century Fox, Chernin Entertainment, TSG Entertainment, 2017.

[18] J. J. Abrams, *Star Trek Into Darkness*. Paramount Pictures, Skydance Media, Bad Robot, 2013.

[19] B. Bird, *The Incredibles*. Pixar Animation Studios, Walt Disney Pictures, 2004.

[20] B. Bird, *Mission: Impossible – Ghost Protocol*. Paramount Pictures, Skydance Media, TC Productions, 2011.

[21] M. Andrews, B. Chapman, and S. Purcell, *Brave*. Walt Disney Pictures, Pixar Animation Studios, 2012.

[22] Interview with W. Files, Jan. 27, 2020.

[23] Interview with W. Sereno, Jan. 17, 2020.

[24] A. Cuarón, *Gravity*. Warner Bros., Esperanto Filmoj, Heyday Films, 2013.

[25] J. Algar *et al.*, *Fantasia*. Walt Disney Productions, Walt Disney Animation Studios, 1941.

[26] Interview with L. Hirschberg, Jan. 13, 2020.

[27] C. Buck and J. Lee, *Frozen*. Walt Disney Animation Studios, Walt Disney Pictures, 2013.

[28] C. Buck and J. Lee, *Frozen II*. Walt Disney Animation Studios, Walt Disney Pictures, 2019.

[29] R. Clements, J. Musker, D. Hall, and C. Williams, *Moana*. Hurwitz Creative, Walt Disney Animation Studios, Walt Disney Pictures, 2016.

[30] B. Howard, R. Moore, and J. Bush, *Zootopia*. Walt Disney Pictures, Walt Disney Animation Studios, 2016.

[31] K. Hall, *Planes*. Prana Studios, Disneytoon Studios, 2013.

[32] Interview with D. Fluhr, Jan. 26, 2020.

[33] Interview with G. Rydstrom, Jan. 30, 2020.

[34] K. Bigelow, *Strange Days*. Lightstorm Entertainment, 1995.

Index

Note: Creative works are filed under their title. Page numbers in *italics* refer to illustrations and those in **bold** refer to tables.

290

Index